Studies in Computational Intelligence 458

Editor-in-Chief

Prof. Janusz Kacprzyk
Systems Research Institute
Polish Academy of Sciences
ul. Newelska 6
01-447 Warsaw
Poland
E-mail: kacprzyk@ibspan.waw.pl

For further volumes:
http://www.springer.com/series/7092

Adam Przepiórkowski, Maciej Piasecki,
Krzysztof Jassem, and Piotr Fuglewicz (Eds.)

Computational Linguistics

Applications

Springer

Editors
Prof. Adam Przepiórkowski
Institute of Computer Science
Polish Academy of Sciences
and University of Warsaw
Warsaw
Poland

Dr. Maciej Piasecki
Department of Artificial Intelligence
Institute of Informatics
Wrocław University of Technology
Wrocław
Poland

Prof. Krzysztof Jassem
Faculty of Mathematics and Computer Science
Information Systems Laboratory
Adam Mickiewicz University
Poznań
Poland

Piotr Fuglewicz
TiP Sp. z o. o.
Katowice
Poland

ISSN 1860-949X e-ISSN 1860-9503
ISBN 978-3-642-42956-9 ISBN 978-3-642-34399-5 (eBook)
DOI 10.1007/978-3-642-34399-5
Springer Heidelberg New York Dordrecht London

Softcover reprint of the hardcover 1st edition 2013

Printed on acid-free paper

Springer is part of Springer Science+Business Media (www.springer.com)

Editorial

Computational Linguistics – Applications (CL-A) is a forum for the dissemination of research findings in the field of applied Computational Linguistics. CL-A started in 2007 as a track of AAIA conference[1]. Since then it has evolved into a workshop (2008) and finally into an independent conference (2010, 2011) organized by Polish Information Processing Society.

A distinguishing feature of CL-A conferences was that the speakers were requested to demonstrate a working tool besides their scientific paper. The demonstrations were evaluated by the audience for interest they aroused.

CL-A authors, who had been scored highly for both paper quality (as marked by CL-A reviewers) and their demonstration (as voted by CL-A audience) were requested to propose chapters for this publication. The authors were given the choice to contribute either extended or updated versions of their original CL-A papers.

All chapters included here have one in common: they refer to either a fully functional system or an advanced working algorithm. All chapters focus on text processing: one chapter deals with recorded text, the remaining ones concern written text.

Part I is devoted to NLP toolkits, i.e. sets of programs designed to process text documents written in various languages. Linden et al present the latest functionalities of HFST (Helsinki Finite-State Technology) – a framework for compiling and applying linguistic descriptions with finite-state methods. Graliński et al introduce PSI-Toolkit – an open source project intended to deliver NLP tools for both experienced language engineers and casual users without any technological background. Broda et al present Fextor - a flexible feature extraction toolkit that can be applied in: word sense disambiguation, recognition of inter-chunk syntactic relations, semantic relations between named entities and anaphora resolution.

Part II deals with extracting information from text documents, i.e. automatic creation of structured information from unstructured texts.

[1] AAIA – Advances in Artificial Intelligence and Applications was first held in 2006 at the International Multiconference on Computer Science and Information Technology in Wisła, Poland.

One way of storing the extracted information is in the form of lexicons. Kurc et al introduce a method for the creation of a stand-alone proper name repository with the aid of a wordnet. Seretan describes a multilingual framework that aims at acquisition of collocations from text corpora. Krstev et al apply an existing multipurpose NLP tool (LeXimir) to a new task of Multi-Word Units processing. Horák and Rambousek describe the structure and usage of PRALED, a specialized lexicographic workstation designed for the development of the Czech lexical database.

The lexicons, whether compiled automatically or manually, need an interface for human readers. An example is EcoLexicon, described by Pilar Leon-Arauz et al. The users of the tool may find information about the global environment by means of a user-friendly visual interface.

In some solutions, the information extracted from text documents is stored in database structures. The chapter written by Atkinson et al presents a number of techniques for multilingual event extraction, where information about the events related to public security is stored in a semantic database.

Another idea for storing extracted knowledge is *metadata*, which describe the content and context of digital data. This idea is followed by Vlachidis et al, who present a method for the automatic generation of metadata describing items in an archaeological digital library.

The last paper of Part II is devoted to automatic extraction of information from text read by a lector. Volskaya et al describe an algorithm for the detection of prominence in Russian texts.

The problem of information extraction is nowadays closely linked with the concept of multilinguality, which is the main theme of Part III. Users expect to obtain the information on the topics of their interest even if the answer is to be found in texts written in a different language than the language of the query. This problem is addressed in the chapter by Saad and Nuernberger who demonstrate the [Mult]iSearcher, an interactive tool for cross-lingual information retrieval. The next three chapters deal with the problem of man-machine co-operation in the task of translation. Burchardt et al give an overview of the state-of-the-art concluding that "human post-editing today seems the only way to go". They also describe taraXŰ, an ongoing joint project between industry and research partners that "is aimed at making MT economically feasible and technically usable". Jaworski's paper describes an algorithm that speeds up the process of translation memory search, the technology used in most Computer Assisted Translation tools. Dudan et al describe a new feature of the Apertium MT system: subject-related translation.

The significance of Computational Linguistics increases along with the progress of man-machine communication. Rapid development of Information Extraction in recent years has been stimulated by the popularity of Google-like querying. The upcoming era of dialogue communication will induce the violent progress of other topics of Applied Computational Linguistics.

Contents

Part I: NLP Toolkits

Part II: Information Extraction

Part I
NLP Toolkits

Using HFST for Creating Computational Linguistic Applications

Krister Lindén, Erik Axelson, Senka Drobac,
Sam Hardwick, Miikka Silfverberg, and Tommi A. Pirinen

Abstract. HFST – Helsinki Finite-State Technology (`http://hfst.sf.net/`) is a framework for compiling and applying linguistic descriptions with finite-state methods. HFST currently collects some of the most important finite-state tools for creating morphologies and spellcheckers into one open-source platform and supports extending and improving the descriptions with weights to accommodate the modeling of statistical information. HFST offers a path from language descriptions to efficient language applications. In this article, we focus on aspects of HFST that are new to the end user, i.e. new tools, new features in existing tools, or new language applications, in addition to some revised algorithms that increase performance.

Keywords: finite-state applications, morphology, tagging, HFST.

1 Introduction

HFST – Helsinki Finite-State Technology (`http://hfst.sf.net/`) is designed for creating and compiling morphologies, which has been documented in, e.g., [14, 11]. In this article we focus on the applications created with HFST and some of their theoretical motivations. HFST contains open-source replicas of xfst, lexc and twolc which are well-known and well-researched tools for morphology building, see [3]. The tools support both parallel and cascaded

Krister Lindén · Erik Axelson · Senka Drobac · Sam Hardwick · Miikka Silfverberg · Tommi A. Pirinen
University of Helsinki
Department of Modern Languages
Unioninkatu 40 A
FI-00014 Helsingin yliopisto, Finland
e-mail: {`krister.linden,erik.axelson,senka.drobac`}`@helsinki.fi`,
{`sam.hardwick,miikka.silfverberg,tommi.pirinen`}`@helsinki.fi`

A. Przepiórkowski et al. (Eds.): *Computational Linguistics*, SCI 458, pp. 3–25, 2013.
DOI: 10.1007/978-3-642-34399-5_1

application of transducers. This is outlined in Section 2 along with some new and previously undocumented extensions to the Xerox tools in HFST.

There are a number of tools for describing morphologies. Many of them start with the item-and-arrangement approach in which an arrangement of sublexicons contain lists of items that may continue in other sublexicons. A formula for compiling such lexical descriptions was documented in [14]. In Section 3, we demonstrate a simplified procedure for how such morphotactic descriptions can be compiled into finite-state lexicons using finite-state operations. To realize the morphological processes, rules may be applied to the finite-state lexicon. In addition, HFST now also offers the capability to train and apply part-of-speech taggers on top of the morphologies using parallel weighted finite-state transducers on text corpora, which is outlined and evaluated in Section 4.

Using compiled morphologies, a number of applications have been created, e.g. spellcheckers for close to 100 languages and hyphenators for approximately 40 different languages. The spellcheckers were derived from open-source dictionaries and integrated with OpenOffice and LibreOffice, e.g. a full-fledged Greenlandic spellchecker, which is a polyagglutinative language, is currently available for OpenOffice via HFST. By adding the tagger capability, we have also created an improved spelling suggestion mechanism for words in context. The spellchecker applications and some of their theoretical underpinnings are described in Section 5.

Finally in Section 6, we discuss some additional applications such as synonym and translation dictionaries as well as a framework for recognizing multi-word expressions for information extraction and how this can be done using finite-state technologies.

2 Building Morphologies

One of the earliest and most important goals of the HFST project has been to provide open-source utilities for compiling full-fledged morphological analyzers, which may be used for constructing spellcheckers, taggers and other language technological utilities of good quality. The Xerox toolkit [3] is among the most widely used frameworks for constructing morphological utilities. The toolkit is used to compile linguistic description into morphological analyzers. More specifically, Xerox tools include the finite-state lexicon compiler lexc, the two-level rule compiler `twolc` and the cascading replace rule compiler xfst. HFST includes the tools **hfst-lexc**, **hfst-twolc** and **hfst-xfst**, which provide full backward compatibility with Xerox tools and augment their functionality.

Lexicon files for the lexicon compiler **hfst-lexc** describe the morphotactics of a language using sub-lexicons containing lists of morphs. A rule component can be used for realizing phonological changes occurring within the morphs or at their boundaries. In this article, we do not introduce new features of **hfst-lexc**. Adding weights to lexicons using **hfst-lexc** is described in [13].

The rule compilers **hfst-xfst** and **hfst-twolc** provide almost the same functionality. Both are used to realize morphophonological variations on a lexicon compiled with **hfst-lexc**. The difference between the tools is that **hfst-xfst** rules are applied in succession gradually altering the lexicon whereas **hfst-twolc** rules are applied as parallel constraints limiting the realizations of morphophonemes used in the **hfst-lexc** lexicon.

2.1 Parallel Rules with Negative Contexts

The rule compiler **hfst-twolc** provides backward compatibility with the Xerox tool twolc, but it augments twolc functionality by providing a new type of rules. In **hfst-twolc**, rules can have negative contexts.

```
! Change x to y after the string "a b" unless "d" or
! "c d" follows.
"Rule with negative context"
x:y <=> a b _     ;
        except
            _ ( c ) d ;

"Rule without negative context"
x:y <=> a b _ [ ? - [ c | d ] | c [ ? - d ] | .#. ] ;
```

Fig. 1 Negative context rule and corresponding traditional two-level rule

In traditional two-level rules, it can be very difficult to express that a certain alternation should not occur in a given context. Such restrictions are required because of rule conflicts, i.e. clashes of two or more contradicting rules. When a rule conflict occurs, the contexts of some of the contradicting rules need to be restricted in order to resolve the conflict.[1]

Sometimes an automated mechanism called conflict resolution of two-level rules [24] can be used to resolve rule conflicts, but conflict resolution works only if the conflicting rules can be ordered into chains of subcases. Often this can be difficult to accomplish, especially for grammar writers who are not especially well acquainted with writing regular expressions.

In twolc syntax, prohibition rules such as `x:y /<= C1 _ C2 ;` can be used to forbid a pair `x:y` in a context where the left context matches the regular expression `C1` and the right context matches `C2`. Unfortunately prohibition rules do not solve the problem of conflicting rules, because they do not

[1] http://www.cis.upenn.edu/~cis639/docs/twolc.html

participate in conflict resolution and simply adding new rules to a two-level grammar does not remove rule conflicts.[2]

Using an extension to the Xerox two-level rule syntax in `hfst-twolc`, it is possible to formulate rules with negative contexts, i.e. contexts that prohibit the triggering of the rule. Figure 1 shows a schematic example of a negative context rule and a traditional Xerox style rule, which has the same effect as the rule with the negative context.

Rules with negative contexts are used in the Kyrgyz morphology developed in the Apertium project[3]. They significantly shorten the two-level grammar[4].

2.2 Cascaded Rules: Explanations and Examples

HFST replace rules provide backward compatibility with XFST replace rules, described in [10, 3]. Although they mostly share the same notation, the HFST replace rules differ in that they were compiled with the `.r-glc.` operator [7] and preference relations described in [29]. This approach makes it possible to more freely define contexts in parallel rules and to easily add new functionalities in future.

We present a general account of using replace rules with the command-line tool `hfst-regexp2fst`. Since these rules mostly follow the behaviour of the XFST rules, for a detailed description of each rule, see The Finite State Morphology book [3].

Simple Rules. A simple right arrow replace rule,

$$A \rightarrow B \tag{1}$$

where A and B are regular expressions, expresses that A in the upper language maps to B in the lower language.

Replace rules can be compiled into a transducer using `hfst-regexp2fst`. This tool takes a regular expresion as input and gives a corresponding transducer written in a binary file as output. To convert transducers from binary format to text (ATT) format, there is an HFST tool `hfst-fst2txt`. Therefore, the upper rule could be compiled as in Figure 2.

```
$ echo "A -> B ;" | hfst-regexp2fst | hfst-fst2txt
```

Fig. 2 Compiling simple replace rule

The regular expression can be read from a file, or saved to a file (Figure 3).

[2] Note that, if prohibition rules would participate in conflict resolution, it would still be challenging to write them in such a manner that conflict resolution could apply.

[3] `http://www.apertium.org/`

[4] Personal communication with Francis Tyers.

```
$ echo "A -> B ;" > regex
$ hfst-regexp2fst -i regex -o transducer
$ hfst-fst2txt transducer
```

Fig. 3 Reading from, and writing to a file

Rules can be applied to any language by using composition. The result of composing a word `ABCD` with the aforementioned rule `A -> B` is the transducer in Figure 4. It has 5 states (0 – 4), and only state 4 is final. The transitions go from a state in the first column to a state in the second column. The input label of a transition is displayed in the third column and the output label – in the fourth. The default output of `hfst-regexp2fst` is weighted. Since weights are not used in replace rules, the weights get the value 0.0 when the rule is compiled. The weights are displayed in the final column.

```
$ echo "A B C D .o. A -> B ;" | hfst-regexp2fst | hfst-fst2txt
0       1       A       B       0.000000
1       2       B       B       0.000000
2       3       C       C       0.000000
3       4       D       D       0.000000
4       0.000000
```

Fig. 4 Composing a rule with a word

Context. Every rule can have a context in which the replacement is made. Here, A will be mapped to B if and only if it is between regular expressions L and R.

$$A \rightarrow B \mid\mid L\_R \qquad (2)$$

Also, multiple context pairs are supported, when separated by comma.

$$A \rightarrow B \mid\mid L_1\_R_1, \ \dots\ , L_i\_R_i \qquad (3)$$

In Figure 5, the rule expression says that `a` in the upper language is mapped to `b` in lower language when between `m` and `n`, or in front of `b`. In the word `manaab`, the first and last `a` are mapped to `b`, because they occur between corresponding contexts, but the second `a` is kept unchanged.

In a replace rule where A is the upper and B the lower language, there are four contextual directions that can be used with replace rules. For example, when the // sign is used as context orientation operator, the left context will be taken from the lower language. It is thus possible for the replace function to write its own context. This is shown in Figure 6.

Other Replace Functions. We have hitherto only used the right arrow replace operator, but there are many other operators that can be used. All the operators listed in Table 2 have their left arrow version, which is the inversion

```
$ echo " m a n a a b .o. a -> b || m _ n , _ b;" |
hfst-regexp2fst | hfst-fst2txt
0       1       m       m       0.000000
1       2       a       b       0.000000
2       3       n       n       0.000000
3       4       a       a       0.000000
4       5       a       b       0.000000
5       6       b       b       0.000000
6       0.000000
```

Fig. 5 Replace rule between two contexts

Table 1 Different context directions in Replace Rules

Operator	Operator description
`||`	both contexts are taken from the upper language
`//`	the left context is taken from the lower language, the right from the upper
`\\`	the left context is taken from the upper language, the right from the lower
`\/`	both contexts are taken from the lower language.

```
$ echo " b a a .o. a -> b // b _ ;" | hfst-regexp2fst | hfst-fst2txt
0       1       b       b       0.000000
1       2       a       b       0.000000
2       3       a       b       0.000000
3       0.000000
```

Fig. 6 Replace rule writes its own context

of the right operator. Furthermore, all the rules can be used with epenthesis [. .] and markup [...] operators (see Figure 7). The epenthesis operator should be used with empty strings to avoid replacing infinitely many epsilons, while the markup operator is used to insert markers around a word.

Table 2 List of right replace operators that can be used in HFST

Right replace operators	Replace function
`->`	Replace
`(->)`	Replace optional
`@->`	Longest match from left to right
`->@`	Longest match from right to left
`@>`	Shortest match from left to right
`>@`	Shortest match from right to left

```
$ echo " a a a .o. [.a*.] @-> b ;" | hfst-regexp2fst > a.fst
$ echo " b a a b .o. a @-> %[ ... %] ;" | hfst-regexp2fst > b.fst
```

Fig. 7 Epenthesis and markup operators

Parallel Rules. Parallel rules are used when there are multiple replacements at the same time. The general parallel replace rule expression consists of individual replace rules separated by two commas.

$$A_1 \rightarrow\ B_1 \mid\mid L_1 \_\ R_1\ ,,\ \ldots\ ,,\ A_i \rightarrow\ B_i \mid\mid L_i \_\ R_i \tag{4}$$

In XFST, all rules in a parallel rule expression have to have the same format, i.e. they need to have the same arrow and the same context layout. In HFST, the constraint that all the rules should have the same arrow is kept, but the context layout can differ freely. Therefore, the rule expression in Figure 8 would not be allowed in XFST, but is valid in HFST.

```
$  echo " a -> b || m _ n ,, c -> d ;" | hfst-regexp2fst | hfst-fst2txt
```

Fig. 8 Parallel rules with different context layouts

3 Morphological Descriptions

Popular formalisms for describing the morphotactics of a language tend to be some variation of the item and arrangement scheme. We build on this to generalize from `lexc` into other item-and-arrangement notations such as Hunmorph and Apertium. This gives us the option of describing morphology in a notation that is compiled into finite-state transducers which we can continue to process with other HFST tools. We demonstrate how compilation could proceed when reducing a `lexc` lexicon into a sequence of finite-state operations. A morphological programming language like `SFST-PL` forgoes a standard lexicon interface like `lexc` and only offers the end user an option to construct the lexicon using finite-state operations.

3.1 Morphotax and Morphological Formulæ

Morphotax is the component dealing with morphological combinations in language description. This concerns word-formation processes, such as affixation and compounding. There are numerous formalisms for describing morphotactics in natural language processing applications, such as hunspell[5] (and its older *spell relatives), the Apertium lttoolbox[6] or the Xerox lexc [3].

[5] `http://hunspell.sf.net`
[6] `http://apertium.sf.net`

The HFST toolset contains parsers and compilers for reading descriptions of morphologies written in these formalisms, which can be used for compiling a finite-state automaton for the other HFST tools to process [20, 14].

Consider a trivial lexicon consisting of the English nouns *cat* and *ax* with the empty string as the singular marker and *s* and *es* as plural markers, respectively, forming a lexicon as outlined in Figure 9 using lexc notation.

```
LEXICON Root
Nouns ;

LEXICON Nouns
cat NumberS ;
ax NumberES ;

LEXICON NumberS
# ;
s # ;

LEXICON NumberES
# ;
es # ;
```

Fig. 9 Lexicon in lexc notation

Many contemporary notations for describing morphotactics tend to use the same item-and-arrangement paradigm for combining sets of morphs or morphemes, i.e. they use sublexicons for lists of morphs that may have continuations in other sublexicons. In its most general form, such a paradigm can be defined using a combination of two components expressing local restrictions: (1) a disjunction of morphs in local lexical context repeated infinitely and (2) a morphotactic filter describing the permitted sequence of sublexicons. Both the disjunction of morphs and the morphotactic filter can be described in finite-state form.

In Figure 10, we demonstrate how the lexicon in Figure 9 is compiled into a loop of disjunctions of any of the morphs in the lexicon. Note that we bracket special symbols with at-signs '@'. In this case, we use special joiner symbols to calculate the morphotactics, and the at-signs give the user a hint that they are not supposed to end up in the final result. The sublexicon symbols like `@NumberES@` are used for lining up the sublexicons, and the symbol `@LEX@` is used as a boundary symbol, which simplifies the filter algorithm[7].

In Figure 11, we show the command-line simulation for creating the morphotactic filter and how this filter is composed with the disjunction of morphs

[7] In [14], the algorithm without a separate boundary symbol requires a term complement of the disjunctive closure of the sublexicon symbols.

```
echo "@Root@ @Nouns@ @LEX@" > strings
echo "@Nouns@ c a t @NumberS@ @LEX@" >> strings
echo "@Nouns@ a x @NumberES@ @LEX@" >> strings
echo "@NumberS@ @END@ @LEX@" >> strings
echo "@NumberS@ s @END@ @LEX@" >> strings
echo "@NumberES@ @END@ @LEX@" >> strings
echo "@NumberES@ e s @END@ @LEX@" >> strings
hfst-strings2fst --has-spaces --disjunct-strings < strings  |
    hfst-repeat -f 1 > bag_of_morphs
```

Fig. 10 Compiling a disjunction of all the morphs in the lexicon

to create the final lexicon[8]. In effect, we create a filter that ensures that each morph, e.g. *es*, only follows the morph that was asking for it on the right, i.e. we require that the same sublexicon symbol, e.g. `@NumberES@`, occurs on both sides of the boundary symbol `@LEX@`. For a full algorithm, see e.g. [14].

```
echo "%@Root%@ [? - %@LEX%@]*"  | hfst-regexp2fst > start
echo "%@END%@ %@LEX%@"  | hfst-regexp2fst > end

echo "%@NumberS%@ %@LEX%@ %@NumberS%@ [? - %@LEX%@]*"  |
    hfst-regexp2fst > cont1
echo "%@NumberES%@ %@LEX%@ %@NumberES%@  [? - %@LEX%@]*"  |
      hfst-regexp2fst > cont2
echo "%@Nouns%@ %@LEX%@ %@Nouns%@  [? - %@LEX%@]*"  |
      hfst-regexp2fst > cont3
hfst-disjunct cont1 cont2 | hfst-disjunct - cont3 |
    hfst-repeat -f 1 > conts
hfst-concatenate start conts |
    hfst-concatenate - end > morphotactics
hfst-compose bag_of_morphs  morphotactics > lexicon
```

Fig. 11 Composing the morphotactics with the disjunction of morphs into a lexicon

Since many morphotactic formalisms tend to be some variation of the item and arrangement scheme, we can build on this to generalize into other notations, since the morphotactics itself does not depend on the description language of the morphotactics. This also makes additions like compound-based weighting of the language-model [13] generally applicable.

For Lexicographers. This approach gives the option to choose their favorite notation, e.g. hfst-lexc, Hunspell or Apertium and then continue with other

[8] Whole process is also available from our SVN in `https://hfst.svn.sf.net/svnroot/hfst/trunk/cla-2012-article/morphtest.bash`

HFST tools. In theory, it would also be possible to use this morphotactic formula to write implementations for any item-and-arrangement morphology, but in practice it's easier to simply use some high-level scripting language to convert a lexical database into e.g. lexc notation.

3.2 Performance

The HFST toolkit has been implemented using three back-end libraries: SFST, OpenFst and foma. The libraries are linked with HFST. Usually one library is chosen and used throughout the compilation of a given morphology. We can thus compare how different back-end libraries perform in the same task.

In Table 3, we show compilation times for different morphologies using different HFST back-ends. The morphologies are OMorFi [19] for Finnish, Morphisto [31] for German, Morph-it [30] for Italian, Swelex [9] for Swedish and TrMorph [6] for Turkish. OMorFi, Morphisto and TrMorph have several rules for inflection and compounding, Morph-it and Swelex are basically word lists. We use HFST version 3.3.4, with SFST, OpenFst and foma as back-ends. The times are averages from runs of 10 compilations.

Table 3 Compilation times for different morphologies with different HFST back-ends. The times are given in minutes and seconds and averaged over 10 compilations. HFST version is 3.3.4.

Back-End	Finnish	German	Italian	Swedish	Turkish
SFST	2:48	2:12	0:30	0:13	0:12
OpenFst	7:52	7:45	2:24	0:49	0:40
foma	1:52	1:33	0:31	0:13	0:05

In Table 4, we show both the current compilation times and the ones that we achieved in an earlier benchmarking [11] for Finnish and German. We also show the back-end versions used.

It can clearly be seen that the performance of HFST with SFST as a back-end has improved: the compilation time of the Finnish morphology has almost halved and the compilation time of the German morphology is only one third of the time at the previous benchmarking. This improvement comes from the newer version of SFST that features more optimized composition and Hopcroft minimization functions. The improved functions were developed by Helmut Schmid in cooperation with the HFST team.

[9] `https://kitwiki.csc.fi/twiki/bin/view/KitWiki/HFSTSwelex`

Table 4 Compilation times for different morphologies with different HFST back-ends and their versions. The times are given in minutes and seconds.

Back-End	version	Finnish	German
SFST	1.4.2	5:02	6:39
	1.4.6	2:48	2:12
OpenFst	1.2.7	6:51	6:28
	1.2.10	7:52	7:45
foma	0.1.14	1:49	1:29
	0.1.16	1:52	1:33

4 Building Taggers

The HFST library includes tools for constructing statistical part-of-speech (POS) taggers which resemble Hidden Markov Models (HMM) from tagged training data. HFST taggers differ from other HMM taggers such as Tnt [4] and Hunpos [8] in that HFST allows combining different estimates for tag probabilities during tagging. It is possible to use e.g. the surrounding word forms in estimating the probability of a given tag. This is more thoroughly explained in [26, 27] The accuracy for basic HMM taggers implemented using HFST tools is comparable to the accuracy of Tnt and Hunpos as demonstrated below.

HMM-type taggers include a lexical model and a tag sequence model. The lexical model is needed for determining probabilities for the co-occurrence of tags and word forms disregarding context, and the tag sequence model is used for determining the probabilities for the co-occurrence of tags of neighboring words. The lexical models in HFST taggers include suffix guessers which can be modified to suit the needs of particular languages. Additionally HFST supports using morphological analyzers in tagging.

4.1 The Structure of HFST Taggers

HMM taggers are statistical models which determine the most likely POS tag from some tag set for each word in a sentence. For determining the most likely tags, the taggers use lexical probabilities $p(w|t)$ for each word w and each tag t together with *transition probabilities* for the tag of a word at a given position given the tags of the preceding words $p(t_i|t_{i-1}, \dots t_{i-n})$ [27]. The integer n determines how many preceding tags are considered when estimating the probability of tags. It is called the *order* of the HMM tagger. Second order HMM taggers are the most common in POS tagging.

HFST taggers extend traditional HMM taggers by allowing modifications to the traditional estimate $p(t_i|t_{i-1}, \dots t_{i-n})$. In addition to preceding tags also succeeding tags and word forms can be used when deciding the probability of a tag at a given position. These estimates are combined using finite-state calculus.

In HFST taggers, the lexical probabilities $p(w|t)$ are given by the lexical component of the tagger and the transition probabilities $p(t_i|t_{i-1}, \ldots t_{i-n})$ are given by the sequence component of the tagger. Both models are trained using tagged training data.

Both components can be modified to suit the needs of a particular language. In [27] it is explained how the sequence model can be modified to include preceding and succeeding words in the estimates of the probabilities of tags, and how suffix guessers can be modified to better suit agglutinative languages. In the present paper we show how a morphological analyzer can be integrated with an HFST tagger to improve the accuracy of the tagger when there are a lot of out-of-vocabulary words.

4.2 The Lexical Model of HFST Taggers

The accuracy of traditional POS taggers suffers greatly because of out of vocabulary (OOV) words, i.e. word forms which were not observed during training of the tagger. OOV words effect the accuracy, since POS taggers generally rely heavily on lexical probabilities $P(word|tag)$ computed for the words occurring in a training corpus.

For languages like English with few productive morphological phenomena, OOV words are not a big problem when there is a lot of training data and the genres of the training data and the data that is tagged are sufficiently similar. When the genres differ considerably there are more OOV words and consequently accuracy is reduced. When building taggers for morphologically rich languages such as Turkish, Finnish or Hungarian, OOV words become a major problem even when no change of genre is involved and even when there is a lot of training data. In agglutinative languages OOV words arise from productive morphological phenomena such as inflection, derivation and compounding.

Usually, e.g. in [4], OOV words are handled using suffix guessers, which combine estimates of all suffixes of the OOV word whose length does not exceed a given threshold. E.g. the POS tag for *ameliorate* can be guessed based on the distribution of tags for the suffixes *-e*, *-te*, ..., *-liorate* if the threshold for the length of suffixes is 7.

HFST taggers offer two improvements for handling of OOV words: (1) the tagger builder can adjust the way in which the probability distribution for different length suffixes are combined to form probability estimates for the OOV word, and (2) the tagger builder has the option to combine taggers with morphological analyzers, whose tag set does not need to equal the tag set of the training data of the tagger.

Adjusting the way suffix estimates are combined is useful e.g. in Finnish, where the suffix guesser proposed by [4] gives poor results. For Finnish, the accuracy of the guesser improves when the only suffix considered is the longest suffix of the word which was seen during training. This is probably a result of the high number of compound words where the final part of the compound is known, but the compound itself is an OOV word. The final part of the compound is generally a very good predictor of the word-class of the compound, so suffixes of the OOV word containing the final part are very informative.

The idea of incorporating a morphological analyzer in a tagger is not new. E.g. [28, 18] use morphological analyzers as part of statistical taggers. The novel aspect of HFST taggers is that they can incorporate morphological analyzers whose morphological description differs from the morphological description used in the training data of the statistical tagger. This is useful, because it allows utilizing ready-made linguistic utilities in tagging without the need to address problems such as differences in the coarseness of the morphological descriptions between the utilities and the training data of the tagger.

In approaches such as [28, 18], the tag set of the morphological analyzer and the statistical tagger have to be the same. When an OOV word is encountered, the morphological analyzer is used to look up the possible analyses for the OOV word. These analyses are used directly by the tagger. In contrast, HFST taggers keep track of *ambiguity classes*, which are the sets of morphological analyses emitted by the morphological analyzer for words in the training data. The taggers use the tag distributions of all training data words in a given ambiguity class to determine the tag distribution of an OOV word in the same ambiguity class.

Given an English morphological analyzer, which emits two analyses *dog+N+SG* and *dog+V+INF* for the word form *dog*, the ambiguity class of *dog* is the set of its analyses {*+N+SG*, *+V+INF*}. In the training data, *dog* might receive Penn Treebank tags like `NN` and `VB`. When the tagger encounters an OOV word in the same ambiguity class as *dog*, such as *man*, it first maps *man* onto its ambiguity class {*+N+SG*, *+V+INF*} using the morphological analyzer. It then uses the tag distribution of all words in the ambiguity class {*+N+SG*, *+V+INF*} to estimate the probabilities $P(man|tag)$ for *man* given each tag (e.g. `NN` and `VB`). Note that this does not require that the tag sets of the training data and the morphological analyzer are the same. It also does not require that their morphological descripitions are equally fine-grained.

4.3 Experiments with HFST Taggers

We demonstrate HFST taggers by constructing POS taggers for Finnish and English. For English we construct a regular second order HMM tagger. For Finnish we construct two taggers: one regular second order HMM tagger and another second order HMM tagger which utilizes the OMorFi morphological analyzer for Finnish [19].

The English training data and test data come from the Penn Treebank. Sections 1 to 18 were used for training and sections 22 to 24 for testing. The Finnish training and test data come from Finnish newspaper text which has been automatically tagged using the Textmorfo parser[10]. The data is described in Table 5.

Table 5 Data used for training and testing the English and Finnish taggers

Language	Training data size (tokens)	Test data size (tokens)	Distinct tags
English	912,344	129,654	45
Finnish	1,027,511	156,572	764

As Table 6 shows, HFST taggers achieve comparable results to the well known second order HMM tagger TNT. The accuracy for unknown words is slightly worse using the HFST tagger, but for known words the accuracies are nearly identical.

Table 6 Results for second order HMM taggers of English. The accuracy figures for TNT can be found in [8].

English Model	Seen	Unseen	All
TNT	96.77%	85.19%	96.46%
Basic HFST	96.68%	80.71%	96.23%

Table 7 Results for Finnish taggers. The model Basic HFST is a regular second order HMM tagger. The model With Morph HFST is augmented with a morphological analyzer.

Finnish Model	Seen	Unseen	All
Basic HFST	97.51%	77.51%	95.23%
With Morph HFST	97.53%	83.65%	95.90%

For Finnish, the accuracy for the basic second order HMM tagger is poorer than for English as seen in Table 7. This is mostly caused by words in the test data that are missing from the training data, i.e. OOV words. In the Finnish test data 11.51% of the words are OOV words. For comparison, only 2.81% of the words in the English test data are OOV words.

To reduce the number of OOV words in the Finnish test data, a morphological analyzer was used as explained above. Using the morphological analyzer, only 2.73% of the words in the test data were OOV words. Consequently a significant increase in total tagging accuracy is observed, see Table 7. The increase is negligible for known words, but significant for unknown words.

[10] `http://www.csc.fi/kielipankki/`

5 Transducer Applications

Automata technology is a general framework for describing language models and phenomena in a wide range of linguistic fields from phonology to morphology, as well as certain areas of syntax and semantics (in POS tagging and machine translation). The practical applications cover spell-checking, as demonstrated in Voikko[11] with bindings to LibreOffice, Gnome desktop, Mozilla and Mac OS X Spell Service, to morphological and syntactic analysis as demonstrated in native HFST tools from HFST downloads[12], and to machine translation as demonstrated in several released language pairs in the machine translation system Apertium[13]. This demonstrates a very important feature of HFST, alluded to in previous chapters, i.e. language models can be described with one tool in one theoretical and practical framework. For example, existing morphological analyzers for the Sámi languages found on the Internet[14], written with the Xerox formalism were converted into a spell-checker in one evening and evaluated in [21] with additional training from likewise freely available Sámi Wikipedia. Similar results of using machine translation dictionaries from the free/libre open source project Apertium to create not only dictionaries, but morphological analyzers and spell-checkers, are also demonstrated on our web page.[15]

Regarding the general applicability of finite-state language models in practical applications it may be noted that we can now generate e.g. spell-checkers for a language that has a machine readable dictionary (such as Manx) or a morphological analyzer (such as Greenlandic, which is not easily implemented in other formalisms). The transformation of a dictionary or analyzer in transducer format to a baseline spell-checker (with a homogeneous Damerau-Levenshtein edit distance as an error model) can be performed using finite-state tools without feedback from linguists or native speakers.

5.1 Spellcheckers

The task of finite-state spell-checking is well researched and documented. It consists mainly of two phases, identifying incorrect word forms and creating suggestions for corrections. For incorrect word forms there are two types of mistakes, non-word spelling errors, such as writing *cta* where *cat* is meant, and real-word spelling errors, such as writing *there* where *their* is intended. The method for finding the former in finite-state systems is simply to apply the dictionary to the text word by word. Any unrecognized string not belonging to the language of the dictionary automaton is a non-word spelling error.

---

[11] `http://voikko.sf.net/`
[12] `http://hfst.sf.net/`
[13] `http://www.apertium.org/`
[14] `http://divvun.no/`
[15] The examples are available at `http://www.ling.helsinki.fi/cgi-bin/omor/omordemo.bash`

For real-word errors a statistical n-gram model or syntactic parser is typically required. To correct a spelling error in a finite-state system, a two-tape automaton modeling the typing errors should be applied to the misspelt string to get set of potential corrections [21]. For practical purposes the error model can also be implemented as a fuzzy finite-state traversal algorithm or similar methods [17]. The result of the correction step is a set of word-forms that are correct in the language of the spell-checking dictionary. Another related task is to rank this set to provide the most likely corrections first. A trivial way to perform such ranking would be to use unigram [21] or n-gram probabilities of the words [15] or word-form analyses [22].

Creating Spellcheckers. The creation of a finite-state spellchecker involves compiling (at least) two automata: a dictionary that contains the correctly spelled word-forms and the error model that can rewrite misspelt word-forms into correctly spelled ones. The former automaton can be as simple as a reference corpus, containing larger quantities of correctly spelled words and (possibly) smaller quantities of misspelt ones [16]. Also more elaborate dictionaries, such as morphological analyzers described in Section 3 can be used, and also trained for spell-checking purposes with reference corpora without any big modifications to the underlying implementation [21].

For the error-model we can trivially construct an automaton corresponding to the Levenshtein edit distance algorithm [17, 23]. For more elaborate error models it is possible to use hunspell algorithms as automata [20] or construct further extensions by hand [21]. Given an aligned error corpus, it is also possible to construct a weighted error model automaton automatically [5].

Checking Strings and Generating Suggestions. String checking is straightforward. It consists of composing the input I with the lexical automaton L and checking whether the output language of the result is empty. If the result is empty, the correction set must be calculated.

A corrected string is a string that can be generated by transforming the input string with the error model and is present in the lexicon. These strings are thus the output language of the composition $I \circ E \circ L$, where E is the error model.

The desired behavior, or result set R, of a spellchecker given input I, a lexicon L and an error model E is thus given by equation 5, where π_2 is the output projection.

$$R = \begin{cases} \pi_2(I \circ L), & \text{if } \pi_2(I \circ L) \neq \emptyset \\ \pi_2(I \circ E \circ L) & \text{otherwise} \end{cases} \tag{5}$$

It is undesirable to compute either of the intermediate compositions $I \circ E$ and $E \circ L$; the former will require futile work (producing strings that are not in the lexicon) and the latter, though possible to precompute, will be very large (see table 8).

Table 8 Size of two lexical transducers composed with Damerau-Levenshtein edit distance transducers, all results minimized

Transducer	states	transitions	SFST file size
Morphalou (French)	77.0K	190.7K	1.7MB
With edit distance 1	9.6K	733.6K	5.7MB
With edit distance 2	18.2K	344.7K	28MB
OMorFi (Finnish)	203.8K	437.9K	4.0M
With edit distance 1	17.7K	16186.6K	120MB
With edit distance 2	30.5K	53738.2K	410MB

To circumvent these problems, a three-way on-line composition of I, E and L was implemented (distributed as `hfst-ospell` under the GPL and Apache licenses). It is along the lines of a more general n-way composition described by Allauzen and Mohri in [1]. The present algorithm is, however, considerably simpler, due to certain implementation details. Firstly, we use an efficient and compact indexing transducer format (`hfst-optimized-lookup`, documented in [25], obviating the need to process the transducers involved into hash tables. Secondly, in the present application it may be guaranteed that no special handling of epsilons is necessary.

We write $I = (Q^I, 0^I, T^I, \Delta^I)$, where Q is the set of states, 0 is the starting state, T is the set of terminal states and Δ is the set of transitions (triples of (state, symbol pair (input and output), state)), and similarity for E and L.

Eliding for the time being weights and flag diacritics, states in the composition transducer are triples of states of the component transducers. Analogously with two-way composition, the starting state is $(0^I, 0^E, 0^L)$ and the edge $\delta = (q_1, (\sigma_1, \sigma_2), q_2)$ exists if edges $(q_1^I, (\sigma_1, \sigma'), q_2^I)$, $(q_1^E, (\sigma', \sigma''), q_2^E)$ and $(q_1^L, (\sigma'', \sigma_2), q_2^L)$ exist in the respective transducers for some $\sigma' \in \Sigma^I \cap \Sigma^E, \sigma'' \in \Sigma^E \cap \Sigma^L$, where $q_n = (q_n^I, q_n^E, q_n^L)$ and $\sigma_1 \neq \epsilon \neq \sigma_2$.

The set of the states of the composition, Q, are the reachable subset (in the sense of having a path from 0 to the state) of $Q^I \times Q^E \times Q^L$.

If in a given state any of the component transducers has an edge involving epsilon and the other symbols match as above, the resulting composed edges go to states such that any state in a transducer to the left of an input epsilon or to the right of an output epsilon is unchanged, eg. if we have $(q_1^I, (\sigma, \sigma'), q^{I'})$, $(q_1^E, (\epsilon, \sigma''), q_2^E)$ and $(q_1^L, (\sigma'', \sigma'''), q_2^L)$, a successor state of q_1 will be (q_1^I, q_2^E, q_2^L).

There is no special handling of cases where an epsilon output and an epsilon input occur simultaneously in consecutively acting transducers. The algorithm is however guaranteed to terminate in the present application as long as epsilon cycles do not occur on the input side of the transducers and I is acyclic. That being the case, the state reached in I is a loop variant that increases (towards termination) whenever E consumes input. For this not to happen for an indefinitely large number of iterations, either E itself would have to have an input epsilon cycle, or indefinitely many states in the composition would have to be created such that the state of E is unchanged. This would require L to have an input epsilon cycle.

The final states are the reachable subset of $T = T^I \times T^E \times T^L$.

It can trivially be verified that this is equivalent to taking the two compositions $(I \circ E) \circ L$.

This three-way composition is computed in a breadth-first manner with a double-ended queue of states. The queue is initialized with the starting state, and the target of every edge is computed and pushed onto the queue. The starting state is then discarded, and the process is repeated with a new state popped from the queue until the queue is empty.

Conceptually, the state space of $E \circ L$, which contains all the misspellings (in the sense of E) of all the entries in the lexicon, is explored in such a way that only the states visited when looking up I are generated.

For this process to be guaranteed to terminate, it is sufficient that none of the component transducers have input-epsilon loops and that the input transducer accepts strings of only finite length. This is because every newly generated state will either have a shorter sequence of edges to traverse in I ("increment q^I") or be closer to requiring an edge in I to be traversed (due to a finite sequence of epsilon edges becoming shorter), establishing a loop variant.

Weights representing the probability of a particular correction being the correct one are a natural extension, and in `hfst-ospell` correspond to multiplication in the tropical semiring (for details see [2]) of the weight each edge traversed. Multiplication in this semiring is the standard addition operation of positive real numbers, which we approximate by addition of `floats`. Each state in the queue is recorded with an accumulated weight, and its successor states have this weight incremented by the sum of the weights of the edges traversed in the component transducers.

The alphabet of I cannot be determined in advance, and in practice is taken to be the set of Unicode code points. To allow the error model to correct unexpected symbols in this large space, `hfst-ospell` uses a special symbol, `@_UNKNOWN_SYMBOL_@` which is taken to be equal to any symbol that is otherwise absent.

Error Model Tool and Optimizations. For the most common case of generating Levenshtein and Damerau-Levenshtein (in which transposition of adjacent symbols constitutes one operation) distance error models, a tool (`editdist.py`) and definition format was developed.

The definition format serves the purposes of minimizing the number of symbols used in the error model (the number of transitions is $O(|\Sigma|^2)$) and introducing weights for edits. Typical morphologies have a number of unusual characters (punctuation, special symbols) or internally used symbols (e.g. flag diacritics) that should be filtered for more efficient correction. This is accomplished by providing a facility for reading the alphabet from a transducer, ignoring any symbols of more than one Unicode code point, and reading further ignorable symbols from a configuration file.

The configuration file allows specifying weights to be added to any edit involving a particular character or a particular edit operation (for example, assigning a low weight to the edit o → ö for an OCR application).

Certain characteristics of the correction task permit efficiency-oriented improvements to error models. A naive Levenshtein error model with edit distance 2 in `ospell` would, when given the word word `arrivew` where the French word `arriver` was intended, do the following useless work, where e.g. a:0 means output epsilon for input a:

```
a:0 0:a r r i v e    [failure]
0:a a:0 r r i v e    [failure]
a r:0 0:r r i v e    [failure]
a 0:r r:0 r i v e    [failure]
...
a r r i v e w:0      [success]
a r r i v e w:z      [success]
a r r i v e w:r      [success]
...
```

When the correctable error is near the end, almost every symbol is deleted and inserted with no effect. This may be circumvented by adding, for each deletion and insertion, a special successor state from which its inverse is absent.

Ranking Suggestions. When applying the error model and the language model to input with spelling errors, the result is typically an ordered set of corrected strings with some probability associated with each correction. After applying the contextless error correction described earlier, it is possible to use context words and their potential analyses to re-rank the corrections [22].

6 Discussion and Future Work

6.1 Synonym and Translation Dictionaries

Other finite-state applications created with HFST include inflecting thesauri and translation dictionaries. These applications have been created from the bilingual parallel Princeton WordNet and FinnWordNet. The creation of FinnWordNet is documented in [12]. FinnWordNet contains roughly 150,000 word meanings in Finnish with their English translations. The synonym dictionaries for Finnish and English and the Finnish-English and English-Finnish translation dictionaries as well as their demos can be found on `http://hfst.sf.net/`.

The inflecting synonym dictionaries were created as a composition of three transducers: (1) a morphological analyzer, (2) a transducer that replaces one word with another while copying the inflection tags and (3) a morphological generator as an inversion of the morphological analyzer. The translation dictionaries only have components (1) and (2). In the future, we intend to take advantage of the weighted transducers to introduce contexts so as to be able to suggest the most likely synonym or the most likely translation in context.

6.2 Extending Transducers for Pattern Matching

Advanced, fast and flexible pattern matching is a major requirement for a variety of tasks in information extraction, such as named entity recognition. An approach to this task was presented by Lauri Karttunen in [9], and an outline for implementing it in HFST is given here.

Some Desiderata for Pattern Matching. Several patterns, including nested matches, should be able to operate in parallel, and it should be possible to impose contextual requirements (rules) on the patterns to be matched. Matching should be efficient in space and time — in particular, it should be possible to avoid long-range dependencies which are awkward for FST transformations. A powerful system should also have a facility for referring to common subpatterns by name.

The pmatch Approach. For a more detailed overview the reader is directed to [9]. Here we focus on the aspects of `pmatch` that necessitate extensions to a FST formalism from the point of view of the implementation.

`pmatch` is presented as a tool for general-purpose pattern matching, tokenizing and parsing. It is given a series of definitions and a specialized regular expression, and it then operates on strings, passing through unmodified any parts of them that fail to match the expression, and applying transformations to any matched parts. If several matches beginning from a common point in the input can be made, matches of less than maximal length are considered invalid.

The expressions may refer to other expressions (possibly combined with each other by operations on regular languages), contextual conditions (left or right side, with negation, `OR` and `AND`) and certain built-in transformation rules. The most interesting of these transformation rules is `EndTag()`, which triggers a wrap-around XML tag to be inserted on either side of the match.

Referencing other regexes (including self-reference) is unrestricted, so the complete system has the power of a recursive transition network (RTN), and matching is therefore context-free.

The crucial extension-demanding features for an HFST utility with similar applications are:

- A distinction between an augmented universal transducer (the top level which echoes all input in the absence of matches) and sub-transducers
- Ignoring non-maximal matching, ie. a left-to-right longest-distance matcher
- An unrestricted system for referencing other transducers by name
- Special handling of `EndTag()` and instructions for context checking
- Reserved symbols for implementing transducer insertion/reference, `EndTag()` and contexts

The referencing system is apparently the only one of these that would be impossible to implement in a strict FST framework; the other extensions suggest compilations to larger, possibly less efficient transducers.

7 Conclusion

HFST—Helsinki Finite-State Technology (`http://hfst.sf.net/`) is a framework for compiling and applying linguistic descriptions with finite-state methods. We have demonstrated how HFST uses finite-state techniques for creating runtime morphologies, taggers, spellcheckers, inflecting synonym dictionaries as well as translation dictionaries using one open-source platform which supports extending the descriptions with statistical information to allow the applications to take advantage of context. HFST offers a path from language descriptions to efficient language applications.

Acknowledgements. We wish to acknowledge the FIN-CLARIN and META-NORD projects for their financial support as well as HFST users for their many constructive suggestions. Additonally, the research leading to some of these results has received funding from the European Commission's 7th Framework Program under grant agreement n° 238405 (CLARA). Miikka Silfverberg's work is funded by LANGNET.

References

1. Allauzen, C., Mohri, M.: N-way composition of weighted finite-state transducers. International Journal of Foundations of Computer Science 20, 613–627 (2009)
2. Allauzen, C., Riley, M., Schalkwyk, J., Skut, W., Mohri, M.: OpenFst: A General and Efficient Weighted Finite-State Transducer Library. In: Holub, J., Žďárek, J. (eds.) CIAA 2007. LNCS, vol. 4783, pp. 11–23. Springer, Heidelberg (2007), `http://www.openfst.org`
3. Beesley, K.R., Karttunen, L.: Finite State Morphology. CSLI publications (2003)

4. Brants, T.: TnT - a statistical part-of-speech tagger. In: Proceedings of the Sixth Applied Natural Language Processing (ANLP 2000), Seattle, WA (2000)
5. Brill, E., Moore, R.C.: An improved error model for noisy channel spelling correction. In: ACL 2000: Proceedings of the 38th Annual Meeting on Association for Computational Linguistics, pp. 286–293. Association for Computational Linguistics, Morristown (2000)
6. Çöltekin, Ç.: A freely available morphological analyzer for Turkish. In: Proceedings of the 7th International Conference on Language Resources and Evaluation, LREC 2010 (2010)
7. Gerdemann, D., van Noord, G.: Transducers from re-write rules with backreferences. In: Proceedings of the EACL Conference, pp. 126–133 (1999)
8. Halácsy, P., Kornai, A., Oravecz, C.: Hunpos—an open source trigram tagger. In: ACL 2007, Prague, Czech Republic (2007)
9. Karttunen, L.: Beyond morphology: Pattern matching with FST. In: Mahlow, C., Piotrowski, M. (eds.) SFCM 2011. CCIS, vol. 100, pp. 1–13. Springer, Heidelberg (2011)
10. Kempe, A., Karttunen, L.: Parallel replacement in finite state calculus. In: The Proceedings of the 16th International Conference on Computational Linguistics, pp. 622–627 (1996)
11. Lindén, K., Axelson, E., Hardwick, S., Pirinen, T.A., Silfverberg, M.: HFST—Framework for Compiling and Applying Morphologies. In: Mahlow, C., Piotrowski, M. (eds.) SFCM 2011. CCIS, vol. 100, pp. 67–85. Springer, Heidelberg (2011)
12. Lindén, K., Carlson, L.: Finnwordnet—wordnet på finska via översättning. LexicoNordica 17 (2010)
13. Lindén, K., Pirinen, T.: Weighting finite-state morphological analyzers using hfst tools. In: FSMNLP 2009 (2009)
14. Lindén, K., Silfverberg, M., Pirinen, T.: HFST Tools for Morphology – An Efficient Open-Source Package for Construction of Morphological Analyzers. In: Mahlow, C., Piotrowski, M. (eds.) SFCM 2009. CCIS, vol. 41, pp. 28–47. Springer, Heidelberg (2009)
15. Mays, E., Damerau, F.J., Mercer, R.L.: Context based spelling correction. Inf. Process. Manage. 27(5), 517–522 (1991)
16. Norvig, P.: How to write a spelling corrector. Web Page (2010), `http://norvig.com/spell-correct.html` (visited February 28, 2010)
17. Oflazer, K.: Error-tolerant finite-state recognition with applications to morphological analysis and spelling correction. Computational Linguistics 22(1), 73–89 (1996)
18. Oravecz, C., Dienes, P.: Efficient stochastic part-of-speech tagging for Hungarian. In: Proceedings of the Third International Conference on Language Resources and Evaluation, Las Palmas, pp. 710–717 (2002)
19. Pirinen, T.: Suomen kielen äärellistilainen automaattinen morfologinen analyysi avoimen lähdekoodin menetelmin. Master's thesis, Helsingin Yliopisto (2008), `http://www.helsinki.fi/~tapirine/gradu/`
20. Pirinen, T.A., Lindén, K.: Creating and weighting hunspell dictionaries as finite-state automata. Investigationes Linguisticae 19 (2010)
21. Pirinen, T.A., Lindén, K.: Finite-state spell-checking with weighted language and error models. In: Proceedings of the Seventh SaLTMiL Workshop on Creation and Use of Basic Lexical Resources for Less-Resourced Languagages, Valletta, Malta, pp. 13–18 (2010)

22. Pirinen, T.A., Silfverberg, M., Lindén, K.: Context-sensitive spelling correction. In: Gelbukh, A. (ed.) International Conference on Intelligent Text Processing and Computational Linguistics, New Delhi, India (2012)
23. Savary, A.: Typographical Nearest-Neighbor Search in a Finite-State Lexicon and Its Application to Spelling Correction. In: Watson, B.W., Wood, D. (eds.) CIAA 2001. LNCS, vol. 2494, pp. 251–260. Springer, Heidelberg (2003)
24. Silfverberg, M., Lindén, K.: Conflict resolution using weighted rules in hfst-twolc. In: Proceedings of the 17th Nordic Conference of Computational Linguistics NODALIDA 2009, pp. 174–181. Nealt (2009)
25. Silfverberg, M., Lindén, K.: Hfst runtime format—a compacted transducer format allowing for fast lookup. In: Watson, B., Courie, D., Cleophas, L., Rautenbach, P. (eds.) FSMNLP 2009 (July 13, 2009), `http://www.ling.helsinki.fi/~klinden/pubs/fsmnlp2009runtime.pdf`
26. Silfverberg, M., Lindén, K.: Part-of-speech tagging using parallel weighted finite-state transducers. In: Proceedings of the 7th International Conference on NLP, IceTAL 2010 (2010)
27. Silfverberg, M., Lindén, K.: Combining statistical models for POS tagging using finite-state calculus. In: Proceedings of the 18th Conference on Computational Linguistics, NODALIDA 2011, pp. 183–190 (2011)
28. Tzoukermann, E., Radev, D.: Using word class for part-of-speech disambiguation. In: Proceedings, Fourth Workshop on Very Large Corpora WVLC 1996, Copenhagen, Denmark (1996)
29. Yli-Jyrä, A.: Transducers from parallel replace rules and modes with generalized lenient composition. In: Finite-State Methods and Natural Language Processing (2008), `http://www.ling.helsinki.fi/users/aylijyra/all/YliJyra-2008b:trafropar:inp.pdf`
30. Zanchetta, E., Baroni, M.: Morph-it! a free corpus-based morphological resource for the Italian language. Corpus Linguistics 1(1) (2005)
31. Zielinski, A., Simon, C.: Morphisto – an open source morphological analyzer for German. In: Proceeding of the 2009 Conference on Finite-State Methods and Natural Language Processing: Post-Proceedings of the 7th International Workshop FSMNLP 2008, pp. 224–231. IOS Press, Amsterdam (2009)

PSI-Toolkit: A Natural Language Processing Pipeline

Filip Graliński, Krzysztof Jassem, and Marcin Junczys-Dowmunt

Abstract. The paper presents the main ideas and the architecture of the open source PSI-Toolkit, a set of linguistic tools being developed within a project financed by the Polish Ministry of Science and Higher Education. The toolkit is intended for experienced language engineers as well as casual users not having any technological background. The former group of users is delivered a set of libraries that may be included in their Perl, Python or Java applications. The needs of the latter group should be satisfied by a user friendly web interface. The main feature of the toolkit is its data structure, the so-called PSI-lattice that assembles annotations delivered by all PSI tools. This cohesive architecture allows the user to invoke a series of processes with one command. The command has the form of a pipeline of instructions resembling shell command pipelines known from Linux-based systems.

1 Introduction

1.1 PSI-Toolkit – General Outline

PSI-Toolkit consists of a set of tools (called processors) that aim at processing natural language texts. There are three types of processors: readers, annotators, and writers. A reader creates the main data structure, the so-called PSI-lattice (see Section 2.), from an external source of information, e.g. from a file (a text file, HTML file, PDF file) or keyboard. An annotator (e.g. a tokeniser, a lemmatiser, a parser) adds new annotations in the form of new edges to the PSI-lattice. Finally, a writer writes back a PSI-lattice to an output device (e.g a file or screen).

Filip Graliński · Krzysztof Jassem · Marcin Junczys-Dowmunt
Faculty of Mathematics and Computer Science
Adam Mickiewicz University
ul. Umultowska 87, 61–614 Poznań, Poland
e-mail: {filipg,jassem,junczys}@amu.edu.pl

A. Przepiórkowski et al. (Eds.): *Computational Linguistics*, SCI 458, pp. 27–39, 2013.
DOI: 10.1007/978-3-642-34399-5_2

1.2 An Overview of NLP Toolkits

The first widely used NLP toolkit was GATE (General Architecture for Text Engineering) [3] designed in the mid-nineties and still being developed. Currently, dozens of open-source toolkits are available; Wikipedia lists over 30 of them.[1] Most toolkits deal with the English language, but toolkits designed for other languages (like TESLA – German [6], Nooj – French [10], Apertium – Spanish [1]) gain popularity. The situation is a bit different for Slavic languages. In spite of the rapid progress in the development of linguistic tools and resources for Slavic languages in the 21st century, attempts at organizing them in the format of cohesive toolkits are still sparse and rare.

Many NLP toolkits are hard to use not only for casual users, but also for experts. It usually takes time to build and configure a NLP toolkit (NLP toolkits are rarely packaged for popular Linux distributions) and to learn how to do anything with a toolkit. Such obstacles are trivial, but they can discourage users. For instance, in Stanford CoreNLP toolkit[2] one needs to edit a specific line in a configuration file to select annotators (modules) and issue a cryptic command with over 210 characters to run them. It is even more difficult to pass additional options for Stanford CoreNLP annotators.

1.3 PSI-Toolkit Assumptions

1.3.1 Free Licence

PSI-Toolkit is released under General Public License (GPL). Most parts of the toolkit (its core and modules created within the PSI-Toolkit project) are also available under Lesser General Public License (LGPL), which, in addition to GPL, allows for linking with proprietary software and further distributing the result under any terms (also for business purposes).

1.3.2 Easy Access via a Web Browser

All PSI-Toolkit tools may be accessed and tested on-line via a web browser interface at `http://psi-toolkit.wmi.amu.edu.pl`. Figure 1 shows an exemplary usage of the toolkit in a web window. The user inputs a text into an edit box (e.g. *Electric Light Orchestra*) and specifies a command as a sequence of processors (whose names are separated with exclamation marks) that should be executed on the text (e.g. `txt-reader ! tp-tokenizer --lang en ! psi-writer`).The processors are run in the order specified in the command (here: read a raw text, tokenise the text according to rules of the English language, write the output in the dedicated PSI format). The PSI output lists every edge of the PSI-lattice (see: 2.1) in a separate line. In Figure 1: line 01: corresponds to the edge spanning

[1] `http://en.wikipedia.org/wiki/List_of_natural_language_processing_toolkits`
[2] `http://nlp.stanford.edu/software/corenlp.shtml`

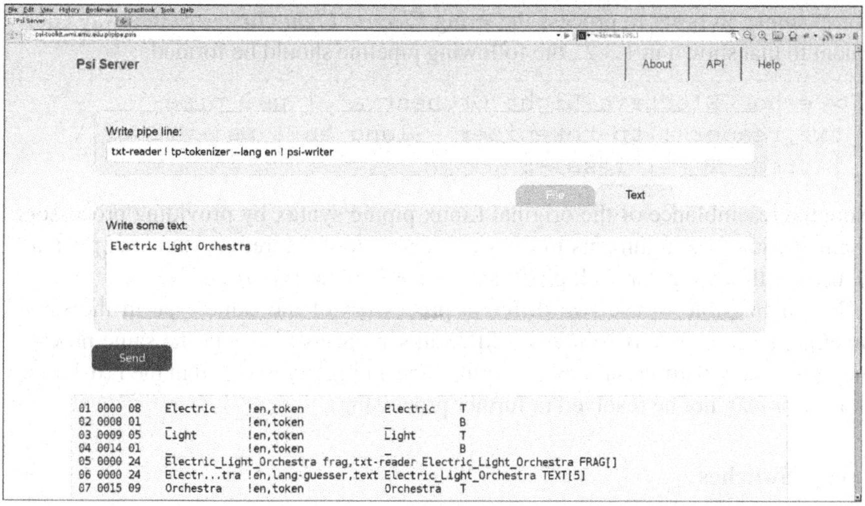

Fig. 1 Web access to PSI-Toolkit (`http://psi-toolkit.wmi.amu.edu.pl`)

over the first 8 characters of the input (start position is equal to 0000, offset is equal to 0008). The type of the edge is "token", the value of the token is *Electric*, the lemma the token belongs to is *Electric*. Line 02 describes a space between the first and the second token of the input. Line 05 corresponds to the edge spanning over the whole input (the edge has been constructed by the `txt-reader`).

The format of the output may be simplified by replacing the `psi-writer` with `simple-writer` in the command line. Combining the simple output format with other PSI-Toolkit processors may result, for instance, in on-line machine translation of the input, yielding a use case similar to that offered by Apertium.

1.3.3 Linux Distribution

The PSI-Toolkit is also distributed in the form of two Linux binaries: `psi-pipe` and `psi-server`. The `psi-pipe` program may be used as a command-line tool on personal computers, whereas `psi-server` allows for the creation of other PSI-Toolkit web pages. Currently, the PSI-Toolkit binaries are distributed in form of easily installable packages for the Arch Linux, Debian and Ubuntu Linux distributions. In our opinion, packaging a NLP toolkit is a simple yet crucial step for making people start using your toolkit.

1.3.4 PSI-Toolkit Pipeline

The PSI-Toolkit command is specified as a pipeline of processors (an example of such a pipeline was given in 1.3.2). If the PSI-Toolkit is used under Ubuntu on a personal computer, the processors should be invoked in a bash-like manner.

For example, in order to process the string *Electric Light Orchestra* in a way equivalent to that shown in 1.3.2., the following pipeline should be formed:

```
> echo "Electric Light Orchestra" | psi_pipe \
txt_reader ! tp-tokenizer --lang en ! psi-writer \
| grep "T_" | less -S
```

Note the resemblance of the original Linux piping syntax by providing processors chained with `!` as arguments to the `psi-pipe` tool. Moreover, `psi-pipe` may be used within a regular shell pipeline (e.g. along with `grep`).

It is admissible to use two different processors of the same type in the same pipeline. For example running two different sentence splitters in the same process may result in two different sentence splits. The ambiguity is stored in the PSI-lattice (it may or may not be resolved in further processing).

1.3.5 Switches

Usage of a PSI-Toolkit processor may be customised with switches (options). Figure 2 shows the list of available switches for the `perl-simple-writer` processor, the "Perl-wrapped" version of the PSI `simple-writer`.

```
Allowed options:
  --linear              skips cross-edges
  --no-alts             skips alternative edges
  --with-blank          does not skip edges with whitespace text
  --tag arg (=token)    basic tag
  --spec arg            specification of higher-order tags
  --with-args           if set, returns text with annotation as
a hash element
```

Fig. 2 Switches of the Perl-simple-writer processor

1.3.6 Java, Perl, Python Libraries

PSI tools are also accessible as libraries of selected programming languages. The user may include the PSI tools in any of the Java, Perl or Python source codes. For instance, in order to tokenise a sentence in Perl, one could write:

```
use PSIToolkit::Simple;

my $runner = PSIToolkit::Simple::PipeRunner->new(
      'tp-tokenizer --lang en ! perl-simple-writer');

my $result = $runner->run_for_perl(
      'To be, or not to be, that is the question');
```

A list of tokens will be returned simply as an array reference.

1.3.7 Natural Languages Processed by PSI-Toolkit

There are no restrictions on languages analysed by the PSI processors (UTF-8 encoding is used uniformly). One PSI pipeline may consist of processors defined for various languages. However, the tools delivered by the authors are oriented mainly towards Polish (some processors are also defined for English). The authors hope that the PSI-Toolkit can bring together most Polish language processing tools in one framework. Currently these tools are dispersed (see `http://clip.ipipan.waw.pl` for an exhaustive list of NLP tools and resources for Polish). The proof of this concept has been implemented on the morphosyntactic level: the external morphological tagger Morfologik[3] has been incorporated into the PSI-Toolkit and can be run in the PSI pipeline instead of or besides (!) the standard morphological analyser of the PSI-Toolkit.

2 PSI-Lattice

2.1 Definition

All PSI-Toolkit processors operate on a common lattice-based data structure called *PSI-lattice*. The term *lattice* refers to a (word) lattice [4] as used in natural language processing rather than to a more general notion of abstract algebra. A PSI lattice is composed of an input (*substrate*) string and *edges* spanning substrings of the substrate string.

Lattice readers read the substrate string (usually from a file) and add some initial edges – usually edges spanning single *symbols* (a symbol is a character occurring as part of a natural-language text) and edges representing mark-up tags of a given format (e.g. the PSI-Toolkit HTML reader would construct edges encapsulating HTML tags). PSI-Toolkit annotators create new PSI-lattice edges based on existing ones and/or the substrate string, e.g. a tokeniser groups symbol edges into token edges, a lemmatiser creates edges representing lemmas and lexemes for each token edge, a parser produces new parse edges based on the lexeme edges and previously added parse edges. Finally, lattice writers do not add any new edges, they just output all or selected PSI-lattice edges in a required format.

A PSI-lattice edge consists of the following elements:

- source and target *vertices* (PSI-lattice vertices are defined as inter-character points),
- *annotation item*,
- *layer tags*,
- *partitions*,
- *score* (weight).

[3] Available at `http://morfologik.blogspot.com`. To our knowledge, this tool has not been described in a published scientific paper.

The annotation item conveys the description of the language unit represented by a given edge. An annotation item is realised as an attribute-value matrix in which two attributes are obligatory: *category* and *text*. The meaning and interpretation of these two attributes varies between the PSI-Toolkit processors, e.g. the category of a token edge is its type (blank, word, punctuation mark etc.) and its text attribute is just the token itself (as a string), whereas for a lexeme edge the part of speech of the given lexeme is used as the category attribute and its identifier (e.g. `long+adj` for the adjective *long*) – as the text attribute. Other attributes are used to describe particular features of the given language unit, e.g. morphosyntactic features such as case, gender, tense, person etc.

Layer tags are used to express some meta-information associated with a given edge, e.g.:

- edge type – whether an edge represents a token, lemma, lexeme, parse etc.,
- the name of the processor that added the edge,
- tagset used in the annotation item.

The important point is that edges with the same annotation item are collapsed into a single edge, even if they have different layer tags (with the exception of plane tags – more on this later). In other words, if a processor produces an edge with the same annotation item and the same source and target vertices as some edge already present in the PSI-lattice, then no new edge is added to the PSI-lattice, the two edges are merged instead, the sum of their layer tags is assigned to the updated edge.

As it is not always reasonable to collapse two edges with the same annotation item (for instance in the context of machine translation it would not make sense to equate a source-language lexeme and its target-language equivalent when they happen to have the same canonical form and the same attributes), so-called *plane (layer) tags* are introduced. Plane tags divide a PSI-lattice into a set of disjoint *planes*, i.e. edges belonging to different PSI-lattice planes (having different plane tags) will not be collapsed even if they share the same annotation item. By convention, plane tags begin with an exclamation mark. Language-code tags specifying the language of a given language unit (e.g. `!pl`, `!en`, `!de` tags) are typical examples of plane tags. When a processor combines some edges into a new edge, the new edge will inherit the plane tags of the subedges by default, unless a list of plane edges was specified explicitly while creating a new edge.

A *partition* specifies which edges were used to create a given edge. For example for a parse edge the partition is a sequence of lexemes (terminals) and subparses (non-terminals) directly combined into the given edge. An edge may have more than one partition, e.g. an edge spanning the expression *Electric Light Orchestra* may result from a parse partitioned into *Electric* + *Light Orchestra* or into *Electric Light* + *Orchestra*, or it could be a lexeme produced by a multi-word unit lexicon (partitioned into tokens in this interpretation). Each partition is assigned the following properties: layer tags, score and (optionally) rule identifier. *Rule identifier* is an arbitrary number the interpretation of which varies between the PSI-Toolkit processors, e.g. for a parser it could be an identifier of a grammar rule (i.e. a partition could be linked to a rule of the parser's grammar this way).

The score (of an edge or a partition) is a floating point value for which the following properties hold:

- the score of an edge is the maximum score of its partitions,
- the score of a partition is the sum of scores of its subedges plus some score for the rule that generated the partition.

For instance, a score for a parse edge/partition could be interpreted as the log probability of the parse and the score of a parse partition could be calculated as the sum of log probabilities of its subedges and the log probability of the grammar rule applied. PSI-lattice scores, however, does not have to be interpreted as (log) probabilities. They might be treated as some kind of weights or penalties (the latter for negative scores).

Taking the maximum score of the partitions (rather than the sum of the partition scores) as the score of an edge might seem controversial from a formal point of view. The reason why we decided to use the maximum value is that the partitions do not have to be independent. For example we would like to run two different lemmatisers on the same words.

2.2 *Motivation and Design Assumptions*

In other toolkits and even more so in the case of independent stand-alone tools the internal data representation differs in general from the input or output data. In consequence, the internal representation is also different between two kinds of tools, e.g. a tokeniser works mostly on raw string data, while a parser works on trees, forests or possibly charts, a statistical machine translation like Moses uses implicit hypothesis graphs etc.

During the conversion of the internal representation to a readable output format usually a lot of information is purposefully discarded. This is the case if only the first-best interpretation for an ambiguous problem is provided, for instance the most probable tokenisation or the best-scored translation. Alternative interpretations can often only be retrieved by analysing log data, the format of which is most certainly not standardised between different tools or toolkits. Providing alternative interpretations to a follow-up tool involves then a lot of manual intervention.

The PSI-lattice is designed to tackle these problems. Firstly, the PSI-lattice is supposed to provide a one-size-fits-all data structure that can be used to contain any annotation generated by various language processing tools without losing information provided by previous processing steps. Secondly, a standardised way to pass around alternate interpretations besides the single-best results is provided. That way, one can easily take advantage of delayed disambiguation, i.e. a higher-order tool can choose among alternatives that a lower-order tool was not able to fully disambiguate. Examples might be a parser that chooses between two part-of-speech tags or a syntax-based machine translation application that translates parse-forests rather than parse trees. Finally the PSI-lattice obsoletes the conversion from an internal data representation to an external output format as it can itself serve for internal data representation, being perfectly capable of representing alternative tokenisations, parse forests or translation ambiguities. This will be illustrated in the next section where a shallow parser directly constructs a PSI-lattice as output data while using it as an intermediate representation of parses obtained so far.

The ease of data interchange between atomic application leads to the second important design guideline of the presented toolkit: simplified combination of tools. Using available toolkits can be a challenge by itself as their special syntax or construction can result in quite a steep learning curve. For instance, the NLTK requires at least a basic knowledge of the Python programming language. For the PSI-Toolkit it has been decided to walk down a path that has already been cut through — in a context much broader than natural language processing.

As far as basic text-processing capabilities are concerned, the popular shells in Linux or other Unix-based operation systems can themselves be considered as capable text-processing toolkits. The rationale behind them is to provide a set of small applications or commands where each tool by itself is limited in functionality in the sense that it can perform only exactly one task. However, each tool is supposed to perform particularly well for the task it has been designed for. Typically, data for a tool is provided on its standard input while the results can be read from its standard output. The power of the command-line comes then from the possibility to combine these tools into pipelines by directing standard output from one tool to the standard input of another tool, thus building up a much more sophisticated application from smaller parts. For instance, one can easily create basic sorted frequency lists of words in a text file combining `sed`, `sort`, `uniq` into a pipeline without the need to write specialised scripts in a more complex programming language like Perl. The operations of each tool in a pipeline can be adjusted by switches — for instance the reverse order of sorting for the `sort` command by `sort -r`.

Due to the popularity of Linux-based systems in the natural language processing community it is safe to assume a high degree of familiarity of the average NLP-researcher with one of the available Linux shells. This assumption leads directly to the second important design decision for our toolkit. Concerning usability it lends itself to exploit this familiarity by simulating the look and feel of the command-line shells in Linux. This includes the construction of a growing set of self-contained single-purpose applications that can be chained into complex pipelines by the use of a familiar looking piping syntax. Also, additional options to any such building-block are provided by switches that again mimic their Linux-shell counterparts.

Every pipeline constructed in such a way constitutes a new stand-alone tool, with well-defined input data, output data and functionality. Formally speaking, a pipeline can be seen as self-contained object similar to a functor, the elements of the pipeline being in turn comparable to combinators that make up the functor. While in Linux shells data is typically passed around in text or plain binary format, the exchange format between the PSI-Toolkit components is the described PSI-lattice.

3 Language Processing with PSI-Lattice

3.1 Segmentation

For both, tokenisation and sentence splitting, the rules created for the Translatica machine translation system[4] were used (the rules had been published under the GNU Lesser General Public Licence). The tokeniser (the processor is called `tp-tokenizer`) uses a Translatica in-house format for "cutting-off" rules (each rule specifies a regular expression describing a token of a given type, only the first matching rule is applied), whereas the sentence splitter (called `srx-segmenter`) uses the SRX (Segmentation Rules eXchange) standard[5].

It is quite straightforward to store alternative segmentations (on both token and sentence level) in a PSI-lattice and to take them into account in the subsequent processing stages (in lemmatisation, parsing etc.). For the time being, both `tp-tokenizer` and `srx-segmenter` produce only one segmentation, as it is not possible to express segmentation non-determinism in either the tokenisation or SRX rules.

Since there are hard-to-disambiguate cases for token/sentence segmentation — e.g. in Polish *gen.* is either an abbreviation for *generał* (= *general*, a military rank) or the word *gen* (= *gene*) at the end of a sentence — i.e. cases in which the decision must be postponed to a later processing stage[6], we are considering enhancing the segmentation rules with some non-determinism. In the case of sentence breaking, however, it would involve extending the widely-used SRX standard. For the time being, segmentation ambiguity could be achieved (at least to some extent) with running segmentation processors twice with slightly different set of rules (e.g. once with *gen.* listed as an abbreviation, once – not listed).

Both `tp-tokenizer` and `srx-segmenter` can be run with an option specifying the maximum length of, respectively, a token and a sentence. In fact, there exist two types of length limits: a soft one and a hard one – in case of the soft limit a token/sentence break is forced only on spaces, whereas exceeding the hard limit always triggers segmentation. Such limits were introduced for practical reasons as a safeguard against extremely long tokens/sentences which may occur when "garbage" (e.g. unrecognised binary data) is fed to PSI-Toolkit (very long tokens or sentences might slow down the subsequent processing to an unacceptable degree).

So far, PSI-Toolkit handles tokenisation and sentence breaking for English, French, German, Italian, Polish, Russian and Spanish.

3.2 Lemmatisation and Lexica

We plan to incorporate as many open source lemmatisers and lexica as possible into PSI-Toolkit. So far, we have created a general framework for adding new

[4] `http://translatica.pl`
[5] `http://www.gala-global.org/oscarStandards/srx/srx20.html`
[6] E.g. In named entity recognition or in parsing, see [2] for discussion of tokenisation ambiguity.

lemmatisers into PSI-Toolkit (now it is relatively easy to add a new lemmatiser on condition that a simple function returning all the morphological interpretations is provided by a given lemmatiser) and incorporated the aforementioned Morfologik lemmatiser for Polish. We are in the process of developing our own finite state library (which will be used not only for lemmatisation, but also for syntax-based machine translation) and adding support for the Stuttgart Finite State Transducer Tools (SFST) [9].

3.3 Shallow Parsing

The PSI-Toolkit includes a shallow parser – Puddle – that can be used to work with any language as long as an appropriate grammar is provided. It started out as a C++ adaptation of the Spejd [8] shallow parser which was a pure Java tool at that time. By now, Puddle has evolved into an independent tool that has been used as a parser for French, Spanish, and Italian in the syntax-based statistical machine translation application Bonsai [7].

The latest version of Puddle has been redesigned to work with the PSI-lattice as an input and output data structure. The parse tree itself is also constructed directly on top of the input lattice. Consecutive iterations work on a PSI-lattice that has been extended by exactly one edge in the previous iteration.

The shallow parsing process of Puddle relies on a set of string matching rules constructed as regular expressions over single characters, words, part-of-speech tags, lemmas and grammatical categories etc. Apart from the matching portion of a rule, it is also possible to define matching patterns for left and right contexts of the main match. The parse tree construction process is linear, matching rules are applied iteratively in a deterministic fashion. The first possible match is chosen and a spanning edge is added to the lattice. No actual search is performed which puts a lot of weight onto the careful design of the parsing grammar. The order of the rules in a grammar determines the choice of rules to be applied to a sentence. The parsing process is finished if no rules can be applied during an iteration.

For the parser to work properly on different types of information, the PSI-lattice already needs to contain edges for tokens, lemmas and morphological properties, so it has to be the result of a pipeline that generated this kind of data. The lattice does not need to be previously disambiguated (although that can be helpful and implemented by adding one or more POS-Taggers to the pipeline) since the parser can also work as a disambiguation tool. Matching rules that require morphological agreement between matched symbols (e.g. between a noun and an adjective) can mark edges that contradict this agreement requirement as `discarded`. However, discarded edges are never quite deleted from the PSI-lattice since they can be of importance in later higher-order processing.

The parser adds a special type of edge to the lattice marked with `parse` tag. The partition of the edge contains information which subedges have been used to construct the new edge. If there are several possible interpretations all of them are added to the lattice. Syntactic heads are marked with additional tags in their respective edges.

4 Work-Flow: An End-to-End Example

In this section we will illustrate an example pipeline for a short Polish phrase *przykładowa analiza* (Eng. *example analysis*). Figure 3 describes the stepwise construction of the PSI-lattice for this phrase. The following commands have been used to create the corresponding lattices:

a) `txt-reader ! dot-writer`
b) `txt-reader ! tp-tokenizer --lang pl ! dot-writer`
c) `txt-reader ! tp-tokenizer --lang pl ! morfologik ! dot-writer`
d) `txt-reader ! tp-tokenizer --lang pl ! morfologik ! puddle --lang pl ! dot-writer`

The common `txt-reader` is used to convert a text string into an unannotated lattice. The characters of that string are the smallest units in the PSI-lattice (Figure 3a). All figures have themselves been generated by another PSI-Toolkit processor, `dot-writer`, that converts PSI-lattices to graphs described in the DOT-language. This can be interpreted for instance by tools from the GraphViz library [5] available at `http://graphviz.org`. The toolkit contains also a writer named `gv-writer` that uses the GraphViz library to generate PDF or SVG files directly.

A Polish tokeniser (Figure 3b) is inserted between the reader and writer processors and adds the first level of annotation to the PSI-lattice. Edges that span characters mark tokens (T) and blanks (B), other symbol types could include punctuation information or HTML mark-up data. Alternative tokenisation results could be included in the PSI-lattice as well, this would in most cases results in crossing edges.

The third level in the PSI-lattice (Figure 3c) is the result of the application of the Morfologik morphological analyser. Different morphological interpretations are added as individual edges (the morphological features have been omitted in the illustration due to space requirements). The first token *przykładowa* has several interpretations as an adjective (`adj`), similarly *analiza* ha been assigned two interpretations as a noun `subst`.

Finally the last level of annotation (Figure 3d) is added by the shallow parser described in the previous section. The ambiguity of the morphological annotation results in an ambiguous parsing result with two parallel adjective phrase edges (`AP`). Together with the following noun this adjective phrases forms a noun phrase (`NP`) that spans the entire input string. To facilitate the analysis for the shallow parser a part-of-speech tagger can be added between the morphological analyser and the parser. Currently, a simple maximum entropy-based tagger processor is being added to the toolkit. For the moment, the shallow parser represents the highest-order processor in the PSI-Toolkit, but deep parsers, machine-translation processors and many other tools will be available in the near future, as well as converters for existing tools like the mentioned Morfologik processor.

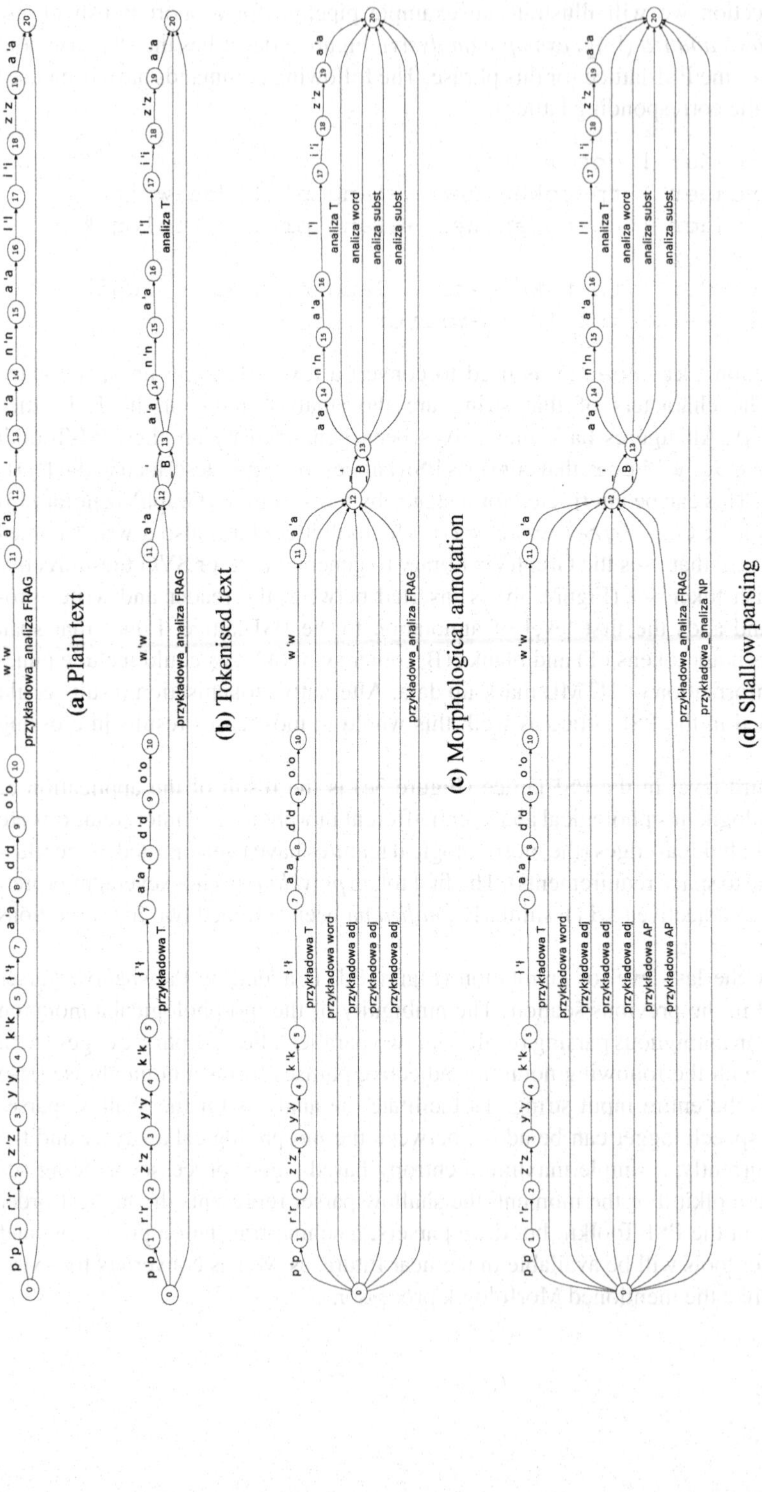

Fig. 3 Stepwise construction of a PSI-lattice

5 Conclusions and Future Work

The paper presents the main ideas and the architecture of the open source PSI-Toolkit. The linguistic annotations are stored in the form of the lattice, which allows for keeping annotations of various types in one structure. The PSI-lattice allows for the delayed disambiguation and the usage of multiple processors of the same type (e.g. a few different sentence splitters) in the same run. The PSI tools may be accessed via the web browser as well as run locally. A processing command should be formed as a pipeline similar to the command pipelines known for Linux shells. The PSI wrappers make it possible to use the PSI tools in Java, Perl or Python applications. The project is under rapid development, ending in April 2013. The list of the processors available in PSI-Toolkit is being systematically increased. At the moment this paper is prepared the list consists of 17 processors. The current list of processors, together with their switches and examples of usage is accessible at `http://psi-toolkit.wmi.amu.edu.pl/help.html`.

Acknowledgements. This paper is based on research funded by the Polish Ministry of Science and Higher Education (Grant No. N N516 480540).

References

1. Carme Armentano-Oller, C., Corbí-Bellot, A.M., Forcada, M.L., Ginestí-Rosell, M., Bonev, B., Ortiz-Rojas, S., Pérez-Ortiz, J.A., Ramírez-Sánchez, G., Sánchez-Martínez, F.: An open-source shallow-transfer machine translation toolbox: consequences of its release and availability. In: OSMaTran: Open-Source Machine Translation, A Workshop at Machine Translation Summit X, pp. 23–30 (September 2005)
2. Chanod, J.P., Tapanainen, P.: A non-deterministic tokeniser for finite-state parsing. In: ECAI 1996 Workshop on Extended Finite State Models of Language, August 11-12 (1996)
3. Cunningham, H., Maynard, D., Bontcheva, K., Tablan, V., Aswani, N., Roberts, I., Gorrell, G., Funk, A., Roberts, A., Damljanovic, D., Heitz, T., Greenwood, M.A., Saggion, H., Petrak, J., Li, Y., Peters, W.: Text Processing with GATE, Version 6 (2011)
4. Dyer, C., Muresan, S., Resnik, P.: Generalizing word lattice translation. In: McKeown, K., Moore, J.D., Teufel, S., Allan, J., Furui, S. (eds.) ACL, The Association for Computer Linguistics, pp. 1012–1020 (2008)
5. Ellson, J., Gansner, E.R., Koutsofios, E., North, S.C., Woodhull, G.: Graphviz and dynagraph static and dynamic graph drawing tools. In: Graph Drawing Software, pp. 127–148. Springer (2003)
6. Hermes, J., Schwiebert, S.: Classification of text processing components: The tesla role system. In: GfKl, pp. 285–294 (2008)
7. Junczys-Dowmunt, M.: It's all about the trees — towards a hybrid syntax-based MT system. In: Proceedings of IMCSIT, pp. 219–226 (2009)
8. Przepiórkowski, A., Buczyński, A.: ♠: Shallow parsing and disambiguation engine. In: Proceedings of the 3rd Language & Technology Conference, Poznań (2007)
9. Schmid, H.: A Programming Language for Finite State Transducers. In: Yli-Jyrä, A., Karttunen, L., Karhumäki, J. (eds.) FSMNLP 2005. LNCS (LNAI), vol. 4002, pp. 308–309. Springer, Heidelberg (2006)
10. Silberztein, M.: Nooj: a linguistic annotation system for corpus processing. In: Proceedings of HLT/EMNLP on Interactive Demonstrations, HLT-Demo 2005, pp. 10–11. Association for Computational Linguistics, Stroudsburg (2005)

Fextor: A Feature Extraction Framework for Natural Language Processing: A Case Study in Word Sense Disambiguation, Relation Recognition and Anaphora Resolution*

Bartosz Broda, Paweł Kędzia, Michał Marcińczuk, Adam Radziszewski, Radosław Ramocki, and Adam Wardyński

Abstract. Feature extraction from text corpora is an important step in Natural Language Processing (NLP), especially for Machine Learning (ML) techniques. Various NLP tasks have many common steps, e.g. low level act of reading a corpus and obtaining text windows from it. Some high-level processing steps might also be shared, e.g. testing for morpho-syntactic constraints between words. An integrated feature extraction framework removes wasteful redundancy and helps in rapid prototyping.

In this paper we present a flexible feature extraction framework called Fextor. We describe assumptions about the feature extraction process and provide general overview of software architecture. This is accompanied by examples of applications in hugely different NLP tasks. Namely, we show the application of Fextor in: word sense disambiguation, recognition of inter-chunk syntactic relations, semantic relations between named entities, as well as anaphora resolution.

Keywords: feature extraction, word sense disambiguation, relation recognition, anaphora resolution, shallow parsing, derivational relations, named entities.

Bartosz Broda · Paweł Kędzia · Michał Marcińczuk · Adam Radziszewski · Radosław Ramocki · Adam Wardyński
Institute of Informatics
Wrocław University of Technology
Wybrzeże Wyspiańskiego 27
50-370 Wrocław, Poland
e-mail: {bartosz.broda,pawel.kedzia,michal.marcinczuk}@pwr.wroc.pl,
{adam.radziszewski,radoslaw.ramocki,adam.wardynski}@pwr.wroc.pl

* This work was financed by the National Centre for Research and Development (NCBiR) project SP/I/1/77065/10.

A. Przepiórkowski et al. (Eds.): *Computational Linguistics*, SCI 458, pp. 41–62, 2013.
DOI: 10.1007/978-3-642-34399-5_3

1 Introduction

Feature extraction process is a preliminary step of Machine Learning (ML) techniques. When working on theoretical aspects, one often uses standardised benchmarking datasets, where features are already provided. E.g., the iris flower dataset [2] allows to put the focus solely on the performance of the algorithm. However, employing ML for real-world problem solving requires the consideration of feature extraction process, first and foremost. In some domains, the nature of features is constrained. For example, when dealing with measurements taken by some device, features can only include data from the sensors of the measuring equipment. On the other hand, there are also domains that provide more freedom when describing the given problem in terms of features.

One of such domains is Natural Language Processing (NLP). In many NLP tasks we deal with textual data from which the features must be extracted, as ML algorithms cannot process text directly. The challenge is to select features that are informative and discriminative. Often an initial set of features is established and some experiments are conducted to improve the set, modyfing, adding or removing features. Having an easy to use and flexible framework for feature extraction can help and speed up the process considerably. This is especially important when dealing with many complex feature types, where being able to describe features in some high-level way is a necessity. Moreover, many steps of feature extraction process are common among different NLP tasks. Reading a corpus, selection of text window around interesting words in text are examples of tasks that can and should be abstracted away while constructing features.

It is hard to find an open source feature extraction framework that could be applied to Slavic languages, esp. to Polish. On the other hand, there are a few NLP frameworks, which contain some form of feature extraction (see Sec. 2). One of our major requirements for feature extraction framework was flexibility. This was motivated by the wide range of NLP research areas we are dealing with: from part-of-speech tagging through word sense disambiguation to anaphora resolution. We have gained some experience with feature extraction while working separately on different problems, thus we had different opinions on what the feature extraction process should look like. Nevertheless, after a few fruitful discussions and brain-storming sessions, common patterns have emerged. We decided to undertake the difficult task of constructing an integrated feature extraction framework called Fextor[1].

We had a few design goals that guided the entire process of constructing the software. The *flexibility* mentioned earlier was very important. Everyone of us needed to solve a different problem, thus without flexibility the undertaking would fail. Opposite to flexibility stands *simplicity*: we joined our efforts to create Fextor in order to reduce total cost of feature extraction. The ability to describe any feature we would need was also required, so

[1] Fextor home page: `http://nlp.pwr.wroc.pl/fextor`

we strived for *expressiveness*. The tools had to work with our standard software libraries, most importantly: `wccl` and `corpus2` [26]. We are constructing practical tools, so we need to process huge corpora in an *efficient* way. We have already constructed an integrated ML framework within LexCSD [7], which enables usage of different classifiers from different software packages, thus the integration with LexCSD[2] was naturally on the list of requirements.

This paper is organised as follows: in the next section we give an overview of existing approaches to feature extraction in NLP. Next, we give an overview of Fextor architecture: we introduce the concepts of *slicers*, *iterators*, *contexts* and *feature generators*. In Section 4 we show some applications. Some of the applications are already developed and some are being under development. The described applications include: word sense disambiguation, recognition of inter-chunk syntactic relations and semantic relations between named entities, anaphora resolution and classification of derivational relations. In Section 5 we summarise the paper and point direction of further works.

2 Related Works

There exist several common frameworks that may be used to solve NLP problems using ML classifiers. Those frameworks usually provide some degree of support for feature extraction.

NLTK [4] is an open source suite containing a large library of Python modules for various NLP tasks. It also includes implementations of various ML algorithms. Although the abundance of practical modules is quite impressive, the user must actually write custom code for feature extraction. What is more, the framework is (somewhat implicitly) targeted at processing English, and hence, no support is given to deal systematically with structured tagsets that are characteristic for inflective languages.

Freeling [20] is another open source suite for language analysis equipped with some ML components. The set of available classifiers is quite limited in comparison with NLTK. What is more, the suite is somewhat loosely coupled, e.g. POS tagging and Word Sense Disambiguation modules are based on completely different APIs. Nevertheless, one component seems particularly interesting for our considerations, namely the *Fex* feature extraction module [27]. The module is used in Freeling for named entity recognition and coreference resolution. Fex is able to process a corpus in a specially prepared input format into a list of feature vectors. Each sentence is processed separately. Features are expressed in a simple domain language. The language allows to extract values of word forms, tags, chunk tags and grammatical roles from different positions in the input corpus (note that this information must be explicitly provided as per-word labels in the input corpus). Fex language expressions may retrieve value of a particular category (tag, form, etc.) from a position relative to the word being processed; it is also possible to retrieve

[2] LexCSD home page: `http://nlp.pwr.wroc.pl/lexcsd`

ranges of values or complex values, composed of values of different categories. Unfortunately, POS tags are treated as atomic symbols and no mechanism is offered to decompose them into values of particular grammatical categories (e.g. case of a noun), not to mention calculation of complex morphosyntactic functions (e.g. tests for agreement between a noun and its adjective modifier, typical for Slavic languages). The language also does not seem to support iteration over larger units, such as existing syntactic annotations or named entities.

WCCL [26] is another open source toolkit for feature generation. The toolkit is targeted at Polish[3] and distinguishes itself with the support for positional tagsets and ability to 'understand' complex tags. As in the case of Fex, WCCL comes with a specialised domain language for writing functional expressions evaluated on annotated sentences with some token highlighted as the centre being processed at the moment. The language is quite expressive, allowing for extraction of simple and complex features. Simple features include orthographic forms, lemmas and grammatical class and values of grammatical categories from tokens. Complex features include the following (the list is not exhaustive):

- test for set-theoretic relations between two sets of strings or sets of symbols defined in the currently used tagset (WCCL supports ambiguity in the input corpus, e.g. some tokens may be assigned more than one tag, resulting in sets of possible values of some grammatical categories);
- constraint satisfaction search, i.e. a given token range is sought for a token satisfying the given predicate and some function is finally evaluated on the token found (e.g. the grammatical case retrieved from a noun being a likely verb object);
- tests for morphological agreement on given grammatical categories, either between two given tokens or between whole token range;
- joining features with help of standard logical connectives and functional-style `if` clauses.

In spite of the mentioned expressiveness of WCCL, there are two limitations particularly important to our considerations. First, as in Fex, WCCL features cannot reach beyond the boundaries of the sentence being processed. This is particularly unfortunate for application in anaphora resolution and word sense disambiguation, where the features typically employed refer to much broader context. The other drawback is that the toolkit is limited to 'token-oriented' feature generation, that is to say, one iterates over subsequent tokens and given functional expressions are evaluated on subsequent tokens along with their local neighbourhood. This is natural for tasks that may be cast as sequence labelling problems, where the sequence being labelled consists of tokens, e.g. morphosyntactic tagging, chunking. Other use cases may require features describing entities other than single tokens,

[3] WCCL has been developed for Polish, although it is able to process any language as long as the tagset conforms to a few additional requirements, cf. [25].

e.g. named entities, syntactic chunks or even pairs of such annotations (to classify inter-annotation relations).

3 Fextor Architecture

Fextor is a Python 2.6 package which consists of several modules: *document* — represents processed document, *reader* — allows to read documents, *iterators* — to iterate over elements of interest in a document, *context* — a trimmed part of the document representing local context, *features* — all available features. Generally, it takes input files as arguments, while details regarding types of examples to take from documents and features to extract from the examples are set in a configuration file. The results are written in CSV format to the standard output or to a selected file. The process of feature generation may be sketched as follows:

- Read the document and make its internal representation;
- Iterate over the `Document` and return pointers to elements of interest (e.g. tokens) where iterator stopped. Type of the `Iterator` determines on which elements the iterator should stop, e.g., `OnlyBaseIterator` can stop only on the words, whose base form occurs in the given list;
- For each `Pointer` from iterator:
  - cut the `Context` — size of the context depends on the slicer used, e.g., n-th tokens from left/right, sentence, document, ...;
  - make the `Pointer` to the element of interest in the `Context` (it may be different than in the document);
  - for each feature specified in the config file, pass pair of `Context` and `Pointer`, apply the feature function and get the resulting feature value;
- Return the results to the user (to the standard output or a file).

The input, be it a single file or a multi-document corpus, is read into an internal representation. Next, the document (or documents) is processed in an appropriate way, as decided by the user in the configuration file. The result of processing are values returned by selected feature extractors. In case of a corpus, the result constitutes all values extracted from all the documents from the corpus.

The class governing the execution flow is `Fextor`. What follows is a functional overview of the architecture, roughly as implemented in the `Fextor` object.

3.1 Fextor I/O

List of input files that constitute a training corpus is provided in the arguments passed to Fextor. The files are read using Corpus2, a C++ module that offers data structures and routines for reading and writing annotated corpora in different formats, supporting configurable tagsets [25].

Two formats are supported so far — *poliqarp* and *document*. The former is a compact, indexed binary representation of a morphosyntactically annotated corpus, as generated by the Poliqarp corpus indexing and querying system [12]. The latter is an XML-based format, so-called *CCL format* (as supported by Corpus2 and WCCL), being a simple extension of XCES with the possibility to annotate chunk-style annotations, their heads and inter-annotation relations. As the relations may hold between annotations from different sentences (typical example being anaphora), while the rest of annotation is limited to sentence boundaries, relations may be read from a separate file. The *document* format also supports batch processing mode, when input files are text files that list the actual XML documents.

The internal representation of input files is handled through the `Document` class. This class holds information about paragraphs, sentences, tokens, relations and annotations from the given document. Tokens in the document are numbered. The document interface allows to get annotations from the document and get all relations from the document. Internal representation of the document is created by a *Reader* that uses Corpus2 classes. Created document is the basis for further processing and for features extraction — this step may be understood as a first phase in the process of feature extraction.

The Fextor output is produced with Python's `csv` module, using `excel` dialect, although with semicolon (`;`) rather than the comma used as the delimeter (more in line with the default CSV file format used by recent spreadsheet software, after which the dialect takes its name). Essentially, the first line is a semicolon-separated list of feature names, and what follows are semicolon-separated lists of extracted values for corresponding features, one example per line.

3.2 Iterators

Once a document is read, it is passed on to an iterator, specified in the Fextor configuration file directly as a Python class. Also, the configuration file can specify parameters that will be passed to the iterator.

The iterators pass through the document and return pointers to examples, one by one as Python generators do. We currently support three types of *pointers*: *token*, *annotation* and *annotation pair*.

`TokenPointer` points to a single token from text. `AnnotationPointer` conversely corresponds to an *annotation*, which is essentially a sequence of tokens belonging to a named *annotation channel*. In our approach, the annotations are organised in independent *channels*; each channel may contain a number of non-overlapping annotations, being sequences of consecutive tokens (as in *chunking* task). For instance, one may have a `chunk_np` channel containing NP chunks, `person_nam` channel containing named entities describing personal names. An annotation may also be given additional properties (key–value string pairs); besides, there is a possibility to annotate annotation's syntactic head (on of its tokens is then highlighted as head). We also provide `AnnotationPairPointer`, which encompasses two selected annotations; this pointer is especially useful when classifying relations holding between two annotations.

Only one iterator, with one type of example is supported in a single Fextor run i.e. features designed for `TokenPointers` won't work for `AnnotationPointers` (for code reuse one could e.g. provide a wrapper that takes head of annotation and invokes some token feature on it).

`EveryTokenIterator` is an example of an iterator which iterates over all tokens in order of their occurrence in the document. Additionally, Fextor offers iteration over annotations from selected channels (`NamedAnnotIterator`) or pairs of annotations as specified with names of the channels (`SentAnnotPairIterator` – limited to sentence boundaris) or with a relation name that should connect the pair (`RelationIterator`).

Fextor outputs extracted features one line per example in the same order as the examples are coming from the selected iterator (in the order of documents coming from `DocumentProvider` i.e. as they were provided in input).

3.3 Slicers and Context

Along with the notion of a pointer, we have *contexts.* A context is a list of tokens, sentences, paragraphs, relations and annotations in the neighborhood of a pointer. It is essentially a fragment of document "cut" around a pointer with the use of a `Slicer` object. In general, slicer is a mechanism used to trim the given context. Size of the returned context depends on the type of slicer used. Sample slicers are:

- `DocumentSlicer` — Returns copy of the document.
- `TokenSlicer` — Context is trimmed to the given range of tokens. N tokens on the right and M on the left site of the given pointer.
- `SentenceSlicer` — Context is trimmed to the boundaries of a sentence that contains the given pointer. Note that the division into sentences is taken intact from the document, no additional sentence splitting is performed.
- `MultiSentenceSlicer` — Context is trimmed to the given range of sentences: N sentences on the right side and M to the left of the given pointer.

A slicer receives a document and a pointer to a fragment in the document. The slicer returns a pair of: context and mapping of the given pointer in document to pointer in the generated context, type of the returned pointer is the same as the used pointer.

The idea is that a feature extraction may be focused on (or limited to) a context around a pointer, so instead of document as a whole, just a pointer and corresponding context is provided (although as mentioned above, context may as well encompass whole document if need be). E.g. WCCL features naturally take a context of a sentence. The type of a slicer required by a feature is specified in the feature definition in the configuration file.

3.4 Feature Generators

The last step of processing is to generate the values of the features for every element returned by iterator. Classes that perform feature generation implement one of the following interfaces:

- `TokenFeatureGenI` if the extraction centers around tokens;
- `AnnotationFeatureGenI` in the case of annotation processing;
- `AnnotationPairFeatureGenI` for pairs of annotations.

As already mentioned, only one type of elements is supported throughout a single Fextor run.

The list of features to be extracted is defined as a `features` parameter in the main extractor section in the configuration file. Each feature has a dedicated configuration section. The section contains name of Python class performing the feature generation and a set of parameters required to initialize and run the generator. This allows to write more generic classes that are subsequently parametrized, e.g. we have classes that directly support WCCL functional expressions. The last element defined in the feature section is the name of context slicer used by the feature generator.

`PosFeature` is a sample implementation of feature generator which extract a set of parts of speech tags from specified context. As it is token feature the generator implements the `TokenFeatureGenI` interface. In case when left or right boundary (specified as relative offset from the current token) is outside the context, then for the missing tokens the '`None`' value is taken. `PosFeature` can perform words filtering according to specified list of stopwords (the `stopwords` parameter). For words present on the list the generator returns empty string. For non-empty sets the generator return a list of part of speech tags separated by semicolon. Below is a sample definition of a feature using the `PosFeature` generator:

```
[wsd_pos]
class      = fextor.features.tokens.PosFeature
slicer     = fextor.contexts.slicer.TokenSlicer(-20, 20)
flex       = adj,adja,adv,aglt, ..., tnum,tsym,winien
tagset     = kipi
stopwords = stopwords.txt
conversion = numeric_set
```

Where: *wsd_pos* — is a unique feature name; *class* — name of Python class used as a generator, *slicer* — type of slicer used to make contexts, here `TokenSlicer` with a range of 20 words from the left and the right side around the pointer; *flex* — a list of POS values of words which are to be processed; *tagset* — the tagset used in the corpus read; *stopwords* — path to a file with stopwords; *conversion* — type of feature value conversion used in Fextor output, as expected by the fextor2lexcsd application. Other `TokenFeatureGenI` features, are: `BaseFeature` — to extract base form of the words, `OrthFeature` — to extract words in form as occurs in the text.

3.5 Interfacing with LexCSD

To facilitate the use of LexCSD [7] which contains a framework for Machine Learning, Fextor package includes `fextor2lexcsd` converter to obtain the LexCSD matrix format out of the CSV format described above.

As described in Sec. 3.1, the Fextor output is a CSV file containing (in successive columns) values generated by the features. These values can be transformed to the new predetermined types of values. Each feature declared in the config file should contain information about the desired type of conversion. The following types of conversion are supported:

- `none` — no conversion is done.
- `numeric` — each generated value has unique number. All occurrence of this value are replaced by this number.
- `binary` — each feature is expanded to a set of new features, each corresponding to its subsequent values. The new features are binary: $F_n^v = 1$ when the original feature f_n has value of v, 0 otherwise. For example, assuming three features `A B C` with the following values:

  ```
  A B C
  -----
  1 3 4
  2 2 4
  ```

  Binary conversion will yield the following result:

  ```
  A:1 A:2 B:2 B:3 C:4
  -------------------
   1   0   0   1   1
   0   1   1   0   1
  ```

- `binary_set` – each generated value is treated as set with values separated by the comma. Then for all values from this set `binary` conversion is made. For example:

```
   A       B
--------------
{1,2,3} {2,3}
 {2,4}   {1}
```

is transformed to:

```
A:1 A:2 A:3 A:4 B:1 B:2 B:3
------------------------
  1  1   1   0   0   1   1
  0  1   0   1   1   0   0
```

- `sparse_binary` – same as `binary`, but value of 0 are not set.
- `sparse_binary_set` – same as `binary_set`, but value of 0 are not set.
- `continuous_vector` – this is an arbitrary fixed size vector of real numbers.

4 Example Applications

Feature extraction is not a goal on its own. Thus, in this section we will show a few applications of Fextor. Some of them are deployed and some are being under development. We have already achieved good results for Word Sense Disambiguation (WSD) and classification of derivational relations between words. The results for inter-chunk relation recognition are promising. The remaining tasks are still under development, so we will confine our discussions to the feature sets we have designed and the role of Fextor in solving the tasks.

4.1 Word Sense Disambiguation

Many words have more than one sense (lexical meaning), but usually only one of them is activated in a given context. The typical example of such words are: *bank* (river bank or financial institution), *ring* (making a call or physical object) or *line* (36 entries in WordNet [11], e.g., queue, cable, shape). The task of Word Sense Disambiguation (WSD) is to choose the right sense for a word in a context. With availability of manually annotated corpora WSD can be described as a classification problem. That is, given the occurrence of ambiguous word in text the classifier has to assign a sense label on the basis of features extracted from surrounding context.

As we had some experience with ad hoc feature extraction for WSD [3, 7] we already had some idea what features we should extract. Also, lots of ideas for WSD features can be found in the literature, especially the important works of [1, 17]. Starting with extraction of bag of words features in vector space model is usually not a bad idea in WSD as this feature type gives the most coverage [1]. In this model the features are represented as a high-dimensional vector, where every dimension corresponds to occurrences of different words in contexts of ambiguous words. The task of feature generator is to mark occurrences of words in a surrounding text window. Words are filtered by their parts of speech as not every part of speech is a good discriminator (e.g., nouns are typically useful, prepositions not so much). The text window is usually large, e.g., 100 words around the target word.

The word order is lost in a bag-of-words representation. Thus another potentially useful feature is a dense vector that holds the information about words occurring on given position relatively to the ambiguous word. Now, the vector's dimensions correspond to the position, and the value represents a numerical identifier of a word occurring on that position. The text window is shorter than in the bag-of-words representation (e.g., 10 words). With even shorter text window (of 2 or 3 words) and some additional statistical filtering one can obtain a good approximation of *collocations*, which where shown to improve WSD performance [18]. Another simple modification of aforementioned feature type leads to another feature: instead of marking occurrence of words in short text window we can mark occurrences of parts of speech.

With the help of WCCL language we can extract more complex features. So far we have experimented with morpho-syntactic constraints similar to the ones used in extraction of Measures of Semantic Relatedness (MSRs) from text corpora [23]. Those features check for a morphological agreement between (ambiguous) noun and an adjective, try to capture a potential predicate for a noun, seek nouns in coordination and check for modification of ambiguous noun by another noun in genitive. So far the preliminary results with employment of morpho-syntactic features are not encouraging. This might be caused by the fact that those operators were written by hand with the goal of the highest precision as possible. Thus the coverage suffered and the features are rather sparse. As so far we have focused only on nouns we haven't tried to employ more sophisticated WCCL features that were employed for construction of verbal MSRs [5].

The currently employed features give state-of-the art results for Polish. That is, results we achieve with help of Fextor are comparable to results achieved in other works [16, 3, 7]. We are working on the extension of already employed feature set. We want to employ a Polish wordnet called Słowosieć [22], MSR directly and additional forms of morpho-syntactic processing using WCCL or more sophisticated parsers [30, 24].

4.2 Recognition of Inter-chunk Syntactic Relations

Another application of Fextor is that of shallow parsing. More specifically, we are working on a simple ML module that recognises selected syntactic relations between syntactic chunks, as defined in the *Polish Corpus of Wrocław University of Technology* (*KPWr*) [29, 6]. There is already a tool suitable for recognition of the chunks, namely the memory-based chunker proposed in [15]. The task described here is that of recognition of partial predicate-argument structure, which is limited to relations holding between VP and NP chunks.

The task may be formulated as a classification problem: given a sentence with two chunks highlighted, classify what type of syntactic relation holds between them. We consider three options: no relation, subject or object. We assume that the input sentence has been tokenised, morphosyntactically tagged and annotated with NP and VP chunks. What is more, each chunk is augmented with information about the location of its syntactic head (always one-token-long). The preliminary experiments described here make use of reference NP/VP annotation, therefore the evaluation presented in this section does not include chunking errors.

In terms of feature extraction, the use case requires iteration over annotation pairs (the annotations being chunks). During training, for each sentence a set of all the possible VP–NP pairs is generated (at least in our baseline model; the set could be limited by some general syntactic constraints). Each such pair, along with the whole sentence as context is employed to generate feature values for classification; the decision class is the desired relation type. The performance phase consists in generating pairs in the same fashion, feature generation and having the trained classifier predict the relation type between the chunks. In terms of Fextor configuration, we employ *sentence annotation pair iterator*, features exploiting WCCL expressions, as well as *annotation pair sentence slicers.* Below are snippets from the configuration file used:

```
[Extractor]
iterator = fextor.iterators.SentAnnotPairIterator
first_channels = chunk_vp
second_channels = chunk_np
features = class_hd1 class_hd2 np_btw ... label

[class_hd1]
class        = fextor.features.pairs.WCCLPairFeatureGen
slicer       = fextor.contexts.slicer.AnnotationPairSentenceSlicer()
operator     = class[$Hd1]
type         = symset
file         =

[class_hd2]
class        = fextor.features.pairs.WCCLPairFeatureGen
slicer       = fextor.contexts.slicer.AnnotationPairSentenceSlicer()
operator     = class[$Hd2]
type         = symset
file         =

[np_btw]
```

```
; the number of chunk_np annots between ann1 and ann2
class       = fextor.features.pairs.DistanceInAnnotations
slicer      = fextor.contexts.slicer.AnnotationPairSentenceSlicer()
channel     = chunk_np
```

The `SentAnnotPairIterator` generates all annotation pairs from each sentence, where first annotation is of `chunk_vp` type, while the second one is a `chunk_np`. Most of the features employed are based on WCCL functions. The role of Fextor here is to set position variables to point to the expected tokens, e.g. `$Hd1` is automatically set to the location of the head of the first annotation (VP in this case); this allows for using concise and expressive functional expressions. E.g., the `class_hd1` feature is responsible for returning the possible values of the grammatical class (POS) of the VP syntactic head. Some of our features are based on functionality that is not available in WCCL at all, e.g. `np_btw` returns the number of NP chunks that occur in the range between the annotations of the considered pair.

A reasonable feature set for this task will include features that characterise both chunks themselves, their local context, as well as the material that comes in-between. As the chunks may be of different length (the same holds for the intervening material), it is necessary to focus on features that refer to particular tokens (at least syntactic heads of both chunks), but also to capture some abstract properties of units bigger than single tokens. The feature set employed here consists of the following items:

1. lemma of heads of both annotations;
2. POS of both heads;
3. grammatical case of the second annotation head (which is an NP);
4. does the second annotation start with a preposition? (the annotation schema of KPWr puts both real NPs and actual PPs into one group called NP);
5. set of POS values collected from all the tokens belonging to the first annotation and the same for the second one;
6. set of POS values collected from all the tokens occupying the range between the annotations;
7. set of values of grammatical case collected from all the tokens occupying the in-between range (some tokens are not specified for case at all, e.g. adverbs and punctuation; such tokens do not contribute to the set);
8. a set of string codes, indicating whether some of important items appeared in the in-between range; this is based on dictionary look-up, where the dictionary was prepared manually and assigns 45 lemmata to one of the following categories: coordinating conjunct, relative pronoun, comma or ambiguous between the two first;
9. how many NP and VP chunks appear in the in-between range;
10. does the NP follow the VP or vice-versa.

Our experiments consisted in using Fextor equipped with the above feature set and configuration file in tandem with the LexCSD ML framework [7]. This allowed us to conveniently test several classifiers, coming from different software packages. Our data set contained 4028 relation instances taken from the part of the KPWr that has been annotated so far. Our experiments are based on 10-fold cross-validation. The results of our preliminary experiments are presented in Table 1. *P*, *R*, *F* columns present respectively values of *precision*, *recall* and *F-measure*. The results labelled TiMBL: k15 correspond to the Tilburg Memory-Based Learning package [9] with $k = 15$ neighbours, using the overlap metric. The results labelled TiMBL: k15,mM are also obtained using TiMBL, but using Modified Value Difference metric. The last row contains results obtained using the LibLINEAR classifier [10] assuming the `C3` regression function with $C = 0.25$.

Table 1 Results for syntactic relation recognition

Classifier	Subject			Object		
	P	R	F	P	R	F
Naive Bayes	61.38%	54.67%	57.33%	58.30%	87.56%	69.87%
Decision Trees (J48)	69.00%	45.38%	53.64%	65.18%	75.34%	69.33%
RIPPER (JRip)	74.57%	45.23%	55.37%	66.37%	74.43%	69.07%
TiMBL: k15	73.44%	58.08%	63.88%	66.98%	81.70%	73.42%
TiMBL: k15,mM	74.67%	58.74%	64.88%	72.72%	76.46%	74.41%
LibLINEAR: S3,C=0.25	73.10%	65.77%	68.93%	69.15%	83.71%	75.50%

The results obtained so far for syntactic relation recognition are not striking but anyway promising. Note that the functionality already implemented in Fextor allows to test more advanced features, e.g., also examining the NPs and VPs appearing beyond the range constituted by both annotations and the in-between material.

4.3 Recognition of Semantic Relations between Named Entities

The next application of Fextor is recognition of semantic relations between named entities. The set of semantic relation categories to be recognised is limited to a predefined set. The task is also limited to relations between named entities present in the same sentence, thus this application is very similar to the task of inter-chunk syntactic relation recognition. The other assumption is that the relations must be supported by some premises stated in sentence — we do not intend to recognise relations supported only by external knowledge. We assume that the named entities will be recognized beforehand with the NER tool presented in [13] extended to 56 categories of proper names.

The recognition of semantic relations between named entities requires iteration over selected pairs of named entities according to a predefined dictionary of pairs of named entity categories. Such enumeration is implemented in the `SentAnnotPairIteratorByDict` iterator. Below is a sample definition of the iterator for *affiliation* relation:

```
[Extractor]
iterator = fextor.iterators.SentAnnotPairIteratorByDict
dict = reference_dict.pkl
relation = affiliation
features = r1 r2 r3 r4 r5 r6 ... context docname relations
```

where `dict` is a path to the dictionary of valid pairs of named entities and `relation` is a name of current relation category. The `context` and `docname` features are used to identify any pair of annotations within set of documents after the classification is done.

In the preliminary experiment we have utilized first-order logic rules obtained by applying Inductive Logic Programming over the data represented as a set of predicates (the description of the predicates used to describe the data is presented in [14]). The feature value is *true* if the rule can be proved and *false* otherwise. The rules are in fact short patterns over the sequence of tokens and token dependency tree around the annotations. The feature generator utilizes external module which transforms every sentence into set of predicates and Yap interpreter for testing the rules.

Below is a sample feature definition which utilizes one of the rules generated for *composition* relation:

```
[r12]
class = fextor.features.ilp.IlpRulesFeature
slicer = fextor.contexts.slicer.AnnotationPairSentenceSlicer()
tagset = nkjp
predicates = predicates.txt
rules = relation(A,B,composition) :-
   annotation_token(B,C), jump_left(D,C), jump_left(E,D),
   token_base(E,word_w), jump_left(F,E), token_base(F,meta_COMMA),
   token_dependency(G,C,adj).
```

We have tested two models of feature space. In the first model, namely *sep-feat*, the feature space for every category of relations contained only rules generated for that category. In the other model, namely *join-feat*, the feature space for every category of relations was the same and contained all features generated for all categories of relations. We have tested a set of basic classifiers available in the LexCSD enviroment which can handle binary features, including: BayesianLogisticRegression, BFTree, ComplementNaiveBayes (CNB), DecisionTable, LMT, NaiveBayes. The classifiers were trained on the training set and evaluated on the tune set[4]. Then, we have selected the classifiers which obtained the best results on the tune set and evaluated them on the testing set.

In Table 2 we present the comparison of results obtained by the hand-crafted rules, ILP rules and classifiers using ILP-rules as features. In case of 5 categories of relations we obtained better results than ILP rules. The highest improvement was obtained for *origin* relation by 20.95 points of F-measure. The other two categories obtained slightly lower results by 1.39 and 0.16 points of F-measure for *composition* and *alias* relation respectively. For the last relation category, i.e. *neighbourhood*, we obtained much worse results.

Table 2 Comparision of results obtained by hand-crafted rules, ILP rules and classifiers using ILP-rules as features

Relation	Rules	ILP	Classifier			
	F [%]	F [%]	Model	Type	F [%]	Change
affiliation	28.90	46.12	sep-feat	DecisionTable	**52.73**	+6.61
alias	24.35	**46.03**	join-feat	BFTree	45.87	-0.16
composition	42.86	**65.63**	sep-feat	NaiveBayes	63.64	-1.39
creator	12.50	32.35	join-feat	CLR+CNB+RT	**41.03**	+8,68
location	26.00	29.58	sep-feat	BLR	**33.46**	+3.88
nationality	30.77	44.44	sep-feat	RandomTree	**54.55**	+10.11
neighbourhood	10.53	**33.33**	sep-feat	DecisionTable	15.38	-17.95
origin	57.14	41.27	join-feat	BFTree	**62.22**	+20.95

We observed that many incorrect relations were recognized in sentences which contain a pair of named entities connected with a relation and a set of irrelevant named entities not connected with any relation. Some of the rules generated by ILP had form of two unlinked patterns matching the context of source and target named entities. Many of the false relations were recognized between named entities appearing far away from each other. We expect that named entities appearing far away from each other are rare to be in any relation. We could eliminate them by introducing some features which will reflect the distance between the named entities.

In the preliminary experiment we have utilized only one type of features. However, the set of features can be extended by features used in the inter-chunk syntactic relations recognition task and also:

1. relative position of named entities (A after B, A before B, but also A inside B, etc.),
2. location of named entities inside chunks (for source and target annotation),
3. syntactic relations for chunks containing the named entities (for source and target annotation),

[4] The training, tune and testing sets are described in [14].

4. wordnet-based generalisation of words (direct and indirect hyperonyms of words),
5. bag of words in *before*, *between* and *after* context (according to [8]),
6. verbs and prepositions closest to named entities (for source and target annotation).

The extended set of features is going to be tested in the future experiments.

4.4 Anaphora Resolution

Anaphora is another example of a relation held between two items in text, namely when one fragment (*anaphor*) points back to a previously mentioned item in the text (*antecedent*). As such, *anaphora resolution* (resolving an anaphor to the expression it refers to) can be treated in similar manner as abovementioned examples concerning syntactic or semantic relations. Specifically, anaphora resolution can be presented as a classification of pairs of expressions, whether given relation holds or not. One important distinction though is that anaphora in general transcends sentence boundaries.

One type of anaphoric relation is direct, identity-of-reference anaphora, where the anaphor and the antecedent have the same referent in the real world. In essence, they are coreferential, and this type of anaphora overlaps with the issue of coreference resolution. A machine learning approach to coreference resolution of noun phrases has been presented in [28] and expanded upon in other publications, such as [19].

The approach represents the problem as a classification task as suggested above, however special care is taken in order to avoid dealing with all possible combinations of suspected anaphor/antecedent pairs. Positive cases are taken from pairs of anaphor and its closest antecedent. To limit the number of negative cases, they are taken only from pairs of actual anaphor with a noun phrase that is not its antecedent, but is textually placed between the anaphor and its real closest antecedent. A set of features describes the pairs and a model is built to predict whether relation holds or not, based on the values of the features.

Such approach should also extend to other anaphoric relations, such as bridging anaphoras, where the relation between anaphor and antecedent is indirect, e.g. anaphor is a part of antecedent or in its possession.

Feature extraction for this process can be naturally facilitated by Fextor. The example selection can be performed by a dedicated iterator, or with a separate preprocessing step that would mark interesting pairs, allowing for a simple iterator.

Initially we are basing our feature selection on [19], focusing on

- lexical features such as textual similarity,
- grammatical features such as agreement on gender and number,
- semantic features such as WordNet semantic class,
- positional features such as distance in tokens or sentences.

It is also conveniently possible to try to employ features selected for classification of other types of linguistic relations as mentioned in other examples herein.

4.5 Classification of Derivational Relations

Classification of derivational relations between words is a task of assigning single class to a pair of words. The first word is called a *derivative* and the second one is referred to as a *derivative base.* In our experiments there were four coarse-grained derivational relations (classes): femininity, inhabitant, markedness and semantic role. Markedness can be subdivided into three subclasses: diminutivity, augmentativity and young being. Respectively semantic role can be subdivided into: agent of hidden predicate and location of hidden predicate. More details about Polish derivational relations and their automatic generation can be found in [21].

There are multiple possible knowledge sources one can use in the task of derivational relation classification, e.g., corpora, wordnet and textual association between derivative and its base form. As Fextor was designed to extract features from corpora by specialized iteration method one additional step had to be introduced to the processing pipeline, unlike in previously described applications. We addressed the problem by construction of an artificial corpus file. The 'fabricated' corpus is made up from only one sentence which contains N tokens. Every token represents one instance of derivational relation, so there are N instances of relations. In addition each token contains special set of properties describing particular instance of relation. Following properties are needed for feature generation:

- derivative and derivative base words,
- relation instance class,
- identifiers of derivative and derivative base synsets in plWordnet,
- every possible morphological analysis of derivative and its base.

To generate features we need to iterate over every token in the artificial corpus.

Fextor was used to generate following set of features:

- plWordnet domain identifiers for derivative and base,
- bag of plWordnet synset identifiers describing semantic classes for derivative and its base,
- suffix of derivative, which is not included in its base form,
- length of derivative's suffix,
- bag of part of speech, grammatical numbers, genders and cases for derivative and its base,
- distributional semantics vector for derivative and its base (the simplest model could be a cooccurrence with any word in fixed sliding window, generated by corpora iteration), vectors are taken from precalculated matrix which is available as a resource for feature generator.

In order to evaluate our approach we used instances of 7 derivational relation subtypes from *plWordNet 1.6*. For our experiment 10-fold cross-validation scheme was applied. Pairs for each relation subtype were randomly divided into 10 subsets. One subset per relation subtype was used for testing in each iteration.

Table 3 presents cross-validation results for coarse-grained and fine-grained class subdivision. We used multiclass *SVM* classifier from *LibLINEAR*.

Table 3 Cross-validation results of multiclass semantic classification (coarse-grained relations are in italic)

Relation	**Precision**	**Recall**	**F-score**
femininity	95.09	98.36	96.70
inhabitant	91.26	87.85	89.52
markedness	89.70	98.16	93.74
diminutivity	89.31	97.17	93.08
augmentativity	70.80	62.02	66.12
young being	53.85	26.92	35.90
semantic role	84.79	93.23	88.81
agent of hidden predicate	86.74	96.09	91.17
location of hidden predicate	83.33	83.33	83.33

4.6 Other Possible Usages

In the above sections we presented Fextor usage scenarios corresponding to NLP tasks we are currently involved in. There is, however, a wide range of other possible applications, where the inclusion of the presented framework could bring practical profits. The possible application areas include the following items:

- Sequence labelling tasks, such as POS tagging, chunking, named entity recognition. Some of the tasks could benefit from features that could not be expressed in WCCL, e.g. bag-of-word features describing local contexts stretching beyond sentence boundaries.
- The `AnnotationPointer` may be used to iterate over any type of annotations already placed in text. The annotations may also correspond to larger units of text, e.g. previously recognised definitions, clauses in sentences or perhaps selected whole sentences annotated as spots of interest. Fextor could be used to classify such stretches of text with respect to various criteria, e.g. sentiment, question type, dialogue act classification.

5 Conclusions

Feature extraction from text corpora is an important step in Natural Language Processing based on Machine Learning. In this paper we have described a flexible approach to feature extraction, which was implemented in the *Fextor* environment. We have shown architecture of the system and a few example applications: anaphora resolution, classification of derivational relations, recognition of inter-chunk syntactic relations, recognition of semantic relations between named entities and word sense disambiguation.

There are a few possibilities for further development. Fextor was designed for Polish, but after a few extension it should be possible to adapt it to other languages — especially from the Slavic family. We also need to finish the deployment of Fextor in recognition of syntactic and semantic relations. The software is being released under GNU GPL licence; we still need to finish writing the documentation, which at the present stage is rather rudimentary.

References

1. Agirre, E., Edmonds, P. (eds.): Word Sense Disambiguation: Algorithms and Applications. Springer (2006)
2. Anderson, E.: The species problem in iris. Annals of the Missouri Botanical Garden 23(3), 457–509 (1936)
3. Baś, D., Broda, B., Piasecki, M.: Towards Word Sense Disambiguation of Polish. In: Proceedings of the International Multiconference on Computer Science and Information Technology — 3rd International Symposium Advances in Artificial Intelligence and Applications (AAIA 2008), pp. 65–71 (2008)
4. Bird, S., Loper, E.: Nltk: The natural language toolkit. In: Proceedings of the ACL Demonstration Session, Barcelona, pp. 214–217. Association for Computational Linguistics (2004)
5. Broda, B., Derwojedowa, M., Piasecki, M., Szpakowicz, S.: Corpus-based Semantic Relatedness for the Construction of Polish WordNet. In: European Language Resources Association (ELRA) (ed.) Proceedings of the Sixth International Language Resources and Evaluation (LREC 2008), Marrakech, Morocco (2008)
6. Broda, B., Marcińczuk, M., Maziarz, M., Radziszewski, A., Wardyński, A.: Kpwr: Towards a free corpus of polish. In: Calzolari, N., Choukri, K., Declerck, T., Doğan, M.U., Maegaard, B., Mariani, J., Odijk, J., Piperidis, S. (eds.) Proceedings of LREC 2012, Istanbul, Turkey. ELRA (2012)
7. Broda, B., Piasecki, M.: Evaluating LexCSD in a Large Scale Experiment. Control and Cybernetics 40(2) (2011)
8. Bunescu, R.C.: Learning for information extraction: from named entity recognition and disambiguation to relation extraction. Ph.d., The University of Texas at Austin (2007)
9. Daelemans, W., Zavrel, J., van der Sloot, K., Van den Bosch, A.: TiMBL: Tilburg Memory Based Learner, version 6.3, reference guide. Technical Report 10-01, ILK (2010)

10. Fan, R.E., Chang, K.W., Hsieh, C.J., Wang, X.R., Lin, C.J.: LIBLINEAR: A library for large linear classification. Journal of Machine Learning Research 9, 1871–1874 (2008)
11. Fellbaum, C., et al.: WordNet: An electronic lexical database. MIT Press, Cambridge (1998)
12. Janus, D., Przepiórkowski, A.: Poliqarp: An open source corpus indexer and search engine with syntactic extensions. In: Proceedings of the 45th Annual Meeting of the ACL on Interactive Poster and Demonstration Sessions, pp. 85–88. Association for Computational Linguistics (2007)
13. Marcińczuk, M., Janicki, M.: Optimizing CRF-Based Model for Proper Name Recognition in Polish Texts. In: Gelbukh, A. (ed.) CICLing 2012, Part I. LNCS, vol. 7181, pp. 258–269. Springer, Heidelberg (2012)
14. Marcińczuk, M., Ptak, M.: Preliminary study on automatic induction of rules for recognition of semantic relations between proper names in polish texts. In: Proceedings of the 15th International Conference on Text, Speech and Dialogue. LNCS (LNAI), Springer (to appear, 2012)
15. Maziarz, M., Radziszewski, A., Wieczorek, J.: Chunking of Polish: guidelines, discussion and experiments with Machine Learning. In: Proceedings of the 5th Language & Technology Conference LTC 2011, Poznań, Poland (2011)
16. Młodzki, R., Przepiórkowski, A.: The wsd development environment. In: Vetulani, Z. (ed.) Proc. 4th Language and Technology Conference, Poznań, Poland (2009)
17. Navigli, R.: Word sense disambiguation: A survey. ACM Comput. Surv. 41(2), 1–69 (2009)
18. Ng, T., Lee, H.: Integrating multiple knowledge sources to disambiguate word senses: An examplar-based approach. In: Proceedings of the Thirty-Fourth Annual Meeting of the Association for Computational Linguistics, pp. 40–47 (1996)
19. Ng, V., Gardent, C.: Improving machine learning approaches to coreference resolution. In: ACL, pp. 104–111 (2002)
20. Padró, L., Collado, M., Reese, S., Lloberes, M., Castellón, I.: FreeLing 2.1: Five years of open-source language processing tools. In: Chair, N.C.C., Choukri, K., Maegaard, B., Mariani, J., Odijk, J., Piperidis, S., Rosner, M., Tapias, D. (eds.) Proceedings of the Seventh International Conference on Language Resources and Evaluation (LREC 2010), Valletta, Malta. European Language Resources Association, ELRA (2010)
21. Piasecki, M., Ramocki, R., Maziarz, M.: Automated Generation of Derivative Relations in the Wordnet Expansion Perspective. In: Proceedings of the 6th Global Wordnet Conference, Matsue, Japan (2012)
22. Piasecki, M., Szpakowicz, S., Broda, B.: A WordNet from the Ground Up. Oficyna wydawnicza Politechniki Wroclawskiej (2009)
23. Piasecki, M., Szpakowicz, S., Broda, B.: Extended Similarity Test for the Evaluation of Semantic Similarity Functions. In: Proceedings of the 3rd Language and Technology Conference, Pozna'n, Poland, October 5-7, pp. 104–108. Wydawnictwo Pozna'nskie Sp. z o.o. (2007)
24. Przepiórkowski, A.: Powierzchniowe przetwarzanie języka polskiego. Akademicka Oficyna Wydawnicza EXIT, Warsaw (2008)
25. Radziszewski, A., Śniatowski, T.: Maca — a configurable tool to integrate Polish morphological data. In: Proceedings of the Second International Workshop on Free/Open-Source Rule-Based Machine Translation (2011)

26. Radziszewski, A., Wardyński, A., Śniatowski, T.: WCCL: A morpho-syntactic feature toolkit. In: Proceedings of the Balto-Slavonic Natural Language Processing Workshop. Springer (2011)
27. Roth, D., Cumby, C., Sammons, M., Yih, W.T.: A relational feature extraction language (fex). Technical report, University of Illinois at Urbana Champaign (2004)
28. Soon, W.M., Chung, D., Lim, D.C.Y., Lim, Y., Ng, H.T.: A machine learning approach to coreference resolution of noun phrases (2001)
29. Radziszewski, A., Marek, M., Wieczorek, J.: Shallow syntactic annotation in the Corpus of Wrocław University of Technology. Cognitive Studies 12 (2012)
30. Wróblewska, A.: Polish dependency bank. Linguistic Issues in Language Technology 7(1) (2012)

Part II
Information Extraction

Automatic Construction of a Dynamic Thesaurus for Proper Names

Roman Kurc, Maciej Piasecki, and Stan Szpakowicz

Abstract. Proper names outnumber all other word classes, but they are under-represented in dictionaries and in electronic resources. We propose an automatic method of associating a stand-alone proper name repository with a wordnet. A variety of sources of lexical-semantic knowledge can be harnessed into this task. Semantic proximity between a proper name and a synset can be based on pattern-driven search and on distributional analyses in large corpora. The sources are heterogeneous and the measures of semantic relatedness vary widely. We propose a flexible method, an adaptation of the algorithm of Activation Area Attachment, which treats each type of sources slightly differently. We reach 80% precision in linking proper names with places in a wordnet even if the targets are highly polysemous.

1 Motivation and Related Work

Proper names (henceforth, PNs) far outnumber other classes of lexical units[1] in texts, but their presence in lexicons, especially semantic lexicons, is spotty. Traditional paper dictionaries obviously cannot do justice to PNs. The class changes very rapidly, there is a lot of *ad hoc* coinage, and many names change their referents.

Roman Kurc · Maciej Piasecki
Institute of Informatics, Wrocław University of Technology, Poland
e-mail: {roman.kurc,maciej.piasecki}@pwr.wroc.pl

Stan Szpakowicz
School of Electrical Engineering and Computer Science, University of Ottawa, Ottawa, Ontario, Canada
e-mail: szpak@eecs.uottawa.ca

Stan Szpakowicz
Institute of Computer Science, Polish Academy of Sciences, Warsaw, Poland

[1] We understand a *lexical unit* to be a lemma paired with a unique sense (lexical meaning). A *lemma* is an arbitrarily selected word form which represents a set of word forms with different values of grammatical categories, but the same meaning.

A. Przepiórkowski et al. (Eds.): *Computational Linguistics*, SCI 458, pp. 65–85, 2013.
DOI: 10.1007/978-3-642-34399-5_4

Electronic resources are another matter, but even there PNs are not served all that well. They do appear in wordnets, notably in *Princeton WordNet* [4], but the coverage is unsystematic and limited: even the largest, *Princeton WordNet*, describes – of necessity – only a small subset of English PNs.

The Polish wordnet project, *plWordNet* [13], has assumed all along that in general PNs will not be represented. An exception has only been made for name places linked by derivation to common nouns which denote inhabitants of such places (just as *londyńczyk* 'Londoner' makes the inclusion of *Londyn* 'London' advisable). This is a fairly systematic phenomenon in the Polish lexical system, and such units were necessary to round out the description of certain derivational relations [20]. Even so, the criterion for the inclusion of a PN was high corpus frequency of the corresponding common noun.

We seek a different solution: a dynamic open lexicon of Polish PNs, which can be expanded on demand, mainly based on large corpora. Its elements are mapped to *plWordNet* synsets by virtual *instance-type* relations. Virtuality means automatic mapping based on knowledge extracted from the existing corpora.

PN identification and classification has been much studied in Information Extraction in the scope of Named Entity Recognition (NER), starting in the late 1980s at Message Understanding Conferences. Still, NER usually puts PNs into a limited number of classes, while mapping to a wordnet requires a much finer classification. Starting with version 2.1, *WordNet* provides a fine-grained description of PNs [14], but for a limited number of PNs. Mann [11] proposed a method of automated construction of a PN ontology. He extracted with 60% precision 113000 PNs from a 1GB corpus, but noted that the expansion of *WordNet* with PNs is a complex task. Sundheim *et al.* [27] studied linking gazetteer elements to *WordNet*, but only to geographical terms. First de Loupy [10] and then Toral [28] proposed the building of a thesaurus of PNs based on *WordNet*; de Loupy added about 130675 PNs "founded on several knowledge bases", but only to 55 synsets, probably using simple heuristics. The PNs, however, were not distinguished among named entities in [10] and the resulting resource is not strictly lexical.

Toral [28] expanded *WordNet* with Named Entities (NEs) automatically extracted from the English Wikipedia.[2] The process has two phases. First, each monosemous *WordNet* noun synset is linked to a Wikipedia category if the synset member is one of the words in that category. In the second phase, NEs described in the Wikipedia by the mapped categories are added to *WordNet* as new synsets linked by the instance relation to the synset identified by the mapping. Whether a Wikipedia entry was a NE was decided by a simple language-dependent rule: PNs and common nouns follow different capitalisation patterns. A Web search determined how often the Wikipedia entries in question were capitalised. The results were compared with an empirically defined threshold set to 91%. Thus, if an entry is capitalised less than 91% of the time, it is classified as a common noun [28]. A manual evaluation on a small sample of 100 NEs gave 93.02% precision, 68.97% recall and 83.33% F-measure for a test set of 100 entries.

[2] `http://en.wikipedia.org`

It must be noted that the key problem of disambiguating polysemous category associations between categories and synsets has been neglected (there were only preliminary experiments), and so has been the problem of an extensive use of knowledge sources other than the Wikipedia. PNs have been also linked to *WordNet* but the description granularity did not improve, because the linking strictly followed the categories. In a project similar to ours, Alfonseca *et al.* [1] construct a function which maps PNs to an ontology. The function decides what ontology area is a proper place to link a new PN. The authors, however, worked with a relatively simple ontology, while we have at our disposal a large, complex wordnet – *plWordNet* 1.3.

Ruiz-Casado *et al.* [25] proposed an automatic mapping of the Wikipedia articles for Simple English, including those describing PNs, to *WordNet*. The mapping, however, depends on the vector-based similarity of the synsets glosses and the article text contents; *plWordNet* does not yet contain glosses, so we cannot apply this method.

Mapping PNs onto wordnet synsets is very similar to the problem of placing new words in the appropriate locations of a wordnet hypernymy structure, which is a key problem in wordnet expansion. Several algorithm of automated wordnet expansion have been proposed – see [26, 17], to mention only those with the highest accuracy for a large wordnet.

We seek a method of automated mapping of PNs onto wordnet synsets by a type-instance relation. Ambiguous PNs, with several different referents (such as people with the same name), can be linked to several synsets. As a result, a kind of semantic lexicon of PNs will be built, providing a fine-grained semantic description of PNs. Such a resource can then be used as a knowledge source in NER in text or in question answering. The algorithm should be based mainly on the analysis of large plain-text corpora, and only supported by knowledge extracted from a collection of structured documents such as the Wikipedia.

The starting point is the algorithm of *Activation Area Attachment* (AAA), proposed in [17], a tool for automatic wordnet expansion. We will discuss it presently, but first we will examine what kind of knowledge sources describing PNs can be extracted from a text corpus and supplemented by structured sources like the Wikipedia. Next, we will analyse the applicability of AAA to the problem of PN mapping from the perspective of the identified knowledge sources. The required modifications of AAA will be proposed, and evaluated experimentally.

2 Sources of Lexical Knowledge for the Mapping of Proper Nouns

The Activation Area Attachment algorithm (AAA) [17], developed for the semi-automated expansion of *plWordNet*, has higher accuracy than other methods proposed for this task; see [3] for an analysis. If, however, it is to work well, it requires the appropriate sources of lexical knowledge (henceforth, the abbreviation KS will stand for "knowledge source"). Both the quality and the quantity (precision and coverage) of KSs are important. There are no gold standards – no manually constructed

KSs with a detailed description of large sets of PNs – so we measured the precision of KSs manually. It is prohibitively labour-intensive to build manually a complete KS for a large set of PNs, the recall of such KSs cannot be estimated and their coverage must be used instead.[3]

There are two paradigms of extracting lexical-semantic relation instances from texts:[4] *pattern-based* and *distributional*. Pattern-based methods, first introduced in [5], deliver results with relatively high precision. In particular, a pattern carefully designed by hand can be applied to low-frequency lemmas, but its coverage tends to be low, because only lemma pairs in specific contexts are extracted. On the other hand, automatically generated patterns, such as those produced by the *Espresso* algorithm [15] or *Estratto* algorithm [8], are sensitive to the frequency of lemma pairs. In such algorithms, the evaluation of patterns and relation instances is based on the statistical analysis of the frequency of both pattern matching locations and lemma pairs.

Measures of Semantic Relatedness (MSRs), a well-established tool of Distributional Semantics, assign a real value to any pair of lemmas, but lemma pairs with high semantic relatedness values represent many different lexico-semantic relations.[5] In the context of applying MSRs to the description of PNs, it is important to note that an MSR's precision drops significantly for lemmas with lower frequencies in the corpus (in our experience, lemmas occurring less than 100-330 times in a billion-word corpus). Almost every single-word occurrence contributes to the results produced by an MSR;[6] in theory, every lemma occurring in a corpus can be described by an MSR constructed from this corpus. Wordnet-expansion algorithms, including AAA, are usually applied to lemmas with frequency above some threshold, in order to ensure good MSR performance for those lemmas. It is therefore typical to assume that an MSR is defined for almost every noun pair, and AAA treats an MSR as the basic KS for wordnet expansion. KSs produced by other methods of semantic relation extraction, such as lemma pairs extracted by patterns, were considered as very valuable, but supplementing the basic information provided by an MSR.

The semantic properties of PNs as lexical units can be characterised by their corpus occurrence in a way similar to other nominal lexical units. Many PNs, however, come from such structured text resources as the Wikipedia. Those PNs are often used only in specific subdomains (for example, company names or names of sportsmen other than stars). This means that they can be infrequent even in a very large

[3] We understand the coverage simply as the size of a KS.

[4] The extracted instances of the target relations may be represented as lemma pairs.

[5] They cannot be directly used for finding locations in the wordnet structure where new lemmas, including PNs, can be attached. MSRs are rather sensitive to lemmatisation errors in the corpus and to accidental associations caused by low-frequency words.

[6] The notable exception are MSRs based on the frequency of co-occurrence of words with other words in specific syntactic relations, for example, a specific adjective as a modifier of a noun. In some contexts such relations can be absent, for example a lemma as a definition, a list of lemmas or other contexts in which there are no proper sentential syntactic structures.

corpus, so it is more difficult to perform corpus-based extraction of knowledge related to such PNs. In particular, because the frequency of many PNs is based on less trustworthy statistical data, it may be too low to guarantee good MSR precision. All this suggest the use of many different KSs if we want to describe effectively such infrequent PNs. That is to say, we must redesign AAA in such a way that MSR is no longer the basic background KS, but is instead treated as one of many KSs. PNs are also sometimes directly defined in text – more often than common nouns. Pattern-based methods, especially tuned for a specific type of text, can mine such definitions.

Our experiments ran on two types of KSs, based on a corpus and on metadata. The latter, introduced to supplement limited coverage provided for PNs by corpus-based KS, have not been used in AAA so far. That is because AAA had been only applied in purely corpus-based wordnet expansion. All knowledge sources were stored as sets of lemma pairs, each lemma pair representing a potential instance of a lexico-semantic relation. Extraction methods based on patterns and meta-data exploration produce directly sets of lemma pairs. In the case of MSR, we produced lemma pairs by linking a lemma x with each lemma from the top k lemmas most semantically related to x. The exact type of lexico-semantic relations represented in a KS depends on the KS type. For example, the case of MSR-based KS lemma pairs can represent several different wordnet relations. It is inevitable that any KS includes a substantial percentage of erroneous lemma pairs, not representing any lexico-semantic relation.

Metadata are also a kind of definitions for PNs, so for example PNs can be classified by metadata into semantic classes. Corpus-based sources were obtained using both *distributional* and *pattern-based* methods. Corpus-based methods were applied to our custom-made joint Polish corpus of about 1.2 billion tokens [18].[7]

For our experiments, we extracted from the joint corpus an MSR for 33577 common nouns from *plWordNet* 1.2 and for a large set of PNs; this set will be presented in Section 4. We applied an MSR construction algorithm introduced in [21]. Noun and PN occurrences were described by several lexico-morpho-syntatic relations: modification by a specific adjective or adjectival participle, coordination with a specific noun, modification by a specific noun in the genitive, and association with a specific verb as a subject-predicate pair. No sufficiently robust shallow parser of Polish is available, so the associations were recognised using a set of manually crafted constraints expressed in JOSKIPI, a language of morpho-syntactic constraints developed for TaKIPI – a tagger for Polish [16]. The associations were recognised at any distance within a sentence. JOSKIPI has a higher expressive power than systems based on regular expressions, such as Word Sketches – see for example [24]. Co-occurrence statistics were next filtered [21], Lin's version of Mutual Information [9] was applied to feature weighting, and the cosine measure for row similarity was calculated.

[7] The corpus includes the IPI PAN Corpus [23], the *Rzeczpospolita* corpus [7], Polish Wikipedia (`http://pl.wikipedia.org`) and a corpus of longer documents in Polish collected from the Internet.

Many less frequent PNs can lack a good description if an MSR is based on lexico-syntactic constraints. For example, morphological analysers often do not recognise PNs, so the grammatical category values cannot be assigned. Yet it is those values that make the constraints work. That is why we also used in the experiments a second kind of MSR, based on simple co-occurrences in a limited text window of a PN and one of thousands of common nouns from *plWordNet* 1.2. The former MSR will be called *constraint-based*, the latter – *window-based.*

Pattern-based sources were prepared using *three manually constructed lexico-syntactic patterns* (text patterns) written also in the JOSKIPI language; for a complete description of those patterns, see [22]:

- ⟨NP, NP, … i inne (*and other*) NP⟩,
- ⟨NP jest (*is a*) NP⟩,
- ⟨NP to (*is a*) NP⟩.

We developed the patterns for the extraction of hypernymic and synonymous word pairs [22].

We also used seven manually designed *structural patterns* discussed in [18] and repeated here in a slightly simplified form:

R_*ToRodzaj_Lnk*: $\texttt{Noun1}_{case\in\{nom,acc\}}$ … (“–” | “to” *is*) … (“rodzaj” *kind* | “typ” *type* | “podtyp” *subtype* | “dziedzina” *domain* | “forma” *form* | “sposób” *manner*)$_{case\in\{nom,acc\}}$ $\texttt{Noun2}$
R_Dash_*Lnk*: $\texttt{Noun1}_{case\in\{nom,acc\}}$ … “–” … $\texttt{Noun2}_{case\in\{nom,acc\}}$ …
R_*ToElement_Lnk*: $\texttt{Noun1}_{case\in\{nom,acc\}}$ … (“–” | “to”) … (“element” *element* | “część” *part* | “fragment” *fragment*) … $\texttt{Noun2}$
R_Dash_*Noun*:[8] $\texttt{Noun1}_{case\in\{nom,acc\}}$ … “–” … $\texttt{Noun2}_{case\in\{nom,acc\}}$
R_After_Parentheses: $\texttt{Noun1}_{case\in nom,acc}$ … “(” … “)” … $\texttt{Noun2}_{case\in\{nom,acc\}}$ — there is no beginning of a link before Noun2
R_In_Parentheses: $\texttt{Noun1}_{case\in\{nom,acc\}}$ “(” not(*verbs* and *punctuation marks*) $\texttt{Noun2}_{case\in\{nom,acc\}}$

These patterns were applied to the first sentence of each Wikipedia article. Their construction was similar to the text patterns, but they also used the specific syntax occurring in the initial parts of Wikipedia articles [18].

KSs based on meta-data were the *Wikipedia categories* extracted from articles entitled with PNs. We used a simple manually constructed pattern to obtain these categories. The categories often had long names, consisting of several words, and were mostly absent from *plWordNet* (e.g., *pisarze urodzeni w Polsce* ‘writers born in Poland’). Categories were also usually in the plural form. We lemmatized them, and extracted from the lemmatized categories one-word and multi-word nominals in the nominative (potential syntactic heads).[9] In that manner, we obtained more general

[8] Noun2 is different than the triggering words of the rules R_*ToRodzaj_Lnk* and R_*ToElement_Lnk*.

[9] 6582 Multi-Word Expression categories out of 168678 categories in general, 204 *unique* Multi-Word Expression categories out of 3325 categories.

and much shorter categories which match lemmas present in *plWordNet.* 168 678 pairs $\langle PN; noun \rangle$ were extracted. We did not extract $\langle PN; PN \rangle$ pairs because – as we checked manually on a large sample – they did not represent an instance-type relation. Instead, they represented lexical relations such as meronymy, while we focused on PN classification by the *instance of* relation. In contrast with [28], we used all categories which could be matched against the wordnet, not only monosemous categories.

We prepared two more KSs based on metadata. We extracted common nouns inside *brackets* in expressions of the form PN (common noun); the bracketed expression is a category name, e.g., *Casablanca (film).* We obtained in total 1202 unique nominals, 842 of which were Multi-Word Expressions. This source consisted of 5628 pairs and it described 3456 of our PNs. We also used about 50 *coarse-grained categories* such as name, surname, location, city, company, organisation etc., assigned to PNs in a large Polish gazetteer [12]. All category names were found among lemmas already described in *plWordNet.* Most of them correspond to synsets from the upper parts of the hypernymy structure. The gazetteer contains $\approx$ 1.4 million Polish PNs collected from the Internet. Only some of them have categories assigned.

In a lemma-based version of AAA, described in Section 3, we used all these heterogeneous KSs extracted from text corpora and semi-structured sources (Wikipedia) [18]. Other than in [26], this version does not require all KSs to be probabilistic. Each source in AAA is said to represent a vote, sometimes a fraction of a vote. All KSs are delivered to the algorithm in the form of lists of lemma pairs $\langle x, y \rangle$ produced by patterns and generated from lemma pairs with high values of semantic relatedness. x is a PN (not present in *plWordNet,* likely a multi-word) and y (nominal lemma from *plWordNet*) is semantically related to x according to the given relation extraction method and the corpora analysed.

3 The Lexical Activation Area Attachment Algorithm

The Lexical Activation Area Attachment Algorithm, presented later in this section, is based on the algorithm of Activation Area Attachment (AAA). AAA takes as an input a lemma which represents one or more lexical units. AAA finds synsets to which those lexical units can be attached as synonyms (synset members) or hyponyms/hypernyms if the appropriate synset does not exist yet.[10] We expect that synsets pointed out by AAA are as close as possible (in path length of the wordnet graph) to the most appropriate locations.

As a result of the uncertainty inherent in KSs, AAA presents each suggested location as an *attachment area* – a hypernymy subgraph with one synset marked as the most likely location. In order to identify attachment areas for a new lemma x, a measure called *semantic fit* is first calculated for each new synset during the first phase of AAA. The calculation is based on all KSs. Semantic fit assigns to each $\langle$lemma, synset$\rangle$ pair a value which shows how well the given lemma fits semantically the

[10] A lexical unit belongs to exactly one synset. For a new lemma, AAA marks a handful of synsets as possible locations.

given synset. This is done with respect to both the *content* (synset members) and the *context* (other synsets linked directly or indirectly by paths across the graph of wordnet relations). As it will be shown later in the algorithm description, the semantic fit calculation is based on the support found for x in all KSs of all kinds. So, the semantic fit is derived from heterogeneous knowledge extracted from corpora. In each attachment area, a synset with the highest value of the semantic fit is marked to represent the whole area. Only a limited set of attachment areas, which represent the highest *semantic fit* to a new lemma x under consideration, are suggested as descriptions of the senses of x.

AAA has been conceived as a tool to support wordnet editors, so the completeness of the lists of proposed lexical units is an important aspect. The algorithm is therefore – mostly intentionally – set to over-generate within the limits of the precision acceptable by editors. AAA has become the key component of the WordnetWeaver system for the semi-automated wordnet development.

AAA relies crucially on KSs for the placement of a lemma in the wordnet. We have made a general assumption: no KS is error-free, but errors can be partially compensated for by collecting information from several KSs across local wordnet sub-graphs (subgraphs of the structures of wordnet relations). We have assumed that the overall error level of KSs suggesting wrong attachment of a lemma is inversely correlated with the distance in the graph of wordnet relations, which can be simply calculated as the number of relation links. For example, direct hypernyms and close hypernyms are very difficult to distinguish by automatic methods. Suppose that synset S' is an indirect hypernym of synset S at the distance 2 (they are only separated by a direct hypernym of S) and an extraction method suggests linking a lemma y to a lemma included in S' (as a hypernym of y). One can then expect that in fact it might be reasonable to link y to S. It is much less likely that a good KS would erroneously suggest an attachment of a lemma to a synset far away from the exact point. At the very least, such a KS should be avoided.

For many KSs, it is also often quite possible to have more indirect connections or even links by other relations, such as meronymy or holonymy. This is especially visible when an MSR produces high values of semantic relatedness for lemmas linked by several wordnet relations. That is why the support for linking x to synset S is collected not only from lemmas in S but also from synsets accessible from S via paths across the wordnet graph. This basic idea was expanded into a model of the semantic fit replication over the wordnet network.

For a synset S, we assume that we should consider not only the semantic fit directly assigned to it after analysing KSs, but also the semantic fit assigned to synsets in the local context of S in the wordnet graph. Some portion of the semantic fit can be *replicated* to S from the synsets linked to S by paths in wordnet. That is to say, fit values are increased by a fraction of the starting synset fit. KSs have errors, and those errors can lead to erroneously spreading the semantic fit over lemmas linked by semantic relations. For example, chunking errors can cause a pattern to extract a pair ⟨*tank*, *vehicle*⟩ instead of *military vehicle* or *armoured vehicle*. By copying part of the semantic fit from a synset S' to S, we follow the nature of KSs (we cannot be sure of the right placement of the fit) and we try to compensate for the errors in KSs. Experiments with AAA [3] have shown that using a local context of the size of up to several links increases AAA's performance in the automated wordnet reconstruction task. The increase takes place in comparison with a method based only on the local semantic fit, or using only directly connected synsets as a context.

The replication of the semantic fit cannot be overestimated, however. A typical KS is constructed with the focus on precision. One can expect that the semantic fit directly assigned to a synset S should be weighted more highly (be more trustworthy) than the semantic fit collected from the corpus. Moreover, as we stray further from S via wordnet relation links, the synsets we encounter become less semantically related to S. This observation is reflected in the mechanism of weighting the indirect semantic fit: typically, the decrease of the semantic fit is correlated with the length of the path via wordnet relation links, where the length is equal to the number of links traversed.

The semantic fit replication model of AAA has been later augmented by the notions of *transmittance* and *impedance*. Transmittance models how much of the semantic fit assigned to a synset S can be replicated to the synsets linked to S by a relation R. Transmittance characterises relations. Impedance models replication of the semantic fit between links of different relations. For example, we can assume that semantic fit is freely replicated across the hypernymy-only paths, mixed holonymy/hypernymy paths are questionable. Both notions received formal interpretation, so they will be revisited in more details after we have presented a formal algorithm description.

AAA was presented in [22, 17, 19] in its different development versions. They all treat synsets as the basic elements of the wordnet structure; synsets mediate the semantic fit originating from lemma pairs described in the KSs. In the first step of AAA, semantic fit is calculated for synsets. The calculation is based on the KS supported for lemmas belonging to these synsets. Next, the synset semantic fit is transmitted between synsets – lexical units are not referred to any more. It was implicitly assumed that the MSR describes a large number of lemmas, and that the synset level is useful in calculating the average over MSR values between a new lemma (not yet included in the wordnet) and synset members.

The situation is different with the PNs actually used (taken from the Wikipedia): they are infrequent and the MSR can be reliably calculated only for selected PN pairs – other KSs are also very incomplete. First experiments with applying AAA to PNs showed that the accuracy for non-singleton synsets decreased, because there is no MSR-based KS plus some others, but very often only one single KS which decides about the synset's semantic fit calculated in the first step of AAA. There is no voting effect of several KSs used together typically in wordnet expansion. This problem can be bypassed by treating a wordnet as a graph of links directly between lexical units. That is to say, synset links are copied to all pairs of lexical units, respectively. The support from KSs is directly replicated along those links, instead of being somehow averaged first at the level of synsets.

We now show a reformulated version of AAA, called *Lexical AAA* (LAAA), which works at the level of lexical units. It takes as an input a new lemma, a wordnet and a set of KSs. LAAA has two phases. In *Phase I*, KSs are used to discover synsets semantically close to a new lemma. For each synset *S*, the semantic fit of a new lemma *x* to *S* is calculated as a sum of:

- *direct semantic fit*: the semantic fit of *x* with the lemmas of *S*,
- *indirect semantic fit*: the semantic fit of *x* replicated from the lexical units belonging to synsets in the local context of *S* (synsets connected to *S* by up to *n* relation links).

In AAA synset semantic fit is replicated. In LAAA, on the other hand, when a context synset includes more than one lexical unit, the support from each of them is replicated independently. This small but significant difference gives more freedom in tracing the support and its nature, and modelling the final semantic fit calculation. The indirect semantic fit, replicated across the paths in the context, is processed by selected types of transformations sensitive to the distance measure as the path length and to the types of links in the path. Those transformations occur in step 2 of Phase I of LAAA on page 75. They will be explained shortly. We aim at finding a place to attach a new lemma by the *instance-of* relation, but errors in KSs can cause KS suggestions to point to hypernyms, meronyms or even lexical units linked by other lexical relations. Links of different types should be considered in semantic fit replication which is intended to compensate for these errors.

In *Phase II*, those synsets which have a semantic fit with *x* above certain threshold are grouped into connected sub-graphs called *activation areas*. A number of activation areas which have the highest fit with *x* are returned by LAAA as descriptions of the suggested lexical units for *x*. We call them *attachment areas*. Each attachment area represents a part of the wordnet structure where *x* should be inserted; it is presented as a whole to linguists in the *WordnetWeaver* system. In each attachment area, one synset with the highest semantic fit – selected to represent the area in the tests – is visually marked in *WordnetWeaver*. The results of evaluation presented in Section 4 refer to synsets so marked. Details are presented below in the specification of the algorithm; earlier descriptions appeared in [22, 17, 19].

The contextual influence represented by the indirect semantic fit is modified by the distance, and it decreases with the distance. We assume that a local synset context is defined by heterogeneous paths consisting of links of all possible relations. For such contexts, weights assigned to paths cannot be based only on the distance. We also account for semantic relations represented by links constituting the path [19]. For each relation, its *transmittance* was expressed as a weight modifying the fit transmitted through links of the given relation. The transmittance values for different relations were estimated heuristically and tested in experiments [19]. The highest value was given to hyponymy (inheritance direction), lower to meronymy and very low to antonymy (not zero, because the transmittance is correlated with errors included in the KSs). Transmittance can be also estimated from statistical correlation between semantic fit values observed in the training data, that is, lemmas already described in a wordnet [19].

The indirect semantic fit is calculated across heterogeneous paths, too, for example a series of hypernymic links ending with a meronymic link. Not every combination of links can be interpreted as a meaningful path to infer the fit for a synset from the linked synsets. For example, an antonymy link inside a hypernymic path seems to be a natural border for replicating evidence between synsets. *Impedance* was introduced to model the phenomena of link combination. The types of contexts for which impedance is zero was motivated linguistically by the meaning of the relations.

Lexical Activation Area Attachment

In the algorithm, $Score[]$ is a table of total semantic fit values calculated for synsets and a new lemma; the functions $T()$ and $I()$ give the values of transmittance and impedance, respectively; the function $path(S_1,S_2)$ returns the best path between the synset S_1 and S_2 (this path gives the smallest decrease in the transmitted semantic fit); $W[]$ is a table of weights for the KSs; K is the set of all KSs.

Phase I Calculation of the fit between a lemma x and synsets.

1. $Score[S] \leftarrow \sum_{k\in K} \sum_{y\in S} W[k] fit(x,y,k)$
2. $Score[S] \leftarrow$
   $Score[S] + \sum_{S'\in cntx(S)} \left[\sum_{k\in K} \sum_{y'\in S'} f_w\left(path(S',S),k,W[k]fit(x,y',k)\right)\right],$
   where
   - $cntx(S) = \{S' \in Synsets : S'$ is accessible from S by a path of up to r links$\}$
   - $f_w(P,k,v) =$
     a. $res \leftarrow T(p_n)I(p_n,p_{n-1})\ldots I(p_1,p_2)T(p_1)v$
     b. $res \leftarrow f_{length}(|P|,k,res)$
     c. `return` res
3. `if` $Score[S] \leq \tau$ `then` $Score[S] \leftarrow 0$

Phase II Identification of the lemma senses: areas and centres.

1. Let $\mathbf{G}$ be a selected subset of the synset relations which includes at least hypernymy and hyponymy.
2. $Activation(x) = \{\mathbf{S} : \mathbf{S} = \{S : S$ is a synset & $Score[S] \geq min\_score\}$ & $\mathbf{S}$ is connected subgraph with respect to $\mathbf{G}\}$.
3. $Attachment(x) \leftarrow$
   $\{$ the top max_att subgraphs $\mathbf{A} \in Activation(x)$ according to $max_{S\in\mathbf{A}}(Score[S])\}$
   $\cup$
   $\{$all $\mathbf{A} \in Activation(x)$ such that $max_{S\in\mathbf{A}}(Score[S]) \geq strong\_fit\}$

In Step 1 of Phase I, the semantic fit between x and each lemma y in the wordnet is calculated as a sum of weights assigned to KSs, where $\langle x,y\rangle$ are in KSs. For each synset S, $Score[S]$ is calculated as a weighted sum from the semantic fit for x and each lemma y in the local context of S. Global weights for the whole KSs (table $W[]$) can be based on the evaluated *accuracy* of the KSs. The accuracy can be assessed by comparing the pairs with the instances of wordnet relations (assuming that the wordnet provides complete coverage) or by a manual assessment of a representative sample. We chose the latter. Local weights – values of $fit()$ – depend on the character of the particular KS, e.g., MSR values. In the experiments presented here we assumed a simplified model. All local weights were set to 1 or 0 (for lemma pairs not supported by the given KS).

A local context is formed by synsets connected with S by a *path* no longer than a given length. This breadth-first type of context search can be changed for another model. A path between two synsets is understood as a sequence of links (relations) to be traversed in order to reach from one synset to the other. Because two synsets can be linked by more than one path in a wordnet (like when there are multiple hypernyms), the *path* function in Step 2 returns the best path for the two synsets. The best path is not always the shortest, but one which results in the smallest decrease of the transmitted fit. This decrease depends directly on the path length but also on the types of links in the path. This can be seen in the algorithm of $f_w()$ in Steps 2.a and 2.b of Phase I on page 75: the fit is first multiplied by the weights corresponding to the transmittance of links and impedance of link-to-link connection: transmittance of p_1, next impedance between p_1 and p_2, transmittance of p_2, ...impedance between p_{n-1} and p_n and finally transmittance of p_n. However, the value is next transformed by f_{length}, taking into account the length $|P|$ of the path and the type of the KS; for example, $f_{length}(l,r) = r/(2*l)$ in a simple but effective implementation used in all experiments presented in this paper.

As noted in Section 2, two groups of KSs were introduced for the task of PN mapping, based on a corpus and on metadata. KSs based on metadata have different semantics than corpus-based sources. They can be thought of as PN classifiers because they usually relate PNs with their remote hypernyms. In order to utilise these long-distance associations in AAA, we would need a very large local context. Such context would be too broad for "ordinary" KSs based on corpus analysis. Thus, in LAAA we introduce a non-uniform size and structure of local context for data obtained from particular KSs. For the fist group, the semantic fit is collected from the distance of at most 2 links, and for the distance d the weight is $1/(2d)$. For KSs based on metadata, longer paths are allowed and paths are comprised of selected semantic relations; the size of the context is limited to 1, except paths built from hyponymy links only (data pertaining to a synset S can influence all its direct and indirect hyponyms).

The contextual influence is modified by distance: it is first decreased with the distance and then stabilized at the level of 0.25 of the original value. Thus, the fit based on metadata marks whole hypernymic subtrees – whole classes.

Phase II first identifies, for a PN x, continuous areas (connected subgraphs) in the graph of the selected wordnet relations in such a way that each synset in an area is assigned some minimal score (*min_score* threshold) to the given PN.[11] For each area, a synset with the local maximum of the score is identified. All subgraphs with the score above certain threshold are preserved and returned as descriptions of the possible types of x.

[11] The relations include at least hypernymy/hyponymy but others are also worth considering.

4 Experiments

PNs are described in a detailed way neither in *plWordNet* (with few exception) nor in any available semantic lexicon, so we needed a manual evaluation. The goal was to analyse the precision of the algorithms in selecting the appropriate synsets and recall measured in terms of the PN senses covered (different types of referents identified for a PN). The method was evaluated using *plWordNet* 1.2 and resources described in Section 2. All lemmas already included in *plWordNet* 1.2 were removed from the test data. We wanted to analyse the combination of KSs extracted from the corpus and those based on metadata, so the experiments were performed only for 119820 PNs covered by Polish Wikipedia (dump on Oct. 20, 2010). Moreover, in order to decrease the computational cost of the extraction of KSs (especially for MSR), all experiments were performed on a randomly selected test set of 10000 PNs.

In the first *baseline* experiment, we used only Wikipedia categories as the KS for mapping PNs from the test set onto *plWordNet* 1.2. In this experiment, summed up in Table 1, a KS based on Wikipedia categories was treated as a corpus-based resource, but we did not account for their characteristics. In the second experiment, pairs ⟨PN, noun lemma⟩ extracted from Wikipedia categories were treated as a KS based on meta-data: the categories provided semantic fit for the whole hypernymy subgraphs – see the description of LAAA. These preliminary experiments concluded with a test where we used MSR as the sole KS.

Table 1 shows the results of the baseline experiments – in all of them we ran the LAAA algorithm. For each PN, up to 5 top-scored suggested attachments, pairs $\langle PN, synset \rangle$, were considered. We took a sample of 400 suggestions (5% statistical significance level [6]). We manually revised the sample marking suggestions as three types: *correct* (marked as T in the table); *super type* or an indirect hypernym (H); a *co-hyponym* (K) — the suggested attachment and the PN share a close hypernym (1-2 links up the hypernymy structure.) In all three cases the results were lower than the 93.02% precision reported in a similar experiment in [28], but that experiment used PNs with categories corresponding to monosemous lemmas in the wordnet. We performed mapping for all, also polysemous, PNs from the test set. We were also unable to take advantage of PNs already described in the wordnet because *plWordNet* contains few PNs.

In the baseline experiment, semantic fit was calculated only from the Wikipedia categories and then replicated along the structure of the context. The categories were used as regular KSs. This resulted in a larger attachment via indirect hypernyms and co-hyponyms (indirect "cousins"). Such results were expected because categories from the Wikipedia represent distant, top-level hypernyms. When they play the role of local sources, a new lemma x is attached higher in the hypernymy tree. There are two possible scenarios. The new lemma is connected either to the appropriate tree or to a tree started by a co-hyponym (possibly an indirect one – a "cousin"). In the first case, we observe more indirect hypernyms and co-hyponyms. In the second case, more false results are reported.

In the second experiment, Wikipedia categories corresponding to polysemous *plWordNet* synsets activate several synsets for every PN. Most of them are located in

Table 1 Baseline experiments in mapping PNs onto *plWordNet*, considering only the top-scoring synset. T – correct instances, K – co-hyponym instances (a suggestion is a co-hyponym of the correct synset), H – hypernyms (a suggestion is a close hypernym of the correct synset).

T [%]	K [%]	H	total [%]
Wiki categories as a corpus-based KS			
76.3	3.5	4.5	84.3
Wiki categories as a metadata-based KS			
82.8	2.3	2.5	87.5
MSR as a corpus-based KS			
25.8	8.8	6.8	41.3

the upper parts of the hypernymy hierarchy and, when the context is small, most of the activated synsets are presented as possible attachments. Many of them describe senses unrelated to the PNs analysed.

The third experiment was conducted with the window-based MSR as the only KS. Even though the MSR is characterized by high recall, pairs in such KS may be random due to the lack of statistical evidence. See the appendix for examples of the lists of 15 or so most semantically related nouns and PNs produced by MSRs.

In this series of experiments, summed up in Table 2, we applied LAAA and all KSs (Section 2). LAAA mostly does not define a clear "cut" between true suggestions and the rest. Even so, for the top-scoring suggestions ("First suggestion correct" in the table), the precision is good, and it demostrates LAAA's ability to disambiguate the PN type based on sparse KSs and the wordnet structure. The class labelled "At least one correct suggestion in top 5" presents LAAA as a tool which supports manual work on the construction of a semantic lexicon of PNs: a list of five suggestion is short enough to be quickly analysed and long enough to include several valid classes for a PN. We can notice in Table 2 that in 99% cases the algorithm produced at least one useful suggestion for a linguists. The use of such a tool can make the work of linguists much more efficient.

For many PNs, the MSRs produce results of lower accuracy, even with the high frequency threshold, but the MSRs combined with other sources actually helped disambiguate meaning and improve precision. A version of PN-dedicated MSR extraction algorithm is required.

It is worth pointing out that the experiments were performed for PNs described by both monosemous and polysemous categories.

A detailed review of the results allowed us to notice some interesting phenomena. Errors in the results of LAAA come from the lack of data in KSs or their similarity: many sources have the same pair $\langle x, y \rangle$. When there are insufficient data, it is typical that all synsets containing lemma y will be chosen. The order of the results will

Table 2 Mapping PNs on *plWordNet* based on the LAAA algorithm and heterogeneous KSs for up to 5 top-scored synsets only. T – correct instances, K – co-hyponym instances, H – hypernyms.

T [%]	K [%]	H	total [%]
First suggestion correct			
82.5	3.8	6.0	92.3
All suggestions correct (up to 5)			
73.5	2.8	4.5	81.0
At least one correct suggestion in top 5			
95.3	1.5	2.2	99.0

depend on the order of synsets in a wordnet. When many sources have only one and the same pair, the situation is again similar. Here is an example:

"The Lost Files" – ⟨*obraz* 'picture', *produkcja* 'production', *film* 'movie'⟩
"The Lost Files" – ⟨*kino* 'cinema', *duży ekran* 'silver screen' , *kinematografia* 'cinematography', *film* 'movie', *dziesiąta muza* 'the 10th Muse' ⟩

On the other hand, we note that information coming from different KSs can disambiguate and enhance the classification provided by Wikipedia categories. For example, *Rudolf Schuster*, a former President of Slovakia, but also a former president (mayor) of the city of Košice, was described by LAAA as:

"Rudolf Schuster" – ⟨*osoba* 'person', *człowiek* 'man', *istota ludzka* 'human being', *jednostka* 'individual'⟩, a hyponym of ⟨*osoba* 'person'⟩
"Rudolf Schuster" – ⟨*prezydent*⟩, a hyponym of ⟨*głowa państwa* 'head of state', *dygnitarz* 'dignitary', *urzędnik wysokiego szczebla* 'high official', *dostojnik* 'dignitary'⟩
"Rudolf Schuster" – ⟨*prezydent*⟩ a hyponym of ⟨*samorządowiec* 'self-government official', *dygnitarz* 'dignitary', *urzędnik wysokiego szczebla* 'high official', *dostojnik* 'dignitary'⟩
"Rudolf Schuster" – ⟨*człowiek rozumny* 'rational man', *człowiek* 'human', *homo sapiens*, *ssak* 'mammal'⟩
"Rudolf Schuster" – ⟨*prezydent* 'president', *przewodniczący* 'chairman', *prezes* 'chairman'⟩, a hyponym of ⟨*przełożony* 'superior', *szef* 'chief', *głowa* 'head', *zwierzchnik* 'superior'⟩

Here is another example:

"Dżuma" – ⟨*dramat* 'drama', *sztuka* 'play'⟩ a hyponym of ⟨*utwór literacki* 'work of literature', *dzieło literackie* 'literary creation'⟩
"dżuma" – ⟨*zaraza* 'plague', *epidemia* 'epidemic'⟩, a hyponym of ⟨*plaga* 'plague', *pomór* 'plague', *mór* 'plague'⟩

"Dżuma" – ⟨*powieść* 'novel'⟩, a hyponym of ⟨*proza* 'prose'⟩
"dżuma" – ⟨*choroba zakaźna* 'contagious disease'⟩ a hyponym of ⟨*dolegliwość* 'ailment', *niedyspozycja* 'indisposition', *przypadłość* 'affliction', *syndrom* 'syndrome', *zachorowanie* 'falling sick', *niemoc* 'malaise', *patologia* 'pathology', *niedomoga* 'failure', *zespół* 'syndrome', *brak zdrowia* 'ill health', *schorzenie* 'disease', *niedomaganie* 'ailment', *choroba* 'sickness'⟩

The design for disambiguation has one drawback. It may be observed in the results for the evaluation class "All suggestions correct (up to 5)". LAAA's suggestions are the best suggestions taken from connected subgraphs. This means that an average suggestion from one subgraph can have a higher $fit(x,S)$ than the best result from the next subgraph. This results in a decrease of precision measured for up to 5 top suggestions, because in LAAA there is no method for the estimation of the number of results (the number of possible senses).

It should be noted here that LAAA cannot decide that two lemmas are similar when they are not identical to the letter. Therefore *'Back to the future'* differs from *'Back to the future (movie)'*. The same problem regards errors in lemmatisation. If y from the pair does not match a lemma in the wordnet, then it cannot be used in the process of attaching x.

The evaluation of *the first suggestion only* showed that LAAA had almost reached the level of precision reported in [28], except that in our case ambiguous PNs were also taken into account.

5 Further Research

In general, a wordnet could be used as a source of information about PNs in text processing. The number of PN, however, is very large and increases all the time, so their extensive description in the wordnet seems infeasible. On the other hand, the task could be restated as dynamic mapping, where only selected group of PNs (for example related to a specific domain) would be attached to the wordnet when necessary. Due to the number of PNs, such an approach would require an automated method.

Our starting point was the AAA algorithm. It was then modified to apply to the specificity of PNs. As a result, we obtained a generalized LAAA method such that the basic AAA could be seen as parametrized LAAA. At the same time, we conducted experiments with KSs. Both AAA and LAAA can use various KSs. In the case of PNs, though, corpora-based methods are limited and provide only sparse knowledge. Thus they have to be supplemented by other sources, for example definition-like descriptions of PNs or metadata. These sources should preferably be selected according to a specific text genre or even text type associated with PNs. In LAAA we ensured that sources acquired from semi-structured resources of high reliability and those extracted by means of statistical analysis can be interpreted in a different manner. More general metadata sources are allowed to work in a general context, while corpus-based sources are used in small, local contexts.

We have achieved the precision of 81% for all correct suggestions out of up to the top 5. The precision was 92.3% for first suggestions (with the highest score per PN). Such results allow us to expect positive effect in practical applications of the mapping. Examples of such applications include

- semi-automated construction of gazetteers with fine-grained semantic classification of PNs,
- support for Named Entity Recognition,
- text classification,
- Information Extraction by enhancing the semantic description of the text.

Nonetheless, KSs like MSR suffer from the infrequent occurrences of PNs. That is why they must be used with great care: even though they have high recall, their precision is usually low. We are now working on a dedicated version of MSR extraction method to make this source more useful for PNs. On the other hand, MSR often groups PNs of the same semantic class, e.g., cities or people of the same profession. This property of MSR can be applied in a bootstrapping procedure of PN mapping using PNs already linked to the wordnet. This is similar to the use of categories of PNs from the wordnet, proposed in [28].

The number of suggestions returned by LAAA is also a subject of our future research. The achieved accuracy for the top-scored suggestions is at a good level, but still we require a method of selecting only high-reliability suggestions and their appropriate number for a given PN. Too low a number would result in missing some meanings, while keeping a constant number of suggestions may lead to overgeneralization and loss in precision. We therefore plan to apply methods for the estimation of the number of senses. Also, methods for the evaluation of mapping proposals must be included in our method in the future.

Acknowledgement. Financed by the Polish National Centre for Research and Development, project SyNaT.

References

1. Alfonseca, E., Manandhar, S.: An Unsupervised Method for General Named Entity Recognition and Automated Concept Discovery. In: Proc. of the 1st ICGW (2002)
2. Bouvry, P., Kłopotek, M.A., Leprévost, F., Marciniak, M., Mykowiecka, A., Rybiński, H. (eds.): SIIS 2011. LNCS, vol. 7053. Springer, Heidelberg (2012)
3. Broda, B., Kurc, R., Piasecki, M., Ramocki, R.: Evaluation method for automated wordnet expansion. In: Bouvry, et al. [2]
4. Fellbaum, C. (ed.): WordNet: An Electronic Lexical Database. MIT Press (1998)
5. Hearst, M.A.: Automatic Acquisition of Hyponyms from Large Text Corpora. In: Proc. 14th International Conference on Computational Linguistics, pp. 539–545 (1992)
6. Israel, G.: Determining Sample Size. Tech. rep., University of Florida (1992)
7. Korpus Rzeczpospolitej, corpus of text from the online edtion of daily "Rzeczpospolita" (2008), http://www.cs.put.poznan.pl/dweiss/rzeczpospolita

8. Kurc, R., Piasecki, M., Szpakowicz, S.: Automatic Acquisition of Wordnet Relations by Distributionally Supported Morphological Patterns Extracted from Polish Corpora. In: Sojka, P., Horák, A., Kopeček, I., Pala, K. (eds.) TSD 2010. LNCS, vol. 6231, pp. 133–141. Springer, Heidelberg (2010)
9. Lin, D.: Automatic retrieval and clustering of similar words. In: Proceedings of the Joint Conference of the International Committee on Computational Linguistics, pp. 768–774. ACL (1998)
10. de Loupy, C., Crestan, E., Lemaire, E.: Proper Nouns Thesaurus for Document Retrieval and Question Answering. Atelier Question-Réponse, TALN (2004)
11. Mann, G.S.: Fine-grained proper noun ontologies for question answering. In: Proc. of the 2002 Workshop on Building and Using Semantic Networks, SEMANET 2002, vol. 11, pp. 1–7. ACL, Stroudsburg (2002)
12. Marcińczuk, M., Piasecki, M.: Statistical Proper Name Recognition in Polish Economic Texts. Control and Cybernetics 40(2), 1–26 (2011)
13. Maziarz, M., Piasecki, M., Szpakowicz, S.: Approaching plWordNet 2.0. In: Proc. the 6th Global Wordnet Conference, Matsue, Japan (January 2012)
14. Miller, G.A., Hristea, F.: WordNet Nouns: Classes and Instances. Computational Linguistics 32(1), 1–3 (2006)
15. Pantel, P., Pennacchiotti, M.: Espresso: Leveraging Generic Patterns for Automatically Harvesting Semantic Relations. In: ACL (ed.) Proc. COLING-ACL 2006, Sydney, pp. 113–120. ACL (2006), www.aclweb.org/anthology/P/P06/P06-1015
16. Piasecki, M.: Polish Tagger TaKIPI: Rule Based Construction and Optimisation. Task Quarterly 11(1-2), 151–167 (2007), www.task.gda.pl/files/quart/TQ2007/01-02/tq111t-g.pdf
17. Piasecki, M., Broda, B., Głąbska, M., Marcińczuk, M., Szpakowicz, S.: Semi-automatic Expansion of Polish WordNet based on Activation-Area Attachment. In: Recent Advances in Intelligent Information Systems, pp. 247–260. EXIT (2009)
18. Piasecki, M., Indyka-Piasecka, A., Kurc, R.: Linguistically Informed Mining Lexical Semantic Relations from Wikipedia Structure. In: Nguyen, N.T., Kim, C.-G., Janiak, A. (eds.) ACIIDS 2011, Part I. LNCS (LNAI), vol. 6591, pp. 297–306. Springer, Heidelberg (2011)
19. Piasecki, M., Kurc, R., Broda, B.: Heterogeneous Knowledge Sources in Graph-Based Expansion of the Polish Wordnet. In: Nguyen, N.T., Kim, C.-G., Janiak, A. (eds.) ACIIDS 2011, Part I. LNCS, vol. 6591, pp. 307–316. Springer, Heidelberg (2011)
20. Piasecki, M., Ramocki, R., Maziarz, M.: Automated Generation of Derivative Relations in the Wordnet Expansion Perspective. In: Proc. 6th Global Wordnet Conference, Matsue, Japan (January 2012)
21. Piasecki, M., Szpakowicz, S., Broda, B.: Automatic Selection of Heterogeneous Syntactic Features in Semantic Similarity of Polish Nouns. In: Matoušek, V., Mautner, P. (eds.) TSD 2007. LNCS (LNAI), vol. 4629, pp. 99–106. Springer, Heidelberg (2007)
22. Piasecki, M., Szpakowicz, S., Broda, B.: A Wordnet from the Ground Up. Wrocław University of Technology Press, Wrocław (2009), www.plwordnet.pwr.wroc.pl/main/content/files/publications/A_Wordnet_from_the_Ground_Up.pdf
23. Przepiórkowski, A.: The IPI PAN Corpus: Preliminary version. Institute of Computer Science PAS (2004)
24. Radziszewski, A., Kilgarriff, A., Lew, R.: Polish Word Sketches. In: Vetulani, Z. (ed.) Human Language Technologies as a Challenge for Computer Science and Linguistics. Proc. 5th Language and Technology Conference, Poznań, Poland, pp. 237–242 (2011)

25. Ruiz-Casado, M., Alfonseca, E., Castells, P.: Automatic Assignment of Wikipedia Encyclopedic Entries to WordNet Synsets. In: Szczepaniak, P.S., Kacprzyk, J., Niewiadomski, A. (eds.) AWIC 2005. LNCS (LNAI), vol. 3528, pp. 380–386. Springer, Heidelberg (2005)
26. Snow, R., Jurafsky, D., Ng., A.Y.: Semantic taxonomy induction from heterogenous evidence. In: COLING 2006 (2006)
27. Sundheim, B.M., Mardis, S., Burger, J.: Gazetteer Linkage to WordNet. In: Proc. of the III IWC (2006)
28. Toral, R.M.A., Monachini, M.: Named Entity WordNet. In: ELRA (ed.) Proc. of the VI LREC 2008, Marrakech, Morocco (2008)

Appendix: Examples of Knowledge Sources

The appendix shows the details of the KSs which underlie the two examples given in Section 4.

KS elements for *Rudolf Schuster*	KS elements for *dżuma*
Window-based MSR: 15 best	
jean-luc dehaene	tyfus ''typhoid'
wiktor czernomyrdin 'Victor Chernomyrdin'	dur brzuszny 'typhoid'
szadli bendżedid 'Chadli bendjedid'	żółta febra 'yellow fever'
dmytro tabacznyk 'Tabachnyk'	krztusiec 'whooping cough'
frank-walter steinmeier	tężec 'tetanus'
eduardo duhalde	dur rzekomy 'paratyphoid fever'
lee myung-bak	framboezja 'framboesia'
michael portillo	dyfteryt 'diphteria'
willy claes	ospa wietrzna 'chicken pox'
zaprzysiężenie 'swearing in'	dyzenteria 'dysentery'
kryzys gabinetowy 'government crisis'	ornitoza 'ornithosis'
peter hollingworth	błonica 'diphteria'
michel barnier	szkarlatyna 'scarlet fever'
michl ebner	księgosusz 'cattle plague'
herman van rompuy	eklampsja 'ecclampsia'
Constraint-based MSR: 15 best	
przemysł gazowniczy'gas industry'	dur brzuszny 'typhoid'
tayloryzm 'taylorism'	tyfus 'typhoid'
celinograd 'Tselinograd'	dyfteryt 'diphteria'
wietrznica (village)	gruźlica 'tuberculosis'
wiktor czernomyrdin 'Victor Chernomyrdin'	czarna śmierć 'black death'
rybniczanin 'inhabitant of Rybnik'	żółta febra 'yellow fever'
kwagga	wąglik 'anthrax'
mówczyni 'speaker$_{female}$	tularemia
dżudok 'judoka'	cholera
ramapitek 'ramapithecus'	czarna ospa 'pox'
wicepremier 'deputy prime minister	malaria
prezydent-elekt 'president-elect'	wścieklizna 'rabies'
nasciturus	choroba zakaźna 'contagious disease'
kluch (surname)	zaraza 'plague'
dyfuzjonizm 'diffusionism'	grypa 'flu'
Constraint-based MSR: bidirectional	
przemysł gazowniczy 'gas industry'	dur brzuszny 'typhoid'
wietrznica (village)	tyfus
wiktor czernomyrdin 'Victor Chernomyrdin'	czarna śmierć 'black death'
mówczyni 'speaker$_{female}$	żółta febra 'yellow fever'
kluch (surname)	wąglik 'anthrax'
	tularemia
	cholera
	czarna ospa 'pox'
	malaria
	wścieklizna 'rabies'
	zaraza 'plague'
	grypa 'flu'

KS elements for *Rudolf Schuster*	KS elements for *dżuma*
Words from parentheses in Wikipedia titles	
	powieść 'novel'
KS based on Wikipedia categories	
człowiek 'human' prezydent 'president' samorządowiec 'self-goverment official'	choroba 'illness'
KS based on structural patterns	
polityk 'politician'	choroba 'illness' śmierć 'death'
KS based on the *to* (is a) text pattern	
burmistrz 'mayor'	kara 'punishment' nazwa 'name' pocałunek 'kiss' symbol 'symbol' synonim 'synonym' termin 'term' wirus 'virus' wróg 'enemy' zło 'evil'
KS based on the *jest* (is a) text pattern	
laureat 'laureate' prezydent 'president' Słowak 'Slovakian'	choroba 'illness' kara 'punishment' klęska 'defeat' książka 'book' plaga 'plague' powieść 'novel' świadek 'witness'
KS based on the *i inne* (and other) text pattern	
	choroba 'illness' klęska 'disaster' plaga 'plague'

A Multilingual Integrated Framework for Processing Lexical Collocations

Violeta Seretan

Abstract. Lexical collocations are typical combinations of words, such as *heavy rain*, *close collaboration*, or *to meet a deadline*. Pervasive in language, they are a key issue for NLP systems since, as other types of multi-word expressions like idioms, they do not allow for word-by-word processing. We present a multilingual framework that lays emphasis on the accurate acquisition of collocational knowledge from corpora and its exploitation in two large-scale applications (parsing and machine translation), as well as for lexicographic support and for reading assistance. The underlying methodology departs from mainstream approaches by relying on deep parsing to cope with the high morphosyntactic flexibility of collocations. We review theoretical claims and contrast them with practical work, showing our efforts to model collocations in an adequate and comprehensive way. Experimental results show the efficiency of our approach and the impact of collocational knowledge on the performance of parsing and machine translation.

1 Introduction

Lexical collocations are conventional and syntactically-motivated combinations of words, such as *heavy rain*, *dark night*, *to strike a balance*, *to meet a requirement*, *to work hard*, or *largely ignored*. Such combinations make up the bulk of the multi-word expressions[1] in a language. They are more numerous than any other type of multi-word expressions (Mel'čuk, 1998), yet they have been less studied and are less well understood than other expressions like compounds (e.g., *by and large*,

Violeta Seretan
Department of Translation Technology, Faculty of Translation and Interpreting,
University of Geneva, 40 bd. du Pont-d'Arve, 1211 Geneva, Switzerland
e-mail: violeta.seretan@unige.ch

[1] *Multi-word expressions* are "idiosyncratic interpretations that cross word boundaries" (Sag et al, 2002). The reader is referred to Baldwin and Kim (2010) for a thorough discussion of the relevance of multi-word expressions for Natural Language Processing (NLP).

A. Przepiórkowski et al. (Eds.): *Computational Linguistics*, SCI 458, pp. 87–108, 2013.
DOI: 10.1007/978-3-642-34399-5_5

cell phone), verb-particle constructions (e.g., *to come across, to get up*), light-verb constructions (e.g., *to take a walk, to give a speech*) or idioms (e.g., *to kick the bucket, to shoot the breeze*).

While there is currently no commonly agreed precise definition for collocations, broadly speaking, they can be understood as "the way words combine in a language to produce natural-sounding speech and writing" (Lea and Runcie, 2002). Like compounds and idioms, they constitute the ready-made units or the building blocks of a language (Sinclair, 1991). But what makes collocations more difficult to describe and process than other expressions is that they are easily confounded with regular, compositional productions in language (e.g., *heavy bag, to meet a friend*). Indeed, collocations are quite similar to regular productions, both from a syntactic and semantic point of view. They are characterised by a high morphosyntactic flexibility and semantic transparency, which make them relatively difficult to recognise, describe, classify, as well as to process in natural language applications. Collocations are considered to belong both to the lexicon and to the grammar of a language, and it is precisely this equivocal status that makes them particularly difficult to account for, both theoretically and practically.

The term *collocation* is itself ambiguous. Since it is understood in different ways in the literature, we need to clarify the understanding adopted in our own work. In accordance with the Meaning Text Theory (Mel'čuk, 1998), we consider that a collocation is a word combination made-up of a headword and a collocate. The headword (or base) is the semantic head of the combination and its meaning is preserved in the meaning of the combination (e.g., *rain* in *heavy rain*). In contrast, the collocate (in this example, *heavy*) is dependent on the base and its meaning can only be interpreted contingent upon the base (here, *heavy* does not refer to weight as in *heavy bag* but to the intensity of the rain).[2]

Collocations represent a major issue to be tackled by all NLP systems in which the proper identification of lexical units is crucial for their performance (for instance, parsing or machine translation systems, among many others). An NLP system unaware of collocations misses important information on the relations established between words, the lack of handling related parts of a sentence in a coherent way resulting into lower performance. The problem of dealing with collocations is exacerbated by the possibility for the component words to occur at a long distance from each other in a sentence. But most of all, the importance of collocations stands in their omnipresence (Mel'čuk, 1998). As many researchers state, collocations are pervasive in all text genres and domains (Kjellmer, 1987; Lea and Runcie, 2002; Mel'čuk, 1998; Stubbs, 1995). Lea and Runcie (2002) state, for instance, that "no piece of natural spoken or written English is totally free of collocation". Erman and Warren (2000) claim that about half of fluent native text is constructed using ready-made units, such as collocations. The number of multi-word

[2] In the Meaning Text Theory, the intensity is expressed by the lexical function *Magn* ($Magn(rain) = heavy$), $Magn(work) = hard$, $Magn(appreciation) = deep$). A lexical function can be defined as the relation established between lexical items on the basis of the meaning to express.

expressions in a lexicon is estimated to be of the same order as that of single words (Jackendoff, 1997).

In this chapter, we present a multilingual integrated framework for processing collocations that we developed over the past several years as part of our collaboration with the Language Technology Laboratory (LATL) of the Department of Linguistics, University of Geneva. LATL has long since undertaken extensive work on developing two large-scale NLP applications, namely, a multilingual syntactic parser, called Fips (Wehrli, 2007) and a rule-based machine translation system, called Its-2 (Wehrli et al, 2009a). The collocation processing framework is closely connected to these applications. Its primary aim is to provide collocational knowledge for these applications, by accurately extracting such knowledge from pre-compiled text corpora or from the Web, and by automatically translating it using parallel corpora. In addition, the framework provides tools for assisting the work of lexicographers compiling lexicons of collocations for these applications. Moreover, it integrates modules dealing specifically with collocation identification during parsing and translation, whose role is to ensure the adequate treatment of collocations in these systems in order to improve their performance. Last but not least, it allows the exploitation of the acquired collocational knowledge in real-life scenarios such as the context-sensitive look-up in electronic dictionaries.

Among the distinguishing aspects of the work presented are (1) the fact that the framework built is, to our knowledge, the most complete collocation processing environment and (2) the fact that the underlying methodology departs from mainstream approached by focusing on a detailed (as opposed to shallow) syntactic analysis of the text. The present chapter spans a considerable amount of work, part of which has been described in detail in separate publications. Its contribution is to offer for the first time a global view of the framework built and to situate it as a whole in the current NLP context. Emphasis is put on synthesizing the findings and experimental results obtained, as well as on discussing their role in gaining a better understanding of the way to adequately deal with the collocation phenomenon in NLP.

The chapter is organised as follows. First, we provide a concise synthesis of the theoretical description of collocations, and discuss the extent to which the features stipulated by theoretical accounts are taken into account in the practical modelling of this phenomenon (Sect. 2). Next, we provide a schematic presentation of the collocation framework built and highlight the directions in which our work has extended the state of the art (Sect. 3). Then we move to the question of the interrelation between parsing and collocations, and show that not only collocations benefit from a syntax-based approach, but, in turn, they can be used to improve the performance of the syntactic parser (Sect. 4). In Sect. 5 we address the question of how collocations are processed in the machine translation system of LATL, whereas in Sect. 6 we detail the manner in which they are represented in the computational lexicons used by the parser and the translation system. Finally, we illustrate two client applications of context-sensitive dictionary look-up (Sect. 7), and conclude by situating our work in the larger NLP context and by identifying directions for further developments (Sect. 8).

2 Collocations: From Theory to Practice

The phenomenon of word collocation has been addressed in the theoretical literature from different angles. In this section, we provide a concise overview of the various accounts provided in order to come up with a set of features that are generally agreed upon, despite the fact that existing collocation characterizations are often contradictory. The presentation is paralleled by a brief survey of the state of the art, focused on the implementation of each feature.

Syntactic status. In the contextualism linguistic current (Firth, 1957; Sinclair, 1991), collocations refer to the combinatorial profile of words, which is taken to be critical for defining their meaning: "One of the meanings of *night* is its collocability with *dark*, and of *dark*, of course, its collocation with *night*"; "You shall know a word by the company it keeps!" (Firth, 1957). The collocation is defined as "the cooccurrence of two or more words within a short space of each other in a text" or as words that show "the tendency to occur together" (Sinclair, 1991). Sinclair (1991) states that the usual measure of proximity is a maximum of four word intervening, and that the co-occurrence is more or less independent of grammatical pattern (e.g., adjective-noun, verb-object and so on).

The contextualism provides a purely statistical account of collocation phenomenon, unconcerned with the syntagmatic dimension of the combinations involved. Early practical work on extracting collocations from corpora has followed this purely statistical approach, which continues to remain popular as it requires little pre-processing resources. Extraction methods generally apply statistical association measures[3] on candidate combinations identified either from plain text or from POS tagged text. Candidates are selected as word pairs co-occurring in a so-called *sliding window*, possibly filtered according to predefined patterns. A detailed description of existing collocation extraction techniques can be found in Seretan (2011).

In contrast to this purely statistical view, collocations are seen as syntactically-motivated combinations from many other perspectives (e.g., lexicographic, pedagogical, lexis-grammar interface, Meaning-Text Theory, metaphoricity, semantic prosody). For instance, Cowie (1978) defines collocation as the "co-occurrence of two or more lexical items as realizations of structural elements within a given syntactic pattern". Hausmann (1989) provides the following list of syntactic patterns characterising collocations: adjective-noun, noun-verb, verb-noun (object), verb-adverb, adjective-adverb, noun-preposition-noun.

On the practical side, it has also been felt necessary to define collocations as syntactically-motivated co-occurrences. As many researchers have stated, ideally a full syntactic analysis of the source corpus would be needed in order to check that the words are used in a single phrase structure, the recent advances in the parsing field now justifying such an approach (Evert, 2004; Krenn, 2000; Pearce, 2002; Smadja, 1993). An increasing amount of work nowadays rely on the syntactic

[3] An *association measure* can be seen as a formula that assigns to a candidate pair a score that is interpreted as the likelihood for the pair to constitute a genuine, conventional combination (as opposed to a spurious, casual one).

pre-processing of corpora; however, shallow parsing is a more popular technique for identifying collocation candidates than full parsing, which remains less used. Our own work has been mainly devoted to the uptake of syntax-based methods in the area of collocation extraction and has proven the advantages of such an approach (Seretan, 2009, 2011; Seretan and Wehrli, 2006).

The distinction between simple co-occurrences and syntactically-motivated co-occurrences is an important one. In the recent literature, the term *collocation* is generally reserved for syntactic combinations, whereas the denomination of *co-occurrence* is used in the broader sense, for the understanding adopted in contextualism.

Length. In addition to stating that collocations are syntactically-motivated combinations, theoretical accounts also attempt to specify the length of the combinations allowed. First of all, most definitions stipulate that collocations are made up of two or more words (see for instance the definitions above provided by Cowie (1978) – "two or more lexical items" – and Sinclair (1991) – "two or more words"). In rare cases, definitions explicitly mention the presence of exactly two lexical components (Hausmann, 1989; Mel'čuk, 1998). However, these components may not necessarily be single words, but multi-word expressions, as in *weapons of mass destruction* or *to strike a right balance* where the second component is itself a multi-word expression: *mass destruction*, *right balance*. In particular, collocations may contain embedded collocations, the recursive nature of collocations having been noted for instance by researchers like Heid (1994). As Sinclair (1991) states, "there is no theoretical restriction to the number of words involved".

The practice stands, however, in stark contrast to this description. The previous work has been focused almost exclusively on identifying binary collocations. One reason for this situation is that association measures are mainly designed for co-occurrence of two elements only; only rarely have they been extended to a higher arity, such as in Villada Moirón (2005). Another reason is that if we combine more than two elements and allow a wider co-occurrence window, we face the problem of combinatorial explosion. A solution adopted was to restrict the investigation to rigid sequences in order to come up with tractable methods, as in Choueka et al (1983), Smadja (1993), or Dias (2003). This is, obviously, a drastic simplification, leading to failure to cope with variability in word order and with insertion of additional linguistic material, as in Example 1(b) below (compare with the sentence in 1(a) that shows the canonical order).

(1) a. Our designs *strike a right balance* of eastern craftsmanship with a western design.
b. (...) the *right balance* a government should *strike* between protecting its citizens' privacy and ensuring their security.

Configuration. Insofar as the syntactic configuration of collocation is concerned, a closed list of relevant patterns is mentioned in Hausmann (1989): adjective-noun, noun-verb, verb-noun (object), verb-adverb, adjective-adverb, noun-preposition-noun. But unlike Hausmann (1989), many researchers – among whom Fontenelle

(1992) and van der Wouden (2001) – consider that there is actually no restriction on the type of syntactic relation holding between the items of a collocation. Lexicographic evidence showed that, indeed, collocations can be defined by a high variety of configurations. As such, the BBI collocation dictionary (Benson et al, 1986) contains a division into *lexical collocations* defined by a dozen of patterns involving content words only, and *grammatical collocations* defined by a much longer list of patterns allowing function words.[4] In principle, there are no constraints on the combination of syntactic categories, even though some combinations are clearly impossible, e.g., noun-adverb or verb-determiner (van der Wouden, 2001).

A look at the patterns considered in practical work reveals a high divergence in the sets of patterns used, with virtually each system considering its own selection. Moreover, there is a marked disagreement with the theoretical and the lexicographic stipulations. Most systems tend to discard grammatical collocations and thus to exclude function words, whereas theoreticians argue for the inclusion of such words as they are important when studying collocations (van der Wouden, 2001). As a matter of fact, there was no principled way in previous work to select an adequate and complete set of patterns. Moreover, it is not unusual for authors to focus on a single pattern, or a very limited list of patterns for collocation extraction: verb-preposition (Blaheta and Johnson, 2001), preposition-noun-preposition, prepositional phrase - verb (Villada Moirón, 2005), verb-object, noun-adjective, verb-adverb (Lü and Zhou, 2004), noun phrase (Bourigault, 1992; Daille, 1994; Jacquemin et al, 1997). This practice limits, however, the understanding that can be gained through corpus study on the phenomenon of collocation, which is a very complex and important one.

3 The Collocation Processing Framework

The previous section has shown the discrepancies that exists between the theoretical and practical accounts on collocations. Over the past several years, we carried out work directed specifically at extending existing methodologies in order to model the collocation phenomenon in a more adequate and a more comprehensive way. In this section we present the collocation processing framework that we created in the context described in Sect. 1.

3.1 General Architecture

By collocation processing, we understand any type of process that is relevant to acquiring and exploiting collocations by computational means. Thus, the extraction of collocations, their representation in a computational lexicon, their automatic translation and their use in various applications are considered as part of the integrated

[4] Content words carry the meaning of a sentence, and include nouns, verbs, adjectives, and most adverbs. Function words include articles, prepositions, conjunctions, auxiliary verbs, and pronouns.

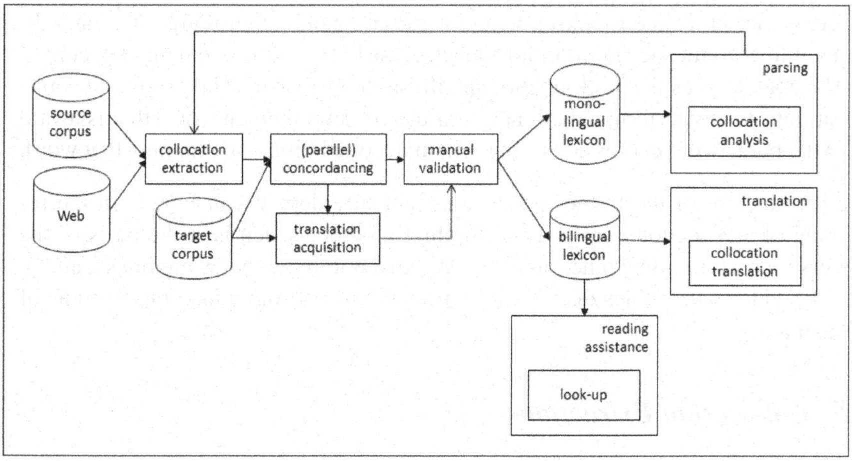

Fig. 1 Architecture of the collocation processing framework

framework built. Integration refers to the possibility to process collocations in a consistent manner, in a process flow that can be pipelined or not, with no need to perform additional work to ensure the interoperability of the modules involved.

The main modules of the framework are:

1. The collocation extraction module, which performs the syntax-based acquisition[5] of collocational knowledge either from pre-compiled corpora or from the Web; the extractor also has two online versions – a first which allows users to upload and process their own corpus, and a second which allows users to consult pre-extracted collocations;
2. The (parallel) concordancing module, which displays collocations in the source context and thus allow users to gain insights on the actual use of the identified collocations; the parallel version displays, in addition, the translation of the source context in the version of the corpus in another language, if available;
3. The translation acquisition module, which detects the target language equivalent of a collocation using parallel corpora, when available;
4. The manual validation module, which supports the work of lexicographers inserting collocations into a lexicon, by parsing the input expression, detecting its morphosyntactic features and displaying possible lexeme readings for the composing words;
5. The collocation analysis module (part of the Fips parser), which ensures that collocation detection and structure attachment decisions are synchronous processes that inform each other;
6. The collocation translation module (part of the Its-2 MT system), which deals with collocations during the translation process, ensuring collocation

[5] Performance evaluation against the sliding window technique is presented in Sect. 3.2.

identification, lexicon look-up and generation of the target representation, taking into account specific morphosyntactic restrictions recorded in the lexicon;
7. The context-sensitive look-up module, illustrated by two slightly different applications, which perform a dictionary search (for an unknown word) that is aware of the multi-word expressions – in particular, collocations – containing that word.

The architecture of the framework is schematically depicted in Fig. 1. In the remaining of this section, we focus on the first module, which is at the basis of the processing performed by other modules. We also briefly overview modules 2 and 3. We will address the issues dealt with by modules 4–7 in the remaining sections of the chapter.

3.2 Collocation Extraction

In our work, the syntactic status of collocations plays a central role in identifying potential candidates from a corpus (see Sect. 2). We define collocation candidates in terms of syntactic relatedness rather than in terms of co-occurrence within a short space in a text. Consider for instance the sentence fragment in Example 2 below.

(2) most voters will find that a candidate for whom they have expressed some preference will have been elected.

Combinations occurring in a 5-word window are extremely noisy, even if POS patterns are used to filter them; for instance, *find-candidate, candidate-expressed, preference elected* are all false positives, and only *voters-find* and *expressed-preference* are true positives. Moreover, the pair *candidate-elected*, which is in fact a collocation, is missed by the window method (false negative).

In contrast, our approach relies on syntactic information provided by the Fips parser (Wehrli, 2007) in order to select collocation candidates. Candidates of predicate-argument type are recovered directly from the argument tables built by the parser. For instance, in the case of the predicate *elected* from Example 2, the argument table contains *candidate* on the direct object position. The task of recovering the verb-object link between *elected* and *candidate* is thus greatly simplified. Contrary to the window method, the syntax-based method succeeds in identifying the pair *candidate-elected* (true positive).

The identification of candidates in predicate-argument relations is a special case. But more generally, the identification algorithm consists of processing each parse tree in a top-down manner. For each lexical head, we find the lexical heads of sibling structures that are in a specific syntactic relation with that head, such as adjective-noun, noun-noun, adverb-verb, etc. Suppose that we are processing the word *balance* in the sentence of Example 1(a), reprinted below as Example 3. Its sibling structures are a left adjectival phrase (*right*) and a right prepositional phrase (*of eastern craftsmanship*). The lexical heads of these structures are *right* and *cratfsmanship*, respectively. Therefore, the adjective-noun pair *right-balance* and the

Table 1 Contingency table for a candidate collocation

	lexeme2	¬ *lexeme2*
lexeme1	*a*	*b*
¬ *lexeme2*	*c*	*d*

noun-preposition-noun pair *balance of craftsmanship* are proposed as collocation candidates at this step of the algorithm.[6]

(3) Our designs strike a right *balance* of eastern craftsmanship with a western design.

After the syntactically-motivated candidates have been selected, we apply the log-likelihood ratio association measure, which ranks candidates in the order of their likelihood to actually constitute collocations. This measure (Dunning, 1993) relies on information on the frequency of lexemes in the source corpus, as well as on frequencies of co-occurrence of lexemes, organised in a so-called contingency table as in Table 1. The log-likelihood ratio score is computed according to the formula shown in Equation 1.

$$\begin{aligned} LLR = \; & 2(a\log a + b\log b + c\log c + d\log d - (a+b)\log(a+b) - (a+c)\log(a+c) \\ & -(b+d)\log(b+d) - (c+d)\log(c+d) + (a+b+c+d)\log(a+b+c+d)) \quad (1) \end{aligned}$$

There are a number of other association measures that are popular in the literature, e.g., χ^2, mutual information, t test, z-score; see Pecina (2005) for a comprehensive list. Our system implements a dozen of such measures; LLR is selected by default since it is argued appropriate for both frequent and infrequent data (Dunning, 1993, 62).

Our collocation extraction approach is a hybrid one, since it combines statistical information with syntactic information in order to detect syntactically valid combinations that are strongly associated. Syntactic pre-processing is necessary in order to account for the morphosyntactic variability of collocations. Example 4 illustrates a few of the numerous morphosyntactic transformation that a collocation like *[to] face a challenge* can undergo. As all examples provided in this chapter, these are attested examples belonging to the corpora we used in our experiments.

(4) a. various global *challenges* that we inevitably have to *face*

b. the *challenge* that was being *faced* by NATO

[6] The list of the most representative configurations considered for English is: adjective-noun (*heavy smoker*), noun-[predicate]-adjective (*effort [be] devoted*), noun-noun (*suicide attack*), noun-preposition-noun (*round of negotiations*), noun-preposition (*inquiry into*), adjective-preposition (*crazy about*), subject-verb (*war breaks*), verb-object (*meet requirement*), verb-preposition-argument (*bring to boil*), verb-preposition (*point out*), adverb-verb (*fully support*), adverb-adjective (*highly important*), and noun-coordination-noun (*nice and warm*).

Table 2 Contingency table for a complex candidate collocation (*to strike a right balance*)

	right balance	¬ *right balance*
strike balance	*a*	*b*
¬ *strike balance*	*c*	*d*

c. we *face* some very demanding *challenges*
d. we are now *facing* entirely new *challenges*

Compared against a syntax-free baseline implementing the sliding window method on POS tags provided by the Fips parser, the syntax-based approach was shown to contribute to a significant increase in performance. We performed several cross-language evaluation experiments involving various data sampling strategies and annotation categories of varied granularity. The extraction precision measured in terms of grammaticality on the 500 top result pairs obtained from French corpora was higher by 20.7% (99% vs 78.3%); in terms of lexicographic interest, it was higher by 8.9% (65.9% vs 57%). When measured using stratified sampling on data in more languages (2,000 pairs in English, French, Spanish, Italian), the precision was 2.7 times higher in terms of grammaticality (88.8% vs 33.2%), 2.5 times higher in terms of lexicographic interest (43.2% vs 17.2%), and 2.6 times higher in terms of collocativity (32.9% vs 12.8%). The positive impact of using syntactic information is consistent with findings obtained by other authors for tasks such as term extraction (Maynard and Ananiadou, 1999), semantic role labelling (Gildea and Palmer, 2002), or semantic similarity computation (Padó and Lapata, 2007).

Our work has contributed not only to the uptake of syntax-based methods for the task of collocation extraction, but also to extending the existing technology in order to deal with the collocation phenomenon in a more comprehensive way. As such, we conducted research in directions that remained less explored by previous work, despite theoretical evidence (see Sect. 2).

One such direction was the extraction of complex collocations, made up of more than two words. We modelled such collocations as recursive constructions involving binary collocations: e.g., *strike a right balance* is seen as a binary verb-object collocation in which the object, *right balance*, is in turn a binary collocation itself. There is no length limitation in our modelling, as an *n*-ary collocation can further be part of a binary combination, yielding a complex collocation of length $n+1$. A score is computed as in the case of binary associations, by treating the component items as single lexemes. For instance, to assign a score to the complex collocation *to strike a right balance*, we consider the contingency table shown in Table 2 (similarly to the case of single lexemes, Table 1).

Another extension direction concerns the selection of configurations considered as relevant for collocations. Our objective was to come up with a method for choosing a set of patterns that is as complete as possible, given that in the existing work this selection is made in a rather arbitrary way. To achieve our objective, we adopted a corpus-based approach. First, we used parse trees generated by the parser in order to record all syntactic configurations that are productive in a language. Then the pairs in each configuration have been ranked according to the log-likelihood ratio measure and those patterns judged as interesting have been selected for inclusion in the list of collocation patterns of a language. This strategy led to the discovery of new patterns such as preposition-noun for English (compare for instance *on page* with **at page*) and adjective-preposition for French (*déterminé comme*, lit. 'determined like').

Collocation extraction experiments have been performed for all the languages currently supported by Fips, i.e., English, French, Spanish, Italian (Seretan and Wehrli, 2009), Greek (Michou and Seretan, 2009), Romanian (Seretan and Wehrli, 2010a) and German. The stand-alone extraction application, FipsCo, is available to researchers upon request. It implements the parsing and extraction from both pre-compiled corpora and from the Web, through Google API search (Seretan et al, 2004). The online version of FipsCo, called FipsCoWeb[7], allows users to upload and process their own corpus, and to visualise the results (Seretan and Wehrli, 2010b). In addition, we developed an online tool for the dictionary-like visualisation of pre-extracted collocations, called FipsCoView[8] (Seretan and Wehrli, 2011). As the Fips parser itself, all these applications are developed in Component Pascal using the BlackBox development environment for Windows.

A screen capture showing the parallel concordancer built around the extractor is shown in Fig. 2. The collocation extraction module is also at the basis of the third module, which deals with the acquisition of translation equivalents for collocations from parallel corpora. Given a source collocation, a mini-corpus of target sentences is built via alignment, from which collocations are extracted on the target language. Then a matching procedure is applied aimed at selecting the target collocation that is likely to correspond to the source collocation. Results obtained on the Europarl corpus (Koehn, 2005) show a performance of 81.6% according to the F-measure (84.1% precision, 79.2% recall) and can be consulted online.[9]

4 Collocations and Parsing

As we shown in Sect. 3, syntactic information is crucial for accurately detecting collocations in corpora. At the same time, collocational information is useful for parsing, as it provides important attachment disambiguation clues. Due to the inherent ambiguities in language, parsing systems are faced with a high number of alternatives, which grows exponentially with the length of the input sentence. Collocation constraints can be used to reduce this number and to guide the parser through the maze of alternatives. Indeed, collocational relations between the words in a sentence have been proven very helpful in selecting the most plausible among all the possible

7 Available at `http://tinyurl.com/FipsCoWeb`

8 Available at `http://tinyurl.com/FipsCoView`

9 `http://www.issco.unige.ch/en/staff/seretan/data/tal/VO.htm`, `http://www.issco.unige.ch/en/staff/seretan/data/tal/AN.htm`, `http://www.issco.unige.ch/en/staff/seretan/data/tal/NPN.htm`

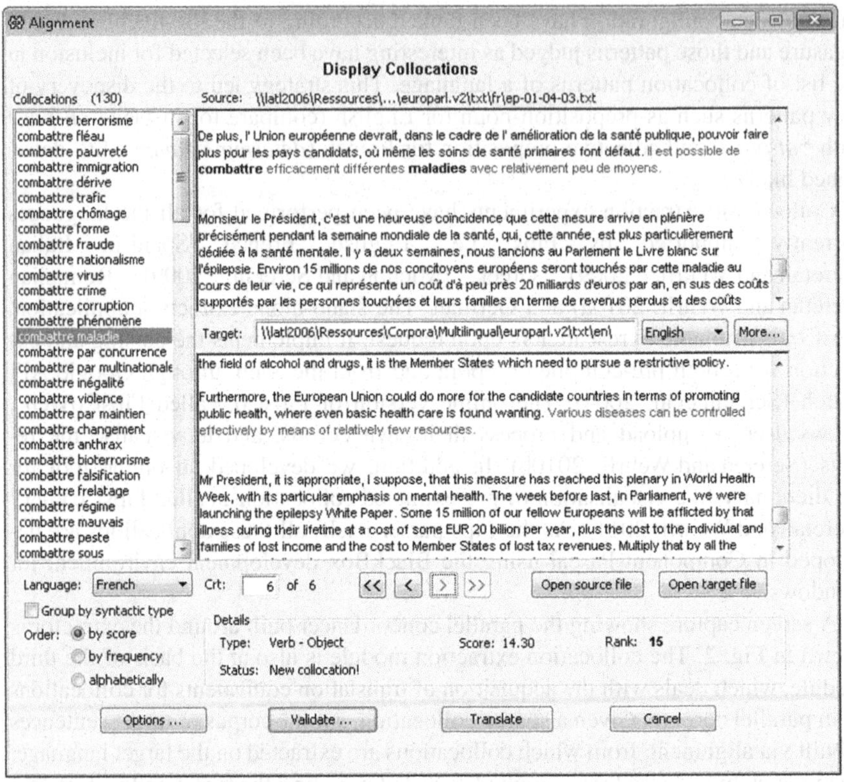

Fig. 2 Screen capture of the parallel concordancer showing collocations extraction results for French, filtered according to type (*combattre* 'to combat' + object)

parse trees of that sentence (Alshawi and Carter, 1994; Hindle and Rooth, 1993). But the problem that arises is that of circularity. Efficient parsing requires reliable corpus-based information, whereas the latter can only be obtained accurately with parsing.

The current literature provides no solution to this circular problem. Just as other types of multi-word expressions, collocations are problematic for parsing because they have to be recognised and treated as a whole, rather than compositionally, i.e., in a word by word fashion. The standard approach is to include multi-word expressions in a "words-with-spaces" pre-processing step. But unlike other expressions that are fixed or semi-fixed, collocations do not allow a "words-with-spaces" treatment, because of their high morphosyntactic flexibility. Collocations are situated at the intersection of lexicon and grammar; therefore, they cannot be accounted for merely by the lexical component of a parsing system, and have to be integrated to the grammatical component as well.

An alternative approach is to identify collocations after the syntactic analysis has been performed, and to output a parse tree in which collocational relations are highlighted between the composing items. The drawback of this alternative is that collocation identification happens too late for the parser. The latter cannot take advantage of collocational information, despite it being a major means of structural disambiguation, along with other types of information like selectional preferences and subcategorization frames.

We argue that collocation identification and syntactic parsing are interrelated processes that must be accounted for simultaneously. We proposed an original approach in which collocations are identified in a sentence as soon as possible during the analysis of that sentence, rather than at the end. To achieve the goal of interconnecting the parsing procedure and the identification of collocations, we have incorporated the collocation identification mechanism within the constituent attachment procedure of the parser Fips.

This parser, like many grammar-based parsers, uses left attachment and right attachment rules to build respectively left subconstituents and right subconstituents. Given the fact that Fips' rules always involve exactly two constituents – see Wehrli (2007) for details – it is easy to add to the attachment mechanism the task of collocation identification. To take a very simple example, when the rule attaching a prenominal adjective to a noun applies, the collocation identification procedure is invoked. It first verifies that both terms bear the lexical feature [+partOfCollocation], which signals that a given word is associated in the parser's lexicon to one or several collocations. Then it searches the collocation lexicon for an adjective-noun collocation with those two terms. If the search is successful, the corresponding parse tree will be given a high priority. Consider the example of *right balance*. The identification of the collocation will immediately relegate any other analyses based on the nominal reading for *right* or on the verbal reading for *balance*.

The collocation identification procedure has been extended to handle complex collocations, which can we view as collocations of collocations. Their recognition is similar to the simpler case. When the two parts have been found, then if they bear the feature [+partOfCollocation], the collocation lexicon is looked up for the complex collocation. If this is found, then – as in the case of binary collocations – an analysis attaching the two parts is given high priority over alternatives. Take for instance the sentence fragment "the right balance a government should strike". The parsers identifies the collocation *right balance*, then the verb *strike*, which are both marked as part of collocation. Since the collocation lexicon contains the entry "to strike a right balance", the parser will attach *right balance* as a direct object to the verb *strike*, rather than pursuing other alternatives.

We assessed the impact of the procedure that interconnects parsing and collocation identification on the performance of both tasks, parsing and collocation identification. We compared the new parser version against the version which does not use collocations for attachment decisions. We found substantial improvement in the precision (95% vs 77.5%) and recall (71% vs 53.5%) of collocation identification, as well as a sensible increase in the coverage of the parser expressed in terms of number of completely parsed sentences (83.3% vs 81.7%). A substantial improvement of performance (13%) has also been observed in a task-based evaluation experiment that was focused on the translations proposed for collocations by the Its-2 translation system (Wehrli et al, 2009a).

5 Collocations and Translation

Collocations are considered a key factor in producing more acceptable machine translation output (Heylen et al, 1994; Orliac and Dillinger, 2003). This is due to their massive presence in language and to their encoding idiomaticity, i.e., the difficulty to predict their components on the basis of the meaning to express. Although collocations are at the first sight semantically transparent and therefore similar to regular constructions, many of them cannot in fact be translated literally. The choice of the "right word" to use in the target language is often a subtle process, with crucial implications on the translation quality. For example, apparently harmless combinations, like *grande attention, grande diversité, grande vitesse* in French would lead to inadequate formulations in English if translated literally: **big attention, *big diversity, *big speed*. The right translations, *great attention, wide range* and *high speed*, illustrate the necessity of using collocations in the target language: the same adjective, *grande* ("big"), is translated in three different ways, depending on the noun it modifies.

Collocations pose the same challenge to machine translation systems as other multi-word expressions do; however, they are more difficult to handle because of their morphosyntactic flexibility. More rigid expressions such as compounds (e.g., *before too long*) and some types of idioms undergoing little variation (e.g., *to shoot the breeze*) can be treated as single words, in a "words-with-spaces" approach. In contrast, collocations allow the inversion of the component words as well as the insertion of additional words between them. Again, they cannot be considered simply as items of the lexicon since they are also found of the intersection of lexicon and syntax. Therefore, a lexical transfer in a "words-with-spaces" approach is not enough. The transfer of collocations is relatively more complex. In the Its-2 rule-based system, it takes place in three main steps, as follows.

Identification. In order to successfully translate a collocation, the first condition that has to be met is to detect it in the source sentence. In Its-2, collocations are identified during the sentence analysis performed by the Fips parser, following the procedure detailed in Sect. 3.2. Each item of the collocation is marked as such in the parse tree. The identification of collocations is possible thanks to the information provided by parser, even in the cases involving complex grammatical transformations, as illustrated in Example 4.

Transfer. Once the source collocation has been identified and its members marked in order to prevent their literal translation, the system looks up the bilingual lexicon for a translation of that collocation. If no translation is found, then it returns a literal translation. Otherwise, it considers the target equivalent – either a simple or a complex lexeme, in particular, a collocation – and proposes it in the target structure. The target structure is built on the basis of the target items obtained. A particular treatment is undergone by those constituents interpreted as predicate arguments, as their structure may in part be determined by the target predicate.

Generation. Finally, morphological and grammatical transformations apply to the obtained target structure, in order to generate the form of the target sentence. The application may be constrained by collocation-specific restrictions recorded in the lexicon, such as the use of a determinerless noun. If no constraints have been stated, then the collocation items undergo exactly the same morphosyntactic processes as regular combinations.

Its-2 is currently available for the following language pairs: English, Italian, Spanish, and German to French, as well as French to English. It uses the monolingual source and target lexicons of the Fips parser, as well as bilingual lexicons which contain pairs of equivalents defined over entries in the monolingual lexicons.

To evaluate the extent to which Its-2 succeeds in proposing a correct translation for the most flexible collocations, we performed an experiment using a test-set of verb-object collocations in English and Italian, occurring in various contexts in a corpus (Wehrli et al, 2009c). We found that Its-2 ranked second in terms of precision on the English test set, when compared against two state-of-the-art competing systems, Google Translate[10] and Systran[11]. On the Italian test set, it ranked first. It was also interesting to note that Its-2 was less affected by the increase in the distance between the component words than the two other systems, a statistical and a rule-based one. This result shows the importance of performing reliable collocation recognition, as a step in translating collocations. A further experiment has confirmed the importance of this step, as the improved collocation identification during parsing has led to a significant increase in the collocation translation precision of 13% (as mentioned in Sect. 4).

Note that we performed a collocation-oriented manual evaluation instead of reporting the results of standard metrics, such as BLEU. The reason is that these metrics underestimate the impact that the substitution of a single word (the collocate) has on the overall sentence quality.[12]

Including explicit collocation knowledge into a MT system might seem useless in the case of modern statistical systems, which model this knowledge internally when building phrase translation tables. Yet, such systems are highly sensitive to the syntactic environment of the source collocations, and often fail to provide accurate translations when additional words are inserted between the collocation components. This is clearly an issue, given the marked collocation flexibility, particularly in languages with a freer word order. Noncontiguous phrases are a challenge for MT, and evidence shows that their inclusion significantly improves the translation accuracy (Bod, 2007).

[10] `http://www.google.com/language_tools`

[11] `http://www.systran.co.uk`

[12] A similar observation can be made in the case of negation: the difference in the BLEU scores does not reflect the huge semantic difference resulting from the change in the polarity of the sentence.

6 Representation in a Computational Lexicon

In Sect. 4 and 5 we have shown that collocations play an important role in large-scale NLP applications such as parsing and machine translation. In this section, we detail the manner in which these expressions are represented in the computational lexicons of these applications. As we mentioned in Sect. 1, collocations may be as numerous as single words in a language. Automatic methods for identifying collocations in text corpora are therefore essential in aiding the manual compilation of collocational resources.

In the framework presented in this chapter, the collocation extraction method discussed in Sect. 3.2 is used to provide the raw material for the compilation of monolingual and bilingual collocation lexicons (see Fig. 1.) The insertion of new entries in these lexicons is performed manually, under the supervision of a lexicographer. The concordancing tools provide lexicographers with good candidates, as well as with statistical information and usage samples selected from corpora.

The information stored in the collocation lexicons for each entry is the following:[13]

- The components of the collocation (these can be either words or collocations);
- The syntactic configuration (e.g., adjective-noun, verb-object, noun-preposition-noun);
- The preposition used, if any (for instance, in the collocation *round of negotiations*, the components are *round* and *negotiation*; the proposition *of* is included for readability);
- The morphosyntactic features which constrain the collocation form (e.g., in the case above, the second noun is in the plural form).

Given that the coverage aimed at is of the order of the coverage of lexicons of single words[14], this represents an immense amount of work required from lexicographers. Therefore, it is reasonable to seek to automate this process as much as possible. Once the lexicographer has entered an input expression, the parser will be invoked and it will detect most of the information needed automatically: the collocation components, the syntactic type, the preposition as well as morphosyntactic features identified given the form entered by the lexicographer. The latter can either validate or modify the parser's choices – for instance, correct the syntactic type, select as a component a lexeme with a different reading, or add more morphosyntactic features.

To illustrate this discussion, consider the entry *weapons of mass destruction*. The parser fills in information such as the following: components of the collocation: *weapon* and *mass destruction* (with the associated entry identification numbers); type: noun-preposition-noun; preposition: *of*, features: [detLessCompl], [pluralColloc].

[13] Here, we describe the case of monolingual entries. Bilingual entries are simply pairs of monolingual entries corresponding to the source and the target language.

[14] Currently, more than 10,000 entries have been added for most of the languages supported by the parser.

It is important to note that the recursive structure used for modelling collocations leads to uniformity in their representation in the lexicon. Each entry is stored in the same way, as an association of two composing lexemes, where a lexeme can be a word or, in turn, a collocation. In this way, we can conveniently represent collocations of unrestricted length. This uniformity enhances the usability of collocation information in client applications.

Since most of the information specified in an entry can be filled in automatically, the effort required for building collocational lexicons is greatly reduced. Part of the lexicographic burden is to manually decide which candidates are worth considering for insertion. Here, again, the manual effort is alleviated by the concordancing module and the possibility to quickly build a list of valid candidates. Compared to the compilation of lexicons of single words, the creation of collocational lexicons is considerably faster in our framework.

7 Applications to Reading Assistance

Parsing and machine translations are the main applications that exploit collocational knowledge in the integrated framework we presented. In addition, there are two other applications concerned with reading assistance (cf. Fig 1). These are Twic and TwicPen (Wehrli et al, 2009b), designed specifically for non-native readers of online and offline material, respectively. The first application is a browser plug-in used to provide contextual translation when the user selects a word in a web page. The second application provides similar help for users scanning a printed text fragment with a hand-held scanner.

The motivation and underlying technology are in both cases the same. Within our increasingly multilingual society, more and more people need to access material in a language that is not their own. In many cases, users have basic knowledge of the target language and can read a text, yet, they may occasionally need to look-up a dictionary when they encounter unknown words. Fortunately, in a digital context the look-up task is considerably alleviated by "intelligent" dictionaries that automatically match the inflected word form to the base word form. Users are no longer expected to know (and look for) the base word form themselves, as when they use paper dictionaries.

The power of context-sensitive search is even more apparent when the electronic dictionary performs a linguistic analysis of the sentence (or fragment) that contains the sought word. That way, the system recognises from the grammatical context the category of the word and is able to narrow down the translations accordingly, thus greatly reducing the noise compared to a simple dictionary lookup. Suppose that a French user reads the sentence in Example 5 below, and is surprised to notice the occurrence of the word *rose*, of which he only knows the nominal and adjectival readings:

(5) The oil price *rose* early in 2011 to just over USD 100.

Through a linguistically-enhanced search, the system will identify the verbal reading that is compatible with the grammatical environment of the word, and will only output the adequate translation, *to rise – s'élever*.

In Twic and TwicPen, a linguistic analysis is performed which allows not only to determine the category of the selected word, but also to select a more appropriate reading by taking into account detailed lexical information available in its lexicons. Thus, in a context like the one in Example 6, the translation proposed is *to rise - se lever, monter*, since the features of the subject trigger the selection of a different reading by the parser.

(6) As soon as the sun *rose*, we started to move to a thick forest.

In addition, a peculiarity of these systems is that they are able to detect whether the sought word is part of a multi-word expression – in particular, a collocation. In that case, they output the translation of the whole expression instead of the translation of the selected word in isolation. Importantly, the underlying methodology allows an expression to be retrieved even if the component items are not adjacent. For example, in the context shown in Example 7, if the selected word is *bridge* or *gap*, the following output is displayed: *to bridge a gap – combler une lacune*. The parser identifies the source collocation and retrieves the target collocation in the bilingual lexicon, thus avoiding to propose the translations for the words considered in isolation, i.e., *to bridge – relier* and *gap – trou, fossé, bèche, écart*.

(7) helping to *bridge* a growing competence *gap* in the manufacturing industry.

8 Conclusion

In this chapter, we dealt with the practical accounts of the phenomenon of collocation, which is pervasive in language. We presented a multilingual integrated framework for collocation processing that we built over the past years at LATL in connection with the development of two large-scale NLP applications, namely, syntactic parsing and rule-based machine translation.

The framework includes modules for acquiring collocational knowledge from corpora; acquiring translation equivalents for source collocations provided that parallel corpora are available; concordancing and manual validation; as well as dedicated modules dealing with the exploitation of collocation knowledge in the parsing and machine translation applications. The approach we used is based on deep syntactic parsing. It departs from mainstream approaches in that it focuses on detailed parsing information, in addition to statistical computation, in order to cope with the high morphosyntactic variability of collocations.

Experiments performed across several languages and in various settings have confirmed the usefulness of incorporating syntactic information in the processing of collocations. This result is in line with findings obtained by related work using syntactic information for coping with other linguistic phenomena in tasks such as term extraction (Maynard and Ananiadou, 1999), semantic role labelling (Gildea and Palmer, 2002), semantic similarity computation (Padó and Lapata, 2007).

The framework built represents an unprecedented environment for modelling collocations in a comprehensive way, in a full processing cycle, from acquisition to exploitation in real-life scenarios. Further developments directions include: extending the framework to other languages that are currently under development in the Fips parser (e.g., Romansh, Japanese, Russian, Hindi, Serbo-Croatian); making the various modules accessible in a high-power distributed computing platform; and adopting data interchange standards to enhance the interoperability with other resources and tools.

Acknowledgement. The work described in this paper has mainly been done while I was affiliated with the Language Technology Laboratory (LATL), University of Geneva. I would like to thank Eric Wehrli for many years of close and fruitful collaboration.

References

Alshawi, H., Carter, D.: Training and scaling preference functions for disambiguation. Computational Linguistics 20(4), 635–648 (1994)

Baldwin, T., Kim, S.N.: Multiword expressions. In: Indurkhya, N., Damerau, F.J. (eds.) Handbook of Natural Language Processing, 2nd edn. CRC Press, Taylor and Francis Group, Boca Raton, FL (2010)

Benson, M., Benson, E., Ilson, R.: The BBI Dictionary of English Word Combinations. John Benjamins, Amsterdam (1986)

Blaheta, D., Johnson, M.: Unsupervised learning of multi-word verbs. In: Proceedings of the ACL Workshop on Collocation: Computational Extraction, Analysis and Exploitation, Toulouse, France, pp. 54–60 (2001)

Bod, R.: Unsupervised syntax-based machine translation: the contribution of discontiguous phrases. In: Proceedings of MT Summit XI, Copenhagen, Denmark, pp. 51–56 (2007)

Bourigault, D.: LEXTER, vers un outil linguistique d'aide à l'acquisition des connaissances. In: Actes des 3èmes Journées d'Acquisition des Connaissances, Dourdan, France (1992)

Choueka, Y., Klein, S., Neuwitz, E.: Automatic retrieval of frequent idiomatic and collocational expressions in a large corpus. Journal of the Association for Literary and Linguistic Computing 4(1), 34–38 (1983)

Cowie, A.P.: The place of illustrative material and collocations in the design of a learner's dictionary. In: Strevens, P. (ed.) Honour of A.S. Hornby, pp. 127–139. Oxford University Press, Oxford (1978)

Daille, B.: Approche mixte pour l'extraction automatique de terminologie: statistiques lexicales et filtres linguistiques. PhD thesis, Université Paris 7 (1994)

Dias, G.: Multiword unit hybrid extraction. In: Proceedings of the ACL Workshop on Multiword Expressions, Sapporo, Japan, pp. 41–48 (2003)

Dunning, T.: Accurate methods for the statistics of surprise and coincidence. Computational Linguistics 19(1), 61–74 (1993)

Erman, B., Warren, B.: The idiom principle and the open choice principle. Text 20(1), 29–62 (2000)

Evert, S.: The statistics of word cooccurrences: Word pairs and collocations. PhD thesis, University of Stuttgart (2004)

Firth, J.R.: Papers in Linguistics 1934-1951. Oxford University Press, Oxford (1957)

Fontenelle, T.: Collocation acquisition from a corpus or from a dictionary: a comparison. In: Proceedings I-II Papers submitted to the 5th EURALEX International Congress on Lexicography in Tampere, pp. 221–228 (1992)
Gildea, D., Palmer, M.: The necessity of parsing for predicate argument recognition. In: Proceedings of 40th Annual Meeting of the Association for Computational Linguistics, Philadelphia, Pennsylvania, USA, pp. 239–246 (2002)
Hausmann, F.J.: Le dictionnaire de collocations. In: Hausmann, F., Reichmann, O., Wiegand, H., Zgusta, L. (eds.) Wörterbücher: Ein internationales Handbuch zur Lexicographie. Dictionaries, Dictionnaires, pp. 1010–1019. de Gruyter, Berlin (1989)
Heid, U.: On ways words work together – research topics in lexical combinatorics. In: Proceedings of the 6th Euralex International Congress on Lexicography (EURALEX 1994), Amsterdam, The Netherlands, pp. 226–257 (1994)
Heylen, D., Maxwell, K.G., Verhagen, M.: Lexical functions and machine translation. In: Proceedings of the 15th International Conference on Computational Linguistics (COLING 1994), Kyoto, Japan, pp. 1240–1244 (1994)
Hindle, D., Rooth, M.: Structural ambiguity and lexical relations. Computational Linguistics 19(1), 103–120 (1993)
Jackendoff, R.: The Architecture of the Language Faculty. MIT Press, Cambridge (1997)
Jacquemin, C., Klavans, J.L., Tzoukermann, E.: Expansion of multi-word terms for indexing and retrieval using morphology and syntax. In: Proceedings of the 35th Annual Meeting on Association for Computational Linguistics, Morristown, NJ, USA, pp. 24–31 (1997)
Kjellmer, G.: Aspects of English collocations. In: Meijs, W. (ed.) Corpus Linguistics and Beyond, Rodopi, Amsterdam, pp. 133–140 (1987)
Koehn, P.: Europarl: A parallel corpus for statistical machine translation. In: Proceedings of the Tenth Machine Translation Summit (MT Summit X), Phuket, Thailand, pp. 79–86 (2005)
Krenn, B.: The Usual Suspects: Data-Oriented Models for Identification and Representation of Lexical Collocations, vol 7. German Research Center for Artificial Intelligence and Saarland University Dissertations in Computational Linguistics and Language Technology, Saarbrücken (2000)
Lea, D., Runcie, M. (eds.): Oxford Collocations Dictionary for Students of English. Oxford University Press, Oxford (2002)
Lü, Y., Zhou, M.: Collocation translation acquisition using monolingual corpora. In: Proceedings of the 42nd Meeting of the Association for Computational Linguistics (ACL 2004), Barcelona, Spain, pp. 167–174 (2004)
Maynard, D., Ananiadou, S.: A linguistic approach to terminological context clustering. In: Proceedings of Natural Language Pacific Rim Symposium (1999)
Mel'čuk, I.: Collocations and lexical functions. In: Cowie, A.P. (ed.) Phraseology. Theory, Analysis, and Applications, pp. 23–53. Claredon Press, Oxford (1998)
Michou, A., Seretan, V.: A tool for multi-word expression extraction in modern Greek using syntactic parsing. In: Proceedings of the Demonstrations Session at EACL 2009, pp. 45–48. Association for Computational Linguistics, Athens (2009)
Orliac, B., Dillinger, M.: Collocation extraction for machine translation. In: Proceedings of Machine Translation Summit IX, New Orleans, Lousiana, USA, pp. 292–298 (2003)
Padó, S., Lapata, M.: Dependency-based construction of semantic space models. Computational Linguistics 33(2), 161–199 (2007)

Pearce, D.: A comparative evaluation of collocation extraction techniques. In: Third International Conference on Language Resources and Evaluation, Las Palmas, Spain, pp. 1530–1536 (2002)

Pecina, P.: An extensive empirical study of collocation extraction methods. In: Proceedings of the ACL Student Research Workshop, Ann Arbor, Michigan, pp. 13–18 (2005)

Sag, I.A., Baldwin, T., Bond, F., Copestake, A., Flickinger, D.: Multiword Expressions: A Pain in the Neck for NLP. In: Gelbukh, A. (ed.) CICLing 2002. LNCS, vol. 2276, pp. 1–15. Springer, Heidelberg (2002)

Seretan, V.: An integrated environment for extracting and translating collocations. In: Mahlberg, M., González-Díaz, V., Smith, C. (eds.) Proceedings of the Corpus Linguistics Conference CL 2009, Liverpool, UK (2009)

Seretan, V.: Syntax-Based Collocation Extraction. Text, Speech and Language Technology. Springer, Dordrecht (2011)

Seretan, V., Wehrli, E.: Accurate collocation extraction using a multilingual parser. In: Proceedings of the 21st International Conference on Computational Linguistics and 44th Annual Meeting of the Association for Computational Linguistics, Sydney, Australia, pp. 953–960 (2006)

Seretan, V., Wehrli, E.: Multilingual collocation extraction with a syntactic parser. Language Resources and Evaluation 43(1), 71–85 (2009)

Seretan, V., Wehrli, E.: Extending a multilingual symbolic parser to Romanian. In: Tufiş, D., Forăscu, C. (eds.) Multilinguality and Interoperability in Language Processing with Emphasis on Romanian. Romanian Academy Publishing House, Bucharest (2010a)

Seretan, V., Wehrli, E.: Tools for syntactic concordancing. In: Proceedings of the International Multiconference on Computer Science and Information Technology, Wisła, Poland, pp. 493–500 (2010b)

Seretan, V., Wehrli, E.: FipsCoView: On-line visualisation of collocations extracted from multilingual parallel corpora. In: Proceedings of the Workshop on Multiword Expressions: from Parsing and Generation to the Real World, Association for Computational Linguistics, Portland, Oregon, USA, pp. 125–127 (2011)

Seretan, V., Nerima, L., Wehrli, E.: Using the Web as a corpus for the syntactic-based collocation identification. In: Proceedings of the 4th International Conference on Language Resources and Evaluation (LREC 2004), Lisbon, Portugal, pp. 1871–1874 (2004)

Sinclair, J.: Corpus, Concordance, Collocation. Oxford University Press, Oxford (1991)

Smadja, F.: Retrieving collocations from text: Xtract. Computational Linguistics 19(1), 143–177 (1993)

Stubbs, M.: Corpus evidence for norms of lexical collocation. In: Cook, G., Seidlhofer, B. (eds.) Principle & Practice in Applied Linguistics. Studies in Honour of H.G. Widdowson. Oxford University Press, Oxford (1995)

Moirón V., Begoña, M.: Data-driven identification of fixed expressions and their modifiability. PhD thesis, University of Groningen (2005)

Wehrli, E.: Fips, a "deep" linguistic multilingual parser. In: ACL 2007 Workshop on Deep Linguistic Processing, Prague, Czech Republic, pp. 120–127 (2007)

Wehrli, E., Nerima, L., Scherrer, Y.: Deep linguistic multilingual translation and bilingual dictionaries. In: Proceedings of the Fourth Workshop on Statistical Machine Translation, pp. 90–94. Association for Computational Linguistics, Athens (2009a)

Wehrli, E., Nerima, L., Seretan, V., Scherrer, Y.: On-line and off-line translation aids for non-native readers. In: Proceedings of the International Multiconference on Computer Science and Information Technology, Mrągowo, Poland, pp. 299–303 (2009b)

Wehrli, E., Seretan, V., Nerima, L., Russo, L.: Collocations in a rule-based MT system: A case study evaluation of their translation adequacy. In: Proceedings of the 13th Annual Meeting of the European Association for Machine Translation, Barcelona, Spain, pp. 128–135 (2009c)

van der Wouden, T.: Collocational behaviour in non content words. In: Proceedings of the ACL Workshop on Collocation: Computational Extraction, Analysis and Exploitation, Toulouse, France, pp. 16–23 (2001)

An Approach to Efficient Processing of Multi-word Units

Cvetana Krstev, Ivan Obradović, Ranka Stanković, and Duško Vitas

Abstract. Efficient processing of Multi-Word Units in the course of development of morphological MWU dictionaries is not easy to achieve, especially when languages with complex morphological structures are concerned, such as Serbian. Manual development of this type of dictionaries is a tedious and extremely slow process. To alleviate this problem we turned to our multipurpose software tool, dubbed LeXimir, in the production of lemmas for e-dictionaries of multi-word units. In addition to that, we developed a procedure aimed at making the production of MWU dictionary lemmas more efficient. This procedure, which strongly relies on our comprehensive e-dictionaries of Serbian simple words, was subsequently implemented as a new functionality LeXimir. In this paper we present our approach, and offer an evaluation of the performance of the new functionality of LeXimir, and hence of our procedure, obtained through two rounds of experiments on various types of data. The paper ends with a brief discussion of some further possible applications of both the procedure and LeXimir in various language processing tasks.

1 Introduction

The term Multi-Word unit (MWU) is used to describe different but related phenomena, such as fixed or semi-fixed phrases, compounds, support verbs, idioms, phrasal verbs, collocations, etc. that act as single units at some level of linguistic analysis

Cvetana Krstev
University of Belgrade — Faculty of Philology, Studentski trg 3, 11000 Belgrade, Serbia
e-mail: cvetana@matf.bg.ac.rs

Ivan Obradović · Ranka Stanković
University of Belgrade — Faculty of Mining and Geology, Djušina 7, 11000 Belgrade, Serbia
e-mail: {ivano,ranka}@rgf.bg.ac.rs

Duško Vitas
University of Belgrade — Faculty of Mathematics, Studentski trg 16, 11000 Belgrade, Serbia
e-mail: vitas@matf.bg.ac.rs

A. Przepiórkowski et al. (Eds.): *Computational Linguistics*, SCI 458, pp. 109–129, 2013.
DOI: 10.1007/978-3-642-34399-5_6

and exhibit reduced syntactic or semantic transparency, non-compositionality, more or less frozen status, high degree of lexicalization and/or high degree of conventionality [3]. MWUs differ in the degree in which these features occur, and are hence positioned between grammar and lexicon.

The NLP community offered various approaches to lexical treatment of multiword units. Since 2003 the workshops on multi-word expressions are being regularly organized in the scope of major events — ACL, EACL, Coling or LREC — not to mention special sessions during other language technology or computational linguistics conferences.[1] On these occasions treatment of MWUs was presented from various points of view showing that significant results were achieved. However, two points need to be stressed. Although much has been done to expand research to less-resourced languages, they are still represented to a lesser degree. The second point is that it seems that the identification and extraction of MWUs has attracted more attention of researchers than their lexical representation. In [18] authors give reasons against the lexical representation (or "a words-with-spaces description") for various types of MWUs, most notably the risk of lexical proliferation and inability to deal with internal MWU modifications. Various approaches to successful lexical representation of MWUs were analyzed in detail by Savary [20].

Slavic languages are analyzed in [17] and arguments are presented why they are in general more difficult for Natural Language Processing (NLP) than Romance and Germanic languages, and which of their features are making them, nevertheless, more suitable for higher levels of processing, like parsing. However, for lexical representation of Serbian MWUs, less favorable features of Slavic languages predominate, most notably its rich morphology, derivation and word order. Nevertheless, when deciding on the approach to process MWUs in Serbian we had to take into consideration that many resources and tools for Serbian are either under-developed or non-existant. One of the exceptions are lexical resources [27].

Morphological electronic dictionaries of Serbian for NLP were successfully developed. Their development follows the methodology and format known as DELAS/DELAF presented for French in [5]. E-dictionaries in the same format have been produced for many other languages. This format can be briefly described in the following way: in a dictionary of lemmas (DELAS) every lemma is described in full detail so that a dictionary of forms containing all necessary grammatical information (DELAF) can be generated from it, and subsequently used in various NLP tasks. Two corpus processing systems that support work with this dictionary format were developed, Unitex [16] and Nooj [24], both of which use finite-state technology as initially introduced in [8]. Serbian e-dictionaries of simple forms have reached a considerable size: they have a total of more than 127,000 lemmas [9] generating close to 4.4 million forms. Unitex distribution includes a large sample from the Serbian e-dictionary which covers a specific text, the Serbian translation of Voltaire's *Candide*.

In order to produce a robust lexical representation of Serbian MWUs we applied two approaches. Productive classes of MWUs, like numerals and various named

[1] Programs and proceedings of these workshops can be found at `http://multiword.sourceforge.net/`

entities that rely on them (e.g. measurement phrases) can best be described by dictionaries in the form of finite-state transducers (FST), and a number of them were produced for Serbian as well [13]. Other contiguous MWUs that are idiosyncratic in nature, namely nouns and adjectives, had to be lexically described in a similar way as simple words. Discontinuous MWUs, like verb phrases, ask for a different approach, and they were not yet described for Serbian.

In the computational lexicography school led by Maurice Gross, the interest in MWUs and the production of morphological dictionaries of compounds has been vivid from the very beginning [7]. Following that direction, dictionaries of MWU lemmas (DELAC) that are provided with information enabling the production of all inflected forms (DELACF) were developed for several languages, including French [4], English [19], Greek [14], Italian [26], and Portuguese [15]. At Unitex official web site[2] a comprehensive list of references related to the production of e-dictionaries of MWUs for these languages is given.

The lexical description of MWUs in the so-called DELAC/DELACF format in practice means that MWU lemmas have to be collected, generated, and inflected.

2 Inflection of MWUs

In order to produce a list of MWU forms in a systematic way, it is necessary to decide what the lemma of all these forms is, what are its additional features, how do its simple word constituents inflect, and what is the inflectional behavior of the MWU as a whole. One can imagine that for some languages this complex procedure can be skipped and a list of MWU forms can be produced from scratch. Serbian is a highly inflectional language and such a shortcut procedure cannot be applied. We will illustrate this with two examples. The nominal MWUs *petokraka zvezda* 'five-pointed star' and *Farenhajtov stepen* 'Fahrenheit degree' consist of an adjective followed by a noun, which in Serbian is the natural order of an adjective and a noun in a MWU. However, these MWUs, together with a few more allow a reverse order as well — *zvezda petokraka* and *stepen Farenhajtov*. Both MWUs can be used in plural form. In Serbian, adjectives and nouns inflect in number and case, while adjective forms also depend on gender, definiteness, comparison, and in some cases animacy. Adjectives and nouns do not inflect freely in a MWU — the values of categories for number, case and gender have to agree. The animacy is important only for the masculine gender nouns in the accusative singular. Since the gender of *zvezda* 'star' is feminine, the animacy is of no relevance for this MWU. This is not the case for *Farenhajtov stepen* since *stepen* is masculine. To obtain the correct accusative singular form *Farenhajtov stepen* it is important to know that *stepen* is inanimate, otherwise the incorrect accusative form *Farenhajtovog stepena* would be obtained. Finally, adjectives *petokrak* 'five-pointed' and *Farenhajtov* 'belonging to Fahrenheit' have no comparative and superlative forms, so they will not be generated. Indefinite adjective forms are rarely used in compounds so they are not generated either.

[2] http://www-igm.univ-mlv.fr/ũnitex/

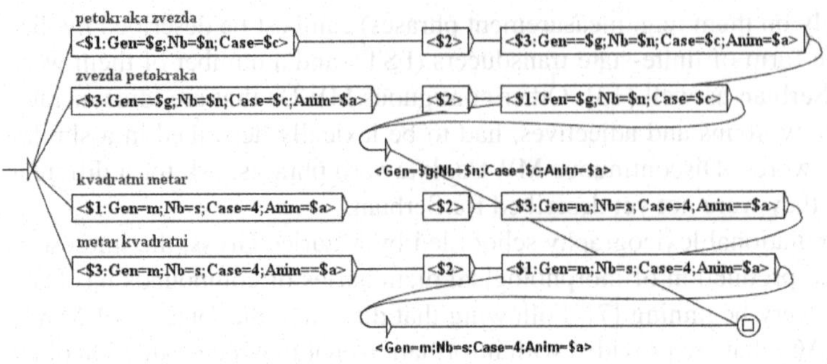

Fig. 1 A simplified transducer for compounds of the type *petokraka zvezda* and *Farenhajtov stepen*

This example illustrates the complexity of capturing all information about one MWU in its DELAC lemma. The most demanding part is to formulate the agreement conditions in a consistent way. A special form of inflectional transducers developed by Savary [21] and implemented in the Multiflex system answers most of these questions. The inflectional graph in Figure 1 illustrates this. A MWU serving as lemma is tokenized and its tokens become values of variables: in the case of *petokraka zvezda* values of variables become $1=*petokraka*, $2=<space>, $3=*zvezda* while in the case of *Farenhajtov stepen* they are $1=*Farenhajtov*, $2=<space>, $3=*stepen*. If a simple pattern of the form <$*i*> appears in the inflectional graph it means that the corresponding token is recopied in all MWU inflectional forms as it is — in our examples the second token, a space, is reproduced in all inflectional forms.

A token pattern can be followed by one or more equations of the type *Grammatical_feature=value*. In that case the specific form of a token is needed. In our example the token <$3:Gen=m;Nb=s;Case=4> from the lower part of the graph means that the masculine gender, singular and accusative form of the third token is needed. However, the gender of the noun *zvezda* from the MWU *petokraka zvezda* is feminine, so this form cannot be produced and the lower paths in the graph will be ignored. They will not be ignored for some other MWUs, like *Farenhajtov stepen*, since the gender of *stepen* is masculine.

Additionally, grammatical-feature equations can contain not only concrete values but also unification variables. A unification variable instantiates to all values of the corresponding grammatical feature. For Serbian, a pattern <$3:Case=$c> means that forms for all cases — 7 different values — will be generated for the third token. The occurrence of the same unification variables in the same path means that their values have to agree. If a pattern <$1:Case=$c> appears in the same path as <$3:Case=$c> it means that when the genitive form of the first token is generated then the genitive form of the third token has to be generated as well, and that will also be the value of the 'Case' feature of the generated MWU form — the output of the transducer.

Finally, a unification variable does not need to instantiate to all values of some grammatical feature. Instead, it can inherit its value from a token itself. In the pattern

<$3:Gen==$g> the variable $g inherits its value from the third token. For *petokraka zvezda* the variable $3 will instantiate to the value *f* since the gender of the third token is feminine, while for *Farenhajtov stepen* it will instantiate to the value *m* — the gender of the token *stepen*. In both cases, the variable $g from the pattern <$1:Gen=$g> occurring in the same path will have to agree with the value inherited from the third token; hence, in the first case it will have the value *f* and in the second case the value *m*.

The two possible orders of the adjective and the noun in the MWU are achieved with two separate paths in the graph, one for the order given by a lemma itself, and the other for the reverse order. Many other variations in MWUs can also be easily described using Multiflex graphs like: the optional use of a hyphen, omission of some of their constituents, their capitalization and/or abbreviation, as well as some limited changes in their syntactic structure [22]. The Multiflex system is incorporated into Unitex, but it was also successfully used for Polish proper names in another environment [23]. For the inflection of Serbian MWUs 104 such transducers were developed — 18 for adjectives and 86 for nouns.

By analogy with entries in a dictionary of simple word lemmas, an entry in a DELAC dictionary consists of a MWU lemma to which a name of an inflectional transducer (similar to the one represented in Figure 1) is assigned. Similarity ends here, because simple word constituents of a MWU lemma also have to be described in a way that enables the production of all needed forms. This leads finally to the following lemma forms:

```
petokraka(petokrak.A6:aefs1g) zvezda(zvezda.N600:fs1q),NC_AXNr
Farenhajtov(Farenhajtov.A1:akms1g) stepen(stepen.N5:ms1q),
  NC_AXNr
```

These DELAC entries enable the production of all MWU forms for DELACF dictionary of forms; forms representing the genitive singular with reverse order of constituents for these two MWUs are:

```
zvezde petokrake,petokraka zvezda.N:fs2q
stepena Farenhajtovog,Farenhajtov stepen.N:ms2q
```

Production of a lemma in the format presented is far too demanding to be done manually because for each MWU one has to provide the following information:

1. What is the lemma? One has to decide that *petokraka zvezda* and *Farenhajtov stepen* are more preferable as lemmas then *zvezda petokraka* and *stepen Farenhajtov*.
2. How does this MWU inflect and which inflectional transducer should be used for it? The two example MWUs have an adjective/noun structure and allow a reverse order of constituents, therefore inflectional transducer NC_AXNr should be used.
3. Which MWU constituents inflect? In our example both constituents inflect which means that inflectional information about them is needed as well.
4. What are DELAS entries of these MWU constituents that enable the generation of all needed forms? These entries for *petokraka zvezda* are `petokrak.A6` and

zvezda.N600, and for *Farenhajtov stepen* they are Farenhajtov.A1 and stepen.N5.

5. What are the values of grammatical features of constituent forms used in the MWU lemma? For the first example they are aefs1g and fs1q, while for the second they are akms1g and ms1q.

A fully manual production of MWU lemmas is, however, not necessary, because possible answers to the above questions that concern MWU constituents can be found in dictionaries of simple words.

3 LeXimir as a Dictionary Management System

Bearing in mind the aforementioned complexity of production of MWU lemmas we have endeavored towards a procedure for automatic production of DELAC entries. The software tool which enabled the implementation of this procedure was LeXimir,[3] a multipurpose tool developed by the University of Belgrade Language Technology Group [12] to support computational linguists in developing, maintaining and exploiting e-dictionaries. LeXimir is written in C#, and operates on the .NET platform. It can run on any personal computer under Windows and supports simultaneous manipulation of various language resources: e-dictionaries, wordnets, and aligned texts.

Implementation of LeXimir followed a modular approach. Namely, there exists a common core of the system, which is coupled with several modules performing different tasks. Modular organization of components provides two obvious benefits. In the first place, it enables the use of various resources in any part of the system, wherever they are needed. Thus, for example, morphological dictionaries can be used for adding additional morphological information to wordnet synsets, whereas both morphological dictionaries and the wordnet can be used in production of concordances for aligned texts. On the other hand, it enables the use of LeXimir's core modules in different scenarios: as a standalone Windows application LeXimir.exe or as a web application VebRanka[4] (previously WS4QE), which is supported by a corresponding web service. The web service accepts and generates data sets in XML format, which are further converted into data structures that can be used for different purposes (string, array, table, etc.). As examples of web service functions we will mention a few characteristic ones: the function that generates inflected forms for a given lemma, the function that returns all synonyms from a given wordnet synset in all inflected forms, and the function that returns synonyms without inflected forms.

As our e-dictionaries are Unitex-based, and Unitex is open source software distributed under the LGPL license, we incorporated its modules in LeXimir for the majority of tasks that involve manipulation of e-dictionaries. For the production

[3] LeXimir is available under CC_BY-NC license. For more information see http://korpus.matf.bg.ac.rs/soft/LeXimir.html

[4] http://hlt.rgf.bg.ac.rs/VebRanka

of MWU DELAC lemmas we used the appropriate Unitex modules for dictionary look-up.

LeXimir provides for concurrent manipulation of several dictionaries of lemmas, both of simple words and MWUs (DELAC), distributed in any number of files. However, the possibility of manipulating dictionaries of word forms is not envisaged, as such files are produced automatically either from DELAS or DELAC by means of appropriate FSTs. Organizing dictionaries in sets of different files is practically motivated. Namely, smaller size files are much easier to manipulate.

With LeXimir's editor for MWUs the user can perform, beside the usual functions — add, insert, copy, change — many more demanding activities. The users can check the correctness of every lemma with the function 'Inflect' that lists all inflected forms of a selected lemma. Another useful function is the extraction of subsets of lemmas based on different criteria: lemmas' beginning, their part of speech (PoS), inflectional class code, syntactic and/or semantic markers or a Boolean combination of these criteria.

Table 1 The DELAC entry management form by LeXimir: to the left is a form for lemma constituents, to the right are two auxiliary tables offered by LeXimir

No	Form	Lemma	FST	GramCat	*petokraka* in DELAF	*zvezda* in DELAF
1	petokraka	petokrak	A6	:aefs1g	petokrak.A:aenp5g	zvezda.N:fp2q
2	zvezda	zvezda	N600	:fs1q	petokrak.A:aenp4g	zvezda.N:fs1q
					petokrak.A:aenw2g	
					petokrak.A:aefs1g	
					petokrak.A:akms4v	
					......	

Table 1 illustrates the table for manual production of a DELAC entry having two constituents: *petokraka* and *zvezda*. The user inserts constituents of the MWU in the column 'Form'. In the next step columns 'Lemma', 'FST' (PoS and inflectional codes of constituents), and 'GramCat' (grammatical codes of constituents) have to be filled. The system automatically offers auxiliary tables with all possible solutions retrieved from DELAS/DELAF dictionaries of simple words. The selection of the correct lemma, FST code and grammatical categories is supported by possible combinations offered in auxiliary tables. In the final step, the user has to fill manually the code of the inflectional transducer for the newly produced MWU lemma, and attach to it the appropriate semantic and other markers. The user can then check the correctness of the new MWU lemma by using the 'Inflect' function that invokes Multiflex to perform the inflection.

The outlined procedure does help in answering the two last questions posed at the end of Section 2. However, answers to questions 2 and 3 have to be provided by the user. Thus, by following this approach not more than 2,800 DELAC entries were produced during three years, which we found very inefficient.

4 A Rule Based Procedure for Inflection of MWUs

4.1 Detection of Inflectional Properties of MWU Lemmas

We have further improved the procedure for production of MWU lemmas when we realized that the answers obtained automatically in support of manual production of MWU lemmas can also help in detection of the syntactic composition of a MWU and therefore indicate the appropriate inflectional transducer. Namely, the MWUs in Serbian have predictable basic structures. For instance, nominal MWUs with two constituents (beside a separator) fall into five basic structures:

- Adjective/noun (both inflect and agree in gender, number and case)
- Noun/noun (both inflect and agree in number and case)
- Noun/noun in the genitive or in the instrumental (only the first noun inflects)
- Word/noun (only the second noun inflects; the first word is usually not a Serbian simple word)
- Noun/adjective (both inflect and agree in gender, number and case)

However, there are 25 different inflectional graphs for the nominal MWUs with two constituents because there are subtleties that have to be taken into consideration besides these basic structures, e.g. can a MWU have plural forms, can a separator be omitted or replaced by another separator, etc. The basic structure, however, determines the general form of a MWU lemma and information that has to be supplied for its constituents.

Table 2 Different interpretations of the sequence *živa rana*

Form	Lemma	Translation	PoS	Relevant grammatical categories
živa	*živ*	'alive'	A	nominative, singular, feminine
živa	*živa*	'mercury'	N	nominative, singular, feminine
rana	*rana*	'wound'	N	nominative, singular, feminine genitive, plural, feminine
rana	*ran*	'early'	A	nominative, singular, feminine

Thus, automatic production of the lemma for *petokraka zvezda* could proceed like this: a look-up in the dictionary of simple word forms determines that *zvezda* can only represent two realizations of the noun lemma *zvezda*, namely in the nominative singular or in the genitive plural. Similarly, it is determined that *petokraka* can be one of 12 different representations of the adjective *petokrak*; however, only one of them agrees with the noun *zvezda*, and that is the singular, feminine gender, nominative case form. Consequently, it can be deduced that only the basic structure adjective/noun applies here.

Of course, not all MWUs are so easy to process. For instance, for the MWU *živa rana* 'open wound' a dictionary look-up offers several possibilities (Table 2).

Thus there are five possible MWU structures: adjective/noun, noun/noun, noun/noun in the genitive, noun/adjective, and adjective/adjective whereas only the first one is correct.

Table 3 Five lemmas offered for the sequence *živa rana*

First constituent	Second constituent	MWU inflectional class
živa(živ.A15:aefs1g)	rana(rana.N600:fs1q)	NC_AXN
živa(živa.N600:fs1q)	rana(rana.N600:fs1q)	NC_NXN
živa(živa.N600:fs1q)	rana	NC_N2X
živa(živa.N600:fs1q)	rana(ran.A17:aefs1g)	NC_NXAr
živa(živ.A15:aefs1g)	rana(ran.A17:aefs1g)	AC_AXA

Based on an analysis illustrated by previous examples, we have developed a new functionality within LeXimir that offers one or more DELAC entries for every MWU presented in its lemma form. As indicated by the example, it relies on information in e-dictionaries of simple words, but also uses a set of manually produced rules to deduce the basic structure of a given MWU, as well as its additional features. For the example *živa rana* this functionality would offer five lemmas; the first one would be selected, the remaining four discarded (Table 3).

In order to design our automated procedure we grouped all inflectional transducers into equivalence classes or super-classes: a super-class consists of all MWUs having the same form of MWU lemma—because they need the same information for the production of inflectional forms—although their syntactic structure need not be the same (see Table 5). This is also reflected in the convention we used for naming the inflectional transducers: A stands for an adjective constituent, N stands for a noun constituent, X stands for a constituent that does not inflect (including a separator), with some additional digits and letters added to differentiate transducers. This is illustrated in Table 4 by four classes (names of inflectional transducers) all belonging to the same AXN super-class and used for the inflection of MWUs consisting of an adjective followed by a noun, where both constituents inflect and must agree in basic grammatical categories.

As we have already mentioned, one super-class need not consist of MWUs having the same syntactic structure. For instance, a super-class N4X consists of three component MWUs for which the first component is a noun that inflects and two remaining components do not inflect. According to our DELAC dictionary MWUs belonging to this super-class may have various syntactic structures, as presented in Table 5.

Also, MWUs having the same syntactic structure need not all belong to the same super-class. Such is the case for MWUs with the syntactic structure noun/noun in the genitive. The plural forms of such MWUs, in the case they exist, can be:

Table 4 Super-class AXN

Class	Example	Translation	Specifics
AXN	*živa rana*	'open wound'	
AXN3	*Pitagorina teorema*	'Pythagorean theorem'	does not inflect in number
AXNF	*serijski ubica*	'serial killer'	second constituent changes gender in plural forms
AXNr	*petokraka zvezda*	'five-pointed star'	allows reverse order

Table 5 Super-class N4X

Example	Translation	Structure
kola hitne pomoći	'first aid car'	noun/adjective in gen./noun in gen.
uskrsenje sina božjeg	'resurrection of the Son of God'	noun/noun in gen./adjective in gen.
menadžment ljudskim resursima	'human resources management'	noun/adjective in instr./noun in instr.
raketa zemlja-vazduh	'air-to-ground missile'	noun/noun in nom./noun in nom.
ugovor o zakupu	'lease contract'	noun/preposition/noun
trgovac na malo	'wholesaler'	noun/preposition/adjective

- only the first component inflects in number, the second component does not inflect, as in *profesor matematike/profesori matematike* 'professor(s) of mathematics' and *red vožnje/redovi vožnje* 'travel schedule(s)';
- both components have to be in the plural form, e.g. *teme ugla/temena uglova* 'angle vertex/angle vertices';
- the second component can be either in the singular form or in the plural form, for instance, *predsednici države/predsednici država* 'presidents of the state/ presidents of states';
- both components can be in the singular and the plural form in all possible combinations, e.g. *analiza dokumenta/analiza dokumenata/analize dokumenta/analize dokumenata* 'document(s) analysis/document(s) analyses'.

Only the MWUs belonging to the first listed group belong to the super-class N2X and they require inflectional information only for the first component. All the other MWUs belong to the super-class NXNg and for them inflectional information is necessary for both components, as illustrated by the example MWU lemmas:

```
profesor(profesor.N2:ms1v) matematike,NC_N2X
teme(teme.N324:ns1q) ugla(ugao.N115a:ms2q),NC_N2X4
```

In order to formulate a strategy for the production of MWU lemmas we analyzed the data available in the existing DELAC dictionary looking for useful information. On the one hand, we identified the additional information assigned to components of MWUs belonging to a particular inflectional class, and on the other, we identified inflectional classes associated with the same additional information.

4.2 *The Rule Design Strategy*

The procedure for automatic construction of a DELAC type dictionary relies on a manually produced set of rules that rely on morphosyntactic and semantic information found in applied dictionaries of simple words (they can be a chosen subset of dictionaries existing for Serbian, e.g. dictionaries of verbs may be omitted). The rule design strategy resulted from the aforementioned expert analysis of available MWU lemmas. The task of the rule based procedure is to automatically generate the complete MWU lemma. However, the strategy and the procedure are independent, and changes in the strategy, in general, do not affect the procedure itself. This approach enabled us to experiment with various rule strategies, and thus the final strategy used is a result of several iterations.

Each rule consists of one set of general conditions (tags `<RuleGenCond>`) and zero to many sets of special conditions (tags `<RuleSpecCond>`). Special conditions are added to general conditions in the processing phase and one such complete set has to be satisfied in full in order to produce a possible solution — a MWU lemma. In that respect each rule behaves as a disjunction of conjunctions. For instance, the rule in Example 1 is applied to two component MWUs as follows: if components satisfy (according to the dictionary of simple words) the specified grammatical conditions, namely, that the first is an adjective in the nominative case and the second component is a noun in the nominative case as well, and these two components agree in gender and animacy, then the additional conditions are checked, and at least one of them needs to be satisfied. In this case it means that one of the following additional conditions must be satisfied: the first component starts with uppercase letter (e.g. *Pariska komuna* 'The Paris Commune'), or both components are already in plural (e.g. *lokalni izbori* 'local elections'), or the second component is a collective noun (e.g. *kandirano voće* 'candied fruits').

Example 1 (XML form of a rule for the class NC_AXN3, super-class NC_AXN — for adjective/noun MWUs that do not inflect in number).

```
<Rule ID="2" CFLX="NC_AXN3" CflxGroup="NC_AXN">
    <RuleGenCond>
        <Word ID="1" POS="A" Flex="true" Case="1" Anim="$a"
              Gen="$g"/>
        <Word ID="2" POS="N" Flex="true" Case="1" Anim="=$a"
              Gen="=$g"/>
    </RuleGenCond>
    <RuleSpecCond ID="1" Example="Pariska komuna">
        <Word ID="1" Num="s" Cond="$PRE"/>
        <Word ID="2" Num="s"/>
    </RuleSpecCond>
    <RuleSpecCond ID="2" Example="lokalni izbori">
        <Word ID="1" Case="1" Num="p"/>
```

```
        <Word ID="2" Case="1" Num="p"/>
    </RuleSpecCond>
    <RuleSpecCond ID="3" Example="kandirano voce">
        <Word ID="1" Case="1" Num="s"/>
        <Word ID="2" Case="1" Num="s"
              SinSem="+VN,+Coll,+HumColl"/>
    </RuleSpecCond>
</Rule>
```

Another rule that applies to three-component MWU adjectives in the form of a simple word adjective followed by the conjunction *kao*, followed by an animate noun, is given in Example 2. An example is the adjective *gladan kao vuk* 'hungry as a wolf'. Adjectives of this type have two plural forms: the noun component can be either in the singular *gladni kao vuk* or in the plural *gladni kao vuci*. This rule has no additional conditions and has no agreement requests.

Example 2 (A rule for the class AC_A3XN2, super-class AC_A3XN).

```
<Rule ID="153" CFLX="AC_A3XN2" CflxGroup="AC_A3XN">
    <RuleGenCond Example="gladan kao vuk">
        <Word ID="1" POS="A" Flex="true" Case="1" Num="s"
              Gen="m"/>
        <Word ID="2" POS="MOT" Flex="false" Cond="=,kao"/>
        <Word ID="3" POS="N,A" Flex="true" Case="1"
              Num="s" Anim="v"/>
    </RuleGenCond>
</Rule>
```

Each rule can check orthographic properties of a processed MWU and/or match its components with applied dictionaries of simple words. Orthographic conditions check separators used between words (a space is presumed by default) and capitalization of components — due to the condition `Cond="$PRE"` in the first set of special conditions in Example 1 this rule is applied only if the first component is written with initial upper-case. Rules can also check whether a component matches a string, e.g. the condition `Cond="=,kao"` in the Example 2 requires that the second component of a MWU is the string *kao* (a conjunction 'as'). The other condition `Sufix="ska,ška,čka"` (Example 5) requires that the suffix of the first component is *-ska*, *-ška* or *-čka* (a comma is used as a disjunction operator).

More interesting are conditions that rely on dictionaries of simple words, and they can offer answers to following questions:

- Does a component exist in dictionaries of simple forms? For instance, due to the condition `POS="!SDIC"` in the set of general conditions in Example 5 this rule applies only if the first MWU component is not in the dictionary of simple forms (it is an "unknown word").
- What are the values of grammatical categories of a MWU component? For instance, the rule in Example 2 applies only if, according to applied dictionaries, the first word is an adjective (`POS="A"`), in the nominative case (`Case="1"`), in the singular (`Num="s"`), and in the masculine gender (`Gen="m"`).

- Do values of a grammatical category agree for two or more components? The rules use unification variables in a similar way as inflectional transducers for MWUs (described in Section 2). For instance, in Example 1 `$g` is one such variable: it receives the value of the gender from the second component (a noun) and has to agree in gender with the first component (an adjective).
- Does a component possess a specific syntactic or semantic feature? In Example 1 the third set of special conditions is applied if the second component is a collective noun or a verbal noun (`SinSem="+VN,+Coll,+HumColl"`).

In general, conditions can be negated by using two different operators: `!` and `~`. The simplest is the condition `!SDIC`, which means that a MWU component does not exist in dictionaries of simple words.[5] The operator `!` is used for atomic values — for instance, the condition `Sep="!-"` requires that a component is NOT followed by a hyphen in a MWU. More often, it is used for agreement conditions. In Example 3 due to the condition `Gen="!$g"` the rule is accepted only if a MWU consists of two nouns having different gender — in our example *leptir* 'butterfly' is masculine and *kravata* 'tie' is feminine.

Example 3 (A rule for the class NC_2XN1, super-class NC_2XN — for two nouns separated by a hyphen and having different gender; the first noun will not inflect).

```
<RuleGenCond Example="leptir-kravata/bow tie">
  <Word ID="1" POS="N" Flex="false" Case="1" Gen="!$g" Num="s"
        Sep="-"/>
  <Word ID="2" POS="N" Flex="true" Case="1" Gen="=$g" Num="s"/>
</RuleGenCond>
```

The operator `~` is the negation of the existence operator, meaning that the subset of word forms from dictionaries of simple words that satisfy other conditions must not contain the element satisfying a given condition. In Example 4 a rule is given that is used for MWUs in which the first component does not inflect. If the separator is not a hyphen this usually happens if the first component is not in dictionaries of simple forms, or is a prefix or an abbreviation. However, the condition `Case="~1"` allows that the first component can also be a noun if it is NOT in the nominative case. Thus abbreviations — like *TEI* in our example — will not be rejected, although *TEI* is a homograph of a dative form of a personal name *Tea*. In this way, some cases of false ambiguity can be resolved.

Example 4 (A rule for the class NC_2XN, super-class NC_2XN — the first component does not inflect; it can be a noun, but not in the nominative case).

```
<Rule ID="17" CFLX="NC_2XN" CflxGroup="NC_2XN">
    <RuleGenCond>
        <Word ID="1" POS="MOT" Flex="false" Sep="!-"/>
        <Word ID="2" POS="N" Flex="true" Case="1" Num="s"/>
    </RuleGenCond>...
    <RuleSpecCond ID="3" Example="TEI zaglavlje/TEI header">
```

[5] Similar notation is used in Unitex for meta-symbols.

```
        <Word ID="1" POS="N" Case="~1"/>
        <Word ID="2"/>
    </RuleSpecCond>...
```

In some rules `set` attributes of a special kind appear. They do not set conditions but rather values for the MWU lemma being generated. That is, instead of obtaining values from applied dictionaries of simple words, they allow rules to set these values themselves. They are thus used for components that do not exist in applied dictionaries ("unknown words"). In Example 5 the first component does not exist in dictionaries (`POS="!SDIC"`), but if it ends with *-ska*, *-ška* or *-čka* it will be treated as an adjective (`setPOS="A"`), with specific grammatical values (`setGramCats="np1gae"`), and a lemma, which can be obtained from the component form by deleting its final character and replacing it with an *i* (`setLemma="[B]i"`). This rule relies on the fact that all MWUs listed in an input file are in a normalized lemma form, that is, in the nominative case.

Example 5 (A rule for the class NC_AXN3, super-class NC_AXN — for proper names for which the first component, a relational adjective, is not in dictionaries of simple words).

```
<Rule ID="14c" CFLX="NC_AXN3" CflxGroup="NC_AXN">
    <RuleGenCond>
        <Word ID="1" POS="!SDIC" Flex="true" Cond="$PRE"
              setPOS="A" setFlexCode="A2"/>
        <Word ID="2" POS="N" Flex="true" Case="1" Num="p"/>
    </RuleGenCond>
    <RuleSpecCond ID="1" Example="Lofotska ostrva">
        <Word ID="1" Sufix="ska,ška,čka" setLemma="[B]i"
              setGramCats="np1gae" />
        <Word ID="2" Gen="n" />
    </RuleSpecCond>...
</Rule>
```

Our rule based strategy presently consists of 117 rules — 97 for nouns and 20 for adjectives. Among them, 38 rules pertain to MWUs with 2 components, 45 rules to MWUs with 3 components, 20 rules to MWUs with 4 components, 9 rules to MWUs with 5 components, and 5 rules to MWUs with 6 and more components.

4.3 Software Implementation

To manipulate the strategy in the form of a XML document our tool LeXimir relies on W3C standard languages Xquery and XSLT supported by .Net. The user interface for automatic production of DELAC lemmas is very straightforward and easy to use. A user can choose a file with a prepared list of MWUs and a file with a strategy, and the results will be presented to him in the form of a table in which the user has only to check the correct solutions upon which a list of DELAC entries is produced.

Table 6 Implementation of the Strategy on a prepared list of MWUs

	CLema	Cflx	No	Cflx_D	Mark	Corr
	Avogadrov broj(broj.N83:ms1q)	2XN3	1			
	Avogadrov broj(broj.N83:ms1q)	2XN	2			
√	Avogadrov(Avogadrov.A1:ms1gak) broj(broj.N83:ms1q)	AXN3	3		OK	
√	Novi(nov.A17:adms1g) Beograd(Beograd.N1001:ms1q)	AXN3	1		OK	
	Novi(nov.A17:adms1g) Beograd(Beograd.N1001:ms1q)	AXN	2			
	Stari(star.A17:adms1g) Grad(grad.N1:ms1q)	AXN3	1			
	Stari(star.A17:adms1g) Grad(grad.N1001:ms1q)	AXN3	2			
√	Stari(star.A17:adms1g) Grad(grad.N81:ms1q)	AXN3	3		OK	
	Stari(star.A17:adms1g) Grad(grad.N1:ms1q)	AXN	4			
	Stari(star.A17:adms1g) Grad(grad.N1001:ms1q)	AXN	5			
	Stari(star.A17:adms1g) Grad(grad.N81:ms1q)	AXN	6			
√	muva(muva.N601:fs1v) zujara(zujara.N601:fs1v)	NXN	1		OK	
	muva(muva.N601:fs1v) zujara	N2X	2			
√	otvorena(otvoren.A17:aenp1g) vrata(vrata.N304:np1q)	AXN3	1		OK	
√	ledeno(leden.A17:aens1g) doba(doba.N338:ns1q)	AXN	1		UOK	AXN3
	ledeno doba(doba.N338:np1q)	2XN3	2			
	ledeno doba(doba.N338:ns1q)	2XN	3			
	petokraka(petokrak.A6:aefs1g) zvezda(zvezda.N600:fs1q)	AXN	1	AXNr	UOK	AXNr
	izmene i dopune UDK		0			

Results for a list of 8 MWUs are given in Table 6[6]. The third option offered by the strategy for the first MWU, *Avogadrov broj* 'Avogadro's number' is the correct solution. It was produced by a rule similar to one presented in Example 5 because the possessive adjective *Avogadrov* is not included in the Serbian DELAS dictionary of adjectives. As for the second MWU, *Novi Beograd* 'New Belgrade (a municipality of Belgrade)', the first of the two options offered by the strategy is the correct solution. For the third MWU, *Stari Grad* 'Old City (a municipality of Belgrade)' the strategy offers as much as 6 options, among which the third represents the correct solution. Such a large number of options offered is due to the fact that the form *grad* can represent as much as three lemmas: city, degree, and hail. Out of the two options offered by the strategy for the fourth MWU, *muva zujara* 'blow fly', the first one is the correct one. As for the 5th MWU *otvorena vrata* 'open door (a meeting of parents with teachers)' only one solution is offered and it is the correct one. Three possible solutions are offered for the 6th MWU, *ledeno doba* 'ice age', and one of them, the first, AXN, is partly correct. Namely, the super-class is properly determined, and hence the lemma form, and what remains is to replace the inflection transducer by AXN3, as this MWU does not have a plural. The correction can be made by the user by stating the new, correct name of the transducer in the last column of this partly correct solution. The 7th MWU, *petokraka zvezda* is already in the dictionary which is evidenced by the fact that the column 'Clfx_D', and the following four columns are already filled. The solution offered by the strategy is almost the same as the one existing in the dictionary, except for the fact that the strategy failed to identify that this MWU allows a reversed order of components, which is a highly exceptional feature. The option of the user interface to detect MWUs already in the dictionary is very useful, as it prevents the introduction of duplicates in the dictionary. In addition

[6] This table excludes many columns that are kept for bookkeeping purposes.

to that, it may alert the user as to the potential shortcomings of the strategy. For the 8th MWU, *izmene i dopune UDK* 'amendments to UDC' no solution is offered — the MWU has an unusual structure for which no prediction was made.

When all options offered by the strategy are reviewed and those for which entries for a DELAC dictionary are to be produced ticked, the system will generate them automatically. Thus, we obtain an automated answer to questions 2 and 3 posed at the end of Section 2. Question 1 is answered by the user, who prepares the list of input lemmas. In some rare cases all rules will fail and a solution — compound lemma — will not be offered to the user. In that cases the user will have to produce a lemma consulting the existing e-dictionary, as illustrated in Table 1.

There are various debugging tools and preference selections at user's disposal. In the strategy development phase the user can compare the results obtained by the use of various strategies on the same MWU input list. The user may also filter the results and obtain only those that differ from the results obtained by the previous version of the strategy. He/she can preview the log file to see which rules were used for a particular MWU and in which order. The user can also see which simple word forms were retrieved from e-dictionaries of simple words and what were their grammatical values.

LeXimir has been successfully used for languages other than Serbian and English, namely, for Bulgarian [11]. The new functionality for production of MWU lemmas is also expected to perform successfully without any modifications for other languages. The prerequisites are that there exists a Unitex module for that language including: a dictionary of simple words, transducers for the inflection of simple words, the automatically produced dictionary of simple word forms, and transducers for the inflection of MWUs. As mentioned before, most of these conditions are satisfied for many languages. However, in order to apply this functionality to a new language it would be necessary to develop a new language-dependent strategy. It is also worth mentioning that the system can be easily modified to work with formats of simple words dictionaries other than those supported by Unitex. To that end, only the dictionary look-up module would have to be changed.

4.4 Procedure Evaluation

In order to evaluate the performance of LeXimir's functionality for automated generation of MWU lemmas we have conducted experiments on two occasions. The first evaluation took place in the first phase of the development of our procedure and strategy, involving three data sets. The first set consisted of 2,571 nouns and 207 adjectives already available in the existing DELAC dictionaries. The MWU lemmas for dictionary entries were (re)produced by LeXimir and then compared to the (correct) dictionary lemmas. The second set of data consisted of 704 common MWUs compiled from several sources, all of them nouns, while the third set consisted of a list of 206 geographic names.

In line with the possibility of a "partly" correct solution that we have recognized in the previous subsections, the evaluation results were classified as follows:

Table 7 Results for two rounds of evaluation

	1st evaluation			2nd evaluation		
	1st set	2nd *set*	3rd *set*	1st set	2nd set	3rd set
1	73.42%	85.92%	57.92%	84.97%	78.01%	93.42%
2	14.72%	10.47%	3.47%	11.17%	10.41%	0.0%
3+4	11.86%	3.61%	38.61%	3.86%	11.58%	6.58%

1. The system produced one or more solutions, among them a solution with the correct lemma, and assigned the correct inflectional class for a given MWU, and thus the overall solution is considered as correct;
2. The system produced one or more solutions, among them a solution with the correct lemma, but failed to assign the correct inflectional class, whereas the assigned super-class was correct — the overall solution was considered as partly correct;
3. The system offered one or more solutions, but they were all rejected as incorrect;
4. The system failed to offer a solution.

In the meantime we have used our system intensively, amended it and refined our strategy, e.g. by adding conditions of type described in Example 4. Then, we have conducted a second round of evaluation using three new data sets. The first two data sets contained MWUs from a terminological dictionary for library and information sciences (LIS): the first one included 519 MWUs of a more general nature, which are used outside this restricted domain, whereas the second included 1,114 MWUs belonging to specific LIS terminology. In addition to that, we used a smaller set of 152 MWU proper names, mostly geographic names and event names.

All results produced by the system in both rounds were validated manually, and are summarized in Table 7 (1 - correct, 2 partly correct, 3+4 - incorrect or no solution). It is obvious from the table that the results varied substantially depending on the type of data used. Results for the first round are discussed in more detail in [10].

When looking at the results of the second evaluation one must also take into account that the size of data sets varies considerably. Although a comparison with the results obtained in the first evaluation would seem natural, it is not easy to draw a conclusion whether we have made a substantial improvement in our strategy from the first evaluation cycle, given the relative heterogeneity of the type and size of data involved. It should, however, be noted that specific terms from the second LIS dictionary data set are often artificial, due to the nature of controlled dictionaries, and thus tend to be longer than average MWUs and consequently closer to free phrases. Hence, we will refrain from a general conclusion and just point out that in the case of relatively comparable sets of geographic names from the first evaluation and proper names from the second, a considerable improvement was reached beyond doubt.

In the second evaluation we also looked at the relation between the number of MWU components and the results obtained, which is presented in Table 8. As it

Table 8 Second evaluation results depending on the MWU length

	2 words	3 words	4 words	5 words	6 words	7 or more
1	83.75%	83.73%	70.08%	64.29%	17.65%	0.0%
2	10.7%	7.39%	13.38%	0.0%	0.0%	0.0%
3+4	5.55%	8.88%	16.54%	35.71%	82.35%	100.00%

was to be expected, the percentage of correct results decreased with the size of the MWU and at the end there were no correct (or even partly correct) solutions in the last case of MWUs with seven or more components.

Although the system in some cases offered as much as eight possible solutions for a single MWU, the correct one, if it existed, was always within the first five, most often the first. This also depended on the size of the MWU, and the larger the MWU the more likely the correct solution to be first.

In general, it is safe to say that the results obtained in both evaluation cycles testify to the fact that our approach yielded a strategy and procedure which can greatly contribute to efficient processing of MWUs.

5 Existing and Further Applications

The outlined procedure is now in everyday use for the production of MWU dictionary entries for Serbian. Due to the new LeXimir functionality the size of the MWU dictionary grew from the initial 2,800 lemmas to existing 9,600 in a relatively short period. We expect this growth rate to be even greater in the forthcoming period, as many new MWU lists are being prepared. LeXimir offers solutions for a list of MWUs practically instantaneously, and an experienced lexicographer needs only to go through the offered solutions to choose the correct ones.

The benefits obtained by including the MWU dictionary in language processing tasks for Serbian are already clearly visible. Besides the benefits that were to be expected, it has been already shown that the MWU dictionary can also be very useful in text disambiguation [1], and further in the parsing process [28]. Similar benefits were achieved for Serbian as well, primarily in named entities recognition tasks [13]. Namely, many simple proper names are ambiguous with other proper names or common nouns, while this is much less the case for MWU proper names. Our MWU morphological dictionary of proper names has already contributed to recognition of named entities with both high precision and recall.

The benefits of our approach do not end here. Our strategy coupled with Unitex and Multiflex modules for inflection and dictionary look-up was successfully used for inflection of free phrases that are not in the MWU dictionary. In these cases, the first option offered by our strategy is always chosen, which is, as we have seen, usually the correct one. This approach has first been tested in VebRanka [12], a web application that enables expansion of queries submitted to the Google search engine. More recently, a similar approach was applied in another web tool, Bibliša, that supports enhanced search of multilingual digital libraries of e-journals [25].[7] The queries initiated by a simple or multiword keyword, in Serbian or English, can be expanded by Bibliša, both semantically and morphologically, using different available monolingual and multilingual resources, such as wordnets and electronic dictionaries, suported by our strategy.

We expect that in the future our efforts will be further justified by solving various tasks involving MWUs in completely new environments. Namely, we have performed a thorough analysis of structures of nominal MWUs in Serbian, and produced an operational tool for detection of the structure that enables conversion of our data — both dictionaries and the strategy — in formats used by other applications. We believe that our strategy rules can be converted, for instance, to rules used by a MWE detection algorithm described in [2]. Namely, various rules in our strategy that produce lemmas belonging to a supper-class N4X could produce rules like those suggested by Arranz: NN [NN|JJ|PREP] [NN], etc. The same goes for rules presented in [6] in the case of the `mewtoolkit` tool used for identification of multi-word expressions. Successful application of these or similar tools to Serbian texts would solve the still remaining problem of detecting new MWUs, with the aim of either incorporating them in our dictionaries or using them in other tasks.

Acknowledgements. This research was supported by the Serbian Ministry of Education and Science under the grant #III 47003.

References

1. Alegria, I., Ansa, O., Artola, X., Ezeiza, N., Nojenola, K., Urizar, R.: Representation and Treatment of Multiword Expressions in Basque. In: Second ACL (ed.) Workshop on Multiword Expressions: Integrating Processing, Barcelona, Spain, pp. 48–55 (2004)
2. Arranz, V., Atserias, J., Castillo, M.: Multiwords and Word Sense Disambiguation. In: Gelbukh, A. (ed.) CICLing 2005. LNCS, vol. 3406, pp. 250–262. Springer, Heidelberg (2005)
3. Calzolari, N., Fillmore, C.J., Grishman, R., Ide, N., Lenci, A., MacLoed, C., Zampolli, A.: Towards best practice for multiword expressions in computational lexicons. In: 3rd LREC 2002, pp. 1934–1940. ELRA, Las Palmas (2002)
4. Courtois, B., Garrigues, M., Gross, G., Gross, M., Jung, R., Mathieu-Colas, M., Silberztein, M., Vivés, R.: Dictionnaire électronique des noms composés delac: les composants NA et NN. Rapport Technique, vol. 55. LADL, Paris (1997)
5. Courtois, B., Silberztein, M.: Dictionnaires électroniques du français. Larousse, Paris (1990)
6. De Araujo, V., Ramisch, C., Villavicencio, A.: Fast and flexible mwe candidate generation with the mwetoolkit. In: Proceedings of the Workshop on Multiword Expressions: from Parsing and Generation to the Real World, pp. 134–136. ACL, Portland (2011)
7. Gross, M.: Lexicon-grammar. the representation of compound words. In: Proceedings of Coling 1986, pp. 1–6 (1986)

[7] `http://hlt.rgf.bg.ac.rs/Biblisha`

8. Gross, M.: The Use of Finite Automata in the Lexical Representation of Natural Language. In: Gross, M., Perrin, D. (eds.) LITP 1987. LNCS, vol. 377, pp. 34–50. Springer, Heidelberg (1989)
9. Krstev, C.: Processing of Serbian — Automata, Texts and Electronic Dictionaries. Faculty of Philology. University of Belgrade, Belgrade (2008)
10. Krstev, C., Stanković, R., Obradović, I., Vitas, D., Utvić, M.: Automatic Construction of a Morphological Dictionary of Multi-Word Units. In: Loftsson, H., Rögnvaldsson, E., Helgadóttir, S. (eds.) IceTAL 2010. LNCS, vol. 6233, pp. 226–237. Springer, Heidelberg (2010)
11. Krstev, C., Stanković, R., Vitas, D., Koeva, S.: E-Connecting Balkan Languages. In: Proc. of the Workshop on Multilingual Resources, Technologies and Evaluation for Central and Eastern European Languages — RANLP 2009, Borovetz, Bulgaria, pp. 23–29 (2009)
12. Krstev, C., Stanković, R., Vitas, D., Obradović, I.: The Usage of Various Lexical Resources and Tools to Improve the Performance of Web Search Engines. In: 6th LREC 2008. ELRA, Marrakech (2008)
13. Krstev, C., Vitas, D., Obradović, I., Utvić, M.: E-dictionaries and finite-state automata for the recognition of named entities. In: Proceedings of the 9th International Workshop on Finite State Methods and Natural Language Processing, pp. 48–56. ACL, Blois (2011)
14. Kyriacopoulou, T., Mrabti, S., Yannacopoulou, A.: Le dictionnaire électronique des noms composés en grec moderne. Lingvisticae Investigationes (2002)
15. Mota, C., Carvalho, P., Ranchhod, E.: Multiword lexical acquisition and dictionary formalization. In: Proceedings of the Workshop Enhancing and Using Electronic Dictionaries, Coling 2004, Geneva, Switzerland, pp. 73–77 (2004)
16. Paumier, S.: Unitex 2.1 User Manual (2011), `http://www-igm.univ-mlv.fr/unitex/UnitexManual2.1.pdf`
17. Przepiórkowski, A.: Slavonic information extraction and partial parsing. In: Proceedings of the Workshop on Balto-Slavonic Natural Language Processing: Information Extraction and Enabling Technologies, ACL 2007, pp. 1–10. ACL, Stroudsburg (2007)
18. Sag, I.A., Baldwin, T., Bond, F., Copestake, A., Flickinger, D.: Multiword Expressions: A Pain in the Neck for NLP. In: Gelbukh, A. (ed.) CICLing 2002. LNCS, vol. 2276, pp. 1–15. Springer, Heidelberg (2002)
19. Savary, A.: Recensement et description des mots composés - méthodes et applications. Ph.D. thesis, Université de Marne-la-Vallée (2000)
20. Savary, A.: Computational Inflection of Multi-Word Units — A Contrastive Study of Lexical Approaches. Linguistic Issues in Language Technologies 1(2) (2008)
21. Savary, A.: Multiflex: A Multilingual Finite-State Tool for Multi-Word Units. In: Maneth, S. (ed.) CIAA 2009. LNCS, vol. 5642, pp. 237–240. Springer, Heidelberg (2009)
22. Savary, A., Krstev, C., Vitas, D.: Inflectional Non-compositionality and Variation of Compounds in French, Polish and Serbian, and Their Automatic Processing. Bulag — Bulletin de Linguistique Appliquée et Générale 32, 73–94 (2007)
23. Savary, A., Rabiega-Wisniewska, J., Wolinski, M.: Inflection of Polish Multi-Word Proper Names with Morfeusz and Multiflex. In: Marciniak, M., Mykowiecka, A. (eds.) Aspects of Natural Language Processing. LNCS, vol. 5070, pp. 111–141. Springer, Heidelberg (2009)
24. Silberztein, M.: Nooj: A Linguistic Annotation System for Corpus Processing. In: Proceedings of HLT/EMNLP on Interactive Demonstrations, HLT-Demo 2005, pp. 10–11 (2005)

25. Stanković, R., Krstev, C., Obradović, I., Trtovac, A., Utvić, M.: A tool for enhanced search of multilingual digital libraries of e-journals. In: 8th LREC 2012. ELRA, Istanbul (2012)
26. Vietri, S., Elia, A., D'Agostino, E.: Lexicongrammar, electronic dictionaries and local grammars in italian. Lingvisticae Investigationes (2004)
27. Vitas, D., Popović, L., Krstev, C., Stanojević, M., Obradović, I.: Languages in the European Information Society — Serbian. META-NET White Paper Series, Berlin (2011)
28. Wehrli, E., Seretan, V., Nerima, L.: Sentence Analysis and Collocation Identification. In: Proc. of the Multiword Expressions: From Theory to Applications — MWE 2010, Beijing, China, pp. 28–36 (2010)

PRALED – A New Kind of Lexicographic Workstation

Aleš Horák and Adam Rambousek

Abstract. This article describes the structure, features and usage of a specialized lexicographic workstation, named PRALED, developed by the Faculty of Informatics, Masaryk University for the purpose of development of new modern lexical database of the Czech language at the Institute of Czech Language, Czech Academy of Sciences.

The PRALED system is based on the Dictionary Editor and Browser (DEB) development platform that is designed for implementations of general dictionary writing systems. The design of the PRALED client and server parts is oriented to fluent editing of very complex dictionary entries. The resulting lexicographic database contains all the morpho-syntactic information of Czech lexical entries in a machine readable form, providing an invaluable resource for both human experts as well as computer applications.

The article describes the DEB platform, as well as some current DEB applications, focusing on the PRALED lexicographic station.

1 Introduction

Preparation of complex lexicographic collections, usually in the form of printed dictionaries, was for long times dependent on large collections of excerpts from literary works that had to be tediously catalogued and organized by lexicographic experts. In the case of the Czech language, [8] and [9] are examples of large Czech dictionaries which were prepared using this technique.

The growing need to handle various lexical resources that take the form of dictionaries, semantic networks, ontologies, valency lexicons, or FrameNets is the cause why researchers seek for software systems that are able to store dictionary-like data

Aleš Horák · Adam Rambousek
NLP Center
Faculty of Informatics, Masaryk University,
Brno, Czech Republic
e-mail: {hales,xrambous}@fi.muni.cz
http://deb.fi.muni.cz

A. Przepiórkowski et al. (Eds.): *Computational Linguistics*, SCI 458, pp. 131–141, 2013.
DOI: 10.1007/978-3-642-34399-5_7

in effective data structures. Many dictionary publishing houses operate large systems with the complex functionality of so called lexicographic stations that manipulate XML [18] and several companies offer dictionary writing programs of different complexity [16] or [2]. However, these and similar tools are not always able to efficiently manipulate resources obtained from data-driven NLP applications.

Reflecting the development of information technologies, the Institute of Czech Language (ICL) in Prague has been working on several digitizing projects. See e.g. `http://bara.ujc.cas.cz/psjc/` for digitized version of [8] with graphical presentations of original (often hand-written) excerpt cards. ICL had set up a special section for the computerization of data. Currently, Czech lexicographers have not only access to the electronic version of the digitized excerpts and previously published printed dictionaries, but also to numerous text and spoken corpora created at the Institute of the Czech National Corpus, the NLP Centre of the Faculty of Informatics, Masaryk University (FI MU) in Brno and the Institute of Formal and Applied Linguistics, Charles University in Prague.

Within the research intent *Creation of a Lexical Database of the Czech Language of the Beginning of the 21st Century* [19] (Academy of Sciences research intent AV0Z90610521), all the available resources (corpora, morphological analyzer, digitized dictionaries, ...) are used in preparing the supportive material for the future new lexicographic description of Czech, which will be finally published as an electronic modern monolingual dictionary of the Czech language named LEXIKON 21.

Before the start of the project, several dictionary writing systems were evaluated by ICL (mainly [16], and [2]). It was decided to develop custom application, so called lexicographic workstation that integrates several linguistic and NLP applications in a single environment, for example corpora query tool, dictionary browsing, and database editing. Custom application fits the needs and requirements of the lexicographic project much better than general dictionary writing system. Furthermore, the goal of the project is not the dictionary developement, but the developement of a complex lexical database. However, to save work and not to build the application completely from the ground, it was decided to use DEB (Dictionary Editor and Browser) platform as a base for the lexical database editor.

The lexicographic work has been divided into several phases – in the years 2005–2008 all the lexicographic resources of ICL have been digitized and, together with FI MU, a new system used as a lexicographic station, named PRALED (short for Prague Lexical Database), has been designed and implemented. In the years 2009–2012 PRALED was used for building a complex database of 100 000 lexical units of various types. PRALED application was also actively developed and updated according to requirement changes during the lexicographic research process. The preparation of LEXIKON 21 will be using the results of this second phase of the ICL's research intent.

2 The DEB Platform for Dictionary Writing Systems

The PRALED lexicographic station is built on the DEB development platform, which allows the system to use many components common to dictionary writing

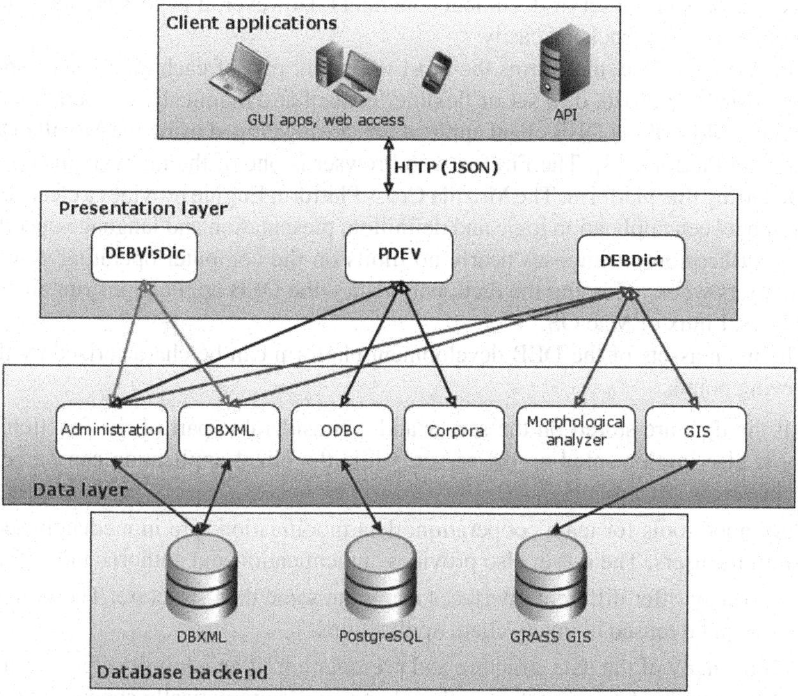

Fig. 1 The DEB platform schema

systems. DEB (Dictionary Editor and Browser, `http://deb.fi.muni.cz/`) is an open-source software platform for the development of applications for viewing, creating, editing and authoring of electronic and printed dictionaries. The platform is developed directly by the PRALED development team (with the authors as team leaders and main developers) at FI MU, thus many new features that have been implemented for PRALED lexicographers are now generally available for all other DEB applications. The DEB platform follows a client-server architecture – see the DEB platform schema in Figure 1. Most of the functionality is provided by the server side and client side offers graphical interfaces for users. The client applications communicate with the server using the standard web HTTP protocol.

The server part is built from small, reusable parts, called servlets, which allow a modular composition of all services. Each servlet provides different functionality such as database access, dictionary search, morphological analysis or a connection to corpora.

The overall design of the DEB platform focusses on modularity. The data stored in a DEB server can use any kind of structural database and combine the results in answers to user queries without the need to use specific query languages for each data source. The main data storage is currently provided by the Oracle Berkeley DB XML [1], which is an open-source native XML database providing XPath and

XQuery access into a set of document containers. However, it is possible to switch to another database backend easily.

The user interface, that forms the most important part of each dictionary application, usually consists of a set of flexible forms that dynamically cooperate with the server. Most of the DEB client applications are developed using the Mozilla Development Platform [3]. The Firefox web browser is one of the many applications created using this platform. The Mozilla Cross Platform Engine provides a clear separation between application logic and definition, presentation and language-specific texts. Furthermore, it imposes nearly no limits on the computer operating system of the users when accessing the dictionary data – the DEB applications run on MS Windows, Linux or Mac OS.

The main assets of the DEB development platform can be characterized by the following points:

- All the data are stored on the server and a considerable part of the functionality is also implemented on the server, while the client application can be very lightweight.
- Very good tools for team cooperation; data modifications are immediately seen by all the users. The server also provides authentication and authorization tools.
- Server may offer different interfaces using the same data structure. These interfaces can be reused by many client applications.
- Homogeneity of the data structure and presentation. If an administrator commits a change in the data presentation, this change will automatically appear in every instance of the client software.
- Integration with external applications.

3 Current DEB Applications

The DEB development platform provides a basis for many different kinds of lexicographic applications. The list of real dictionary systems that was developed on the DEB platform currently contains the following applications:

- DEBDict, a general multiple-dictionary browser
- DEBVisDic, wordnet editor and browser
- DEBTEDI, multilingual terminological dictionary of art terms
- Cornetto, editor and browser of Dutch lexical-semantic database
- Global Wordnet Grid, publicly accessible multilingual wordnet dictionary
- PRALED, complex application for building new Czech lexical database
- KYOTO, backend for wordnet and ontology storage in EU-FP7 project
- PDEV (CPA), Pattern Dictionary of English Verbs, tightly connected with corpora
- Family Names in UK, web editor for Comprehensive Dictionary of English Surnames

The first two applications are widely used with hundreds of users all over the world and with participation in various national and multilingual research projects. In the following paragraphs, we will provide more details about DEBDict and DEBVisDic as well as PDEV and the Dictionary of English Surnames, which are the most interesting (besides PRALED) from the lexicographic point of view. The whole next section will then be devoted to PRALED.

3.1 DEBDict

A DEB application with many of active users is a general dictionary and lexical database browser named DEBDict, which is available at `http://deb.fi.muni.cz/debdict/`. It is designed for all users who need to work with various versions of machine readable dictionaries to obtain the necessary lexical information and it allows to work with any number of electronic dictionaries.

At the DEB server at FI MU, DEBDict offers access to all relevant dictionaries of the Czech language. Thanks to the features of the DEB platform, DEBDict can check user's access rights and thus provide access to selected dictionaries intended for a specific group of people. For example, if the dictionary copyright does not allow public distribution, the access to the dictionary data may be limited to members of a research team.

3.2 DEBVisDic

The specific task of building a lexical semantic network in the form of the Princeton WordNet [4] requires special tools. During the Balkanet project [13], a wordnet browser and editor VisDic was developed by FI MU. VisDic was used for creating several national wordnets. Since 2005, it was replaced by DEBVisDic, a new system based on the DEB development platform.

The DEBVisDic client application is split to the core and the individual modules for each wordnet. This way, it is possible to define different data structure, workflow, or include data from external sources per each of the (national) wordnets. For example, verbs in the Czech wordnet are connected to the verb valency lexicon VerbaLex [11].

Besides the data for the user interface, the DEBVisDic server part provides also application programming interface (API) that is usable by external applications or web services (e.g. the OntoTagger tool from the KYOTO project).

DEBVisDic has been used as a basis for several multilingual projects: the Global Wordnet Grid [12] – aiming to gather freely available wordnets of many languages, Cornetto [14] – Dutch lexical semantic database, and KYOTO [22] – European project building a multilingual knowledge extraction system.

3.3 PDEV – Pattern Dictionary of English Verbs

The *Corpus Pattern Analysis* (CPA) [6] is a new technique for mapping meaning onto words in text. It is currently being used to build a *Pattern Dictionary of English Verbs* (PDEV), which will be a fundamental resource for use in computational linguistics, language teaching, and cognitive science.

The verbs in PDEV are analysed by the verb patterns with links to example sentences from corpora. Lexicographers, when creating new entries in PDEV, divide the concordances of the verb to several groups and create a *verb pattern* for each group, describing the subject, the objects, adverbials and other pattern elements.

The PDEV application is currently also used in projects building Czech, Italian and Spanish pattern dictionaries. All of them share the same base application with custom modifications for each language. The PDEV application is tightly connected with the Sketch Engine. The users can easily display and edit corpus examples while editing the pattern.

3.4 Family Names in the United Kingdom

A joint project with the University of the West of England called *Family Names in United Kingdom* started in May 2010 and is set to create the largest ever database of the UK's family surnames.

The editor of the surname database is a web-based application, which extremely simplifies the installation process for users. All the surnames are divided to groups of surnames connected by variant spelling or references – the lexicographers then work with these groups. The application supports several functions to make the editing easier, for example automatic opening of new surnames added to the group, or moving the explanation between surnames, references are checked and fixed automatically. Other functionalities are designed to help the editors – e.g. quick inserting of special symbols (Greek alphabet etc.), reference documents searching, or templates for frequently used texts.

4 PRALED

The Prague Lexical Database application (called PRALED) is developed in close cooperation with linguists and users from the Institute of the Czech Language. The design of PRALED is based on the DEB platform.

Since the beginning of the project, the application and the user interface is continuously updated according to the changes in the research data and the needs of the lexicographic team. Thanks to the design of the DEB platform and the Mozilla Development Platform, it is possible to prepare prototypes of new versions in short periods of time.

The design of the PRALED client and server parts is oriented to fluent editing of very *complex dictionary entries*. The resulting lexicographic database contains all the morpho-syntactic information of Czech lexical entries in a *machine readable form*, providing an invaluable resource for both human experts as well as computer applications.

The PRALED users can be divided into two groups: the ICL researchers are able to view and create entries, whereas others (usually reviewers) can only view the finished entries. During the editing phase, 25 linguists were using the application, each of them creating or extending over 200 entries per day. During the last year of the project (2011), over 10 reviewers were evaluating completed entries for the purpose of the project final report. As of July 2012, several researchers and reviewers are evaluating the data in the preparation of the continuation project.

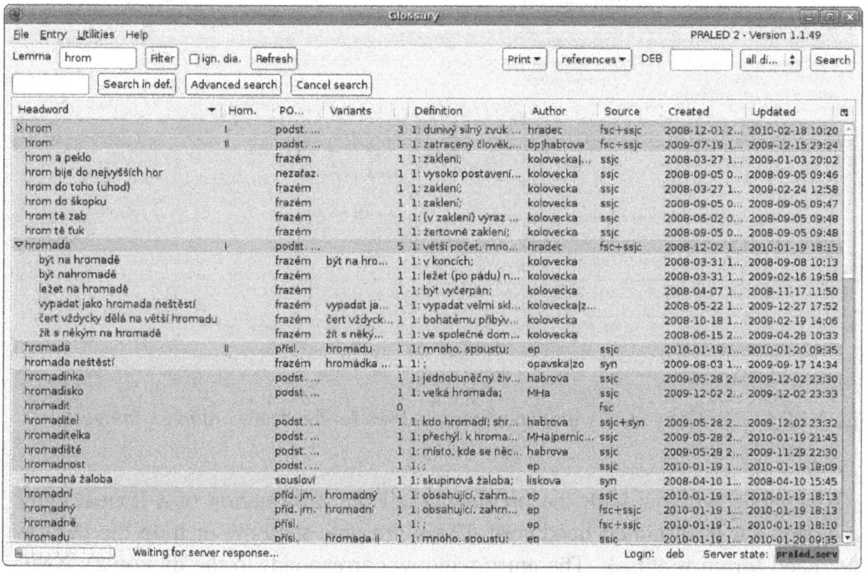

Fig. 2 PRALED: The List window

The client application consists of two parts – the *entry listing*, and the *complex editing form*. After successful login to the application, the entry listing window is displayed (see Figure 2). The dictionary is organized by headwords, or lexical entries – a single word, or multiple word expression, that form a dictionary entry with several meanings. With the basic filtering, a user can search for entries by headword, or by a piece of text from the definition. In the advanced search it is possible to freely combine any criteria from the entry data (for example, entries edited by a selected author in January 2011). The client application then translates the user selection to an XPath query that is executed on the server.

The list of selected entries is provided by the server in the RDF format. Thanks to this format, the user can view the results sorted by different fields (headword, author, last change date, etc.) and it is also possible to nest linked entries together (collocations are linked to the main headword). The resulting list can be printed in several output formats. To distinguish the entries in the list for user, the rows have different colours according to the entry type (single word, collocation, abbreviation).

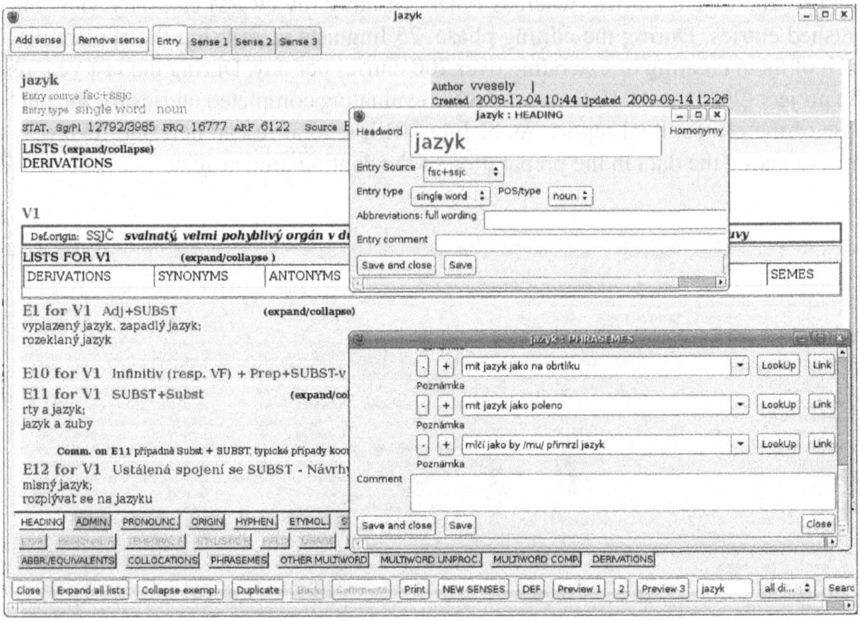

Fig. 3 PRALED: Entry view with two editor windows for the headword *jazyk* (*tongue*)

A separate window with the preview of all the information of a lexical entry is opened for each edited headword. Users are able to show or hide the data as they need for current task. The entry window corresponds to the dictionary XML structure and is divided to general entry information and specific information for each sense. Linguists in ICL usually concentrate on one feature of the word and several people together work on each entry. To make this task easier, users can select which information they want to edit (for example, morphology) and a separate tool is opened to edit just the part of entry structure. Figure 3 shows the entry editing window for the word *jazyk* (*tongue*) with the general information at the top – about the word itself (frequency, PoS, grammar etc.), the editors, and possible derivated words or collocations. Next section displays the data for each sense of the word (user can also select to edit only one sense) – definition, usage evidence from corpora, synonyms etc. The application handles team cooperation and ensures that several users do not overwrite changes of one another.

The PRALED client application currently allows to edit all the structural elements of the complex dictionary entry:

- part of speech and detailed information about the headword type
- orthoepy (spelling)
- pronunciation, hyphenation

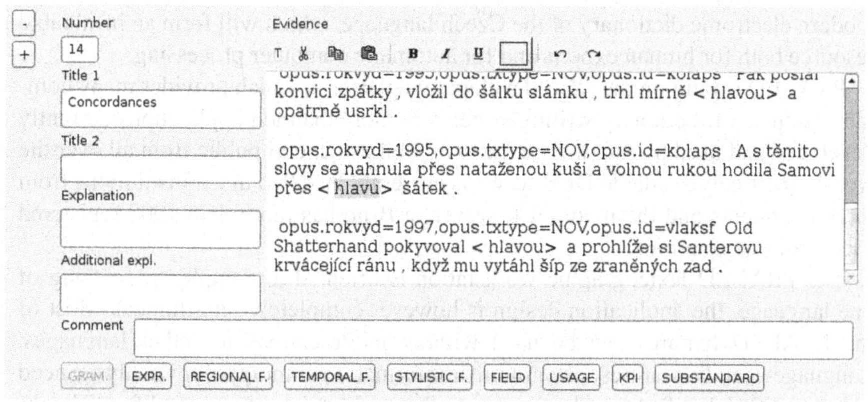

Fig. 4 PRALED: Corpora concordances for an entry

- morphological properties (for the given POS) with the possibility to get these information from the morphological analyzer
- etymology and word origin
- statistical information, automatically inserted from corpora
- linked entries: abbreviations, collocations, phrases, components, synonyms, antonyms, hyponyms
- meaning explanation
- domain, temporal and spatial properties
- examples and corpora concordance

It is possible to link entries together, for example to refer to dialectic variants, related collocations, phrases or hyponyms. Users can easily open linked entries while editing the entry, they can also select whether to open the entries in the preview or edit mode.

To add the word usage evidence from the corpora, PRALED is connected with the Czech National Corpus [15]. Linguists are able to select several examples from the corpora and insert them to the edited entry, see Figure 4. Editor may also easily check several Czech dictionaries and compare the information about the lexical entry from different sources.

Demonstration and more information about the project background are available at `http://deb.fi.muni.cz/praled/` and `http://lexiko.ujc.cas.cz/`.

5 Conclusions

We have described the design and implementation of a system for development of complex lexicographic database, denoted as PRALED lexicographic workstation. The system is actively used by tens of linguistic experts for preparation of a new modern electronic dictionary of the Czech language, which will form an invaluable resource both for human experts and for automatic computer processing.

PRALED is built on the DEB development platform, which provides many common features of dictionary writing systems in more than ten applications currently developed and used by linguistic experts as well as general public from all over the world. The freely available DEB server is currently installed in ten institutions from three continents and the main DEB server in Brno has more than 1000 registered users from 19 countries.

The PRALED lexicographic workstation is oriented to complex processing of one language, the application design is however completely multilingual. Most of the PRALED features can be used without modifications for other languages. Language-specific features, like grammar information or morphology analysis, need to be modified, but fortunately changes to the application are easy and fast. Thanks to the modular design of the DEB platform, generic functions from PRALED, such as e.g. the Sketch Engine interface, are also used as servlet modules in other applications that are currently being developed and used for lexicographic work with tens of languages ranging through nearly all continents.

Acknowledgements. This work has been partly supported by the Ministry of Education of CR within the Center of basic research LC536 and in the National Research Programme II project 2C06009 and by the Czech Science Foundation under the projects P401/10/0792 and 102/09/1842.

References

1. Chaudhri, A.B., Rashid, A., Zicari, R. (eds.): XML Data Management: Native XML and XML-Enabled Database Systems. Addison Wesley Professional (2003)
2. Erlandsen, J.: iLex - new DWS. In: Third International Workshop on Dictionary Writing Systems: Program and List of Accepted Abstracts. Faculty of Informatics. Masaryk University, Brno (2004), `http://www.emp.dk/ilexweb`
3. Feldt, K.: Programming Firefox: Building Rich Internet Applications with XUL. O'Reilly (2007)
4. Fellbaum, C. (ed.): WordNet: An Electronic Lexical Database. MIT Press (1998)
5. Fillmore, C., Baker, C., Sato, H.: Framenet as a 'Net'. In: Proceedings of Language Resources and Evaluation Conference (LREC 2004), vol. 4, pp. 1091–1094. ELRA, Lisbon (2004)
6. Hanks, P.: Corpus pattern analysis. In: Proceedings of the Eleventh EURALEX International Congress, Universite de Bretagne-Sud, Lorient (2004)
7. Hanks, P., Hodges, F.: A Dictionary of Surnames. Oxford University Press, Oxford Oxfordshire (1988)
8. Havránek, B., et al. (eds.): Příruční slovník jazyka českého (Reference Dictionary of Czech Language, PSJČ). Státní pedagogické nakladatelství/SPN, Praha (1935-1957); electronic version, created in the Institute of Czech Language, Czech Academy of Sciences Prague in cooperation with Faculty of Informatics, Masaryk University Brno
9. Havránek, B., et al.: Slovník spisovného jazyka českého (Dictionary of Written Czech, SSJČ), 1st edn. Academia, Praha (1960)

10. Herbst, T., Uhrig, P.: Erlangen Valency Patternbank (2009), http://www.patternbank.uni-erlangen.de/
11. Hlaváčková, D., Horák, A.: VerbaLex – New Comprehensive Lexicon of Verb Valencies for Czech. In: Proceedings of the Slovko Conference, Bratislava, Slovakia (2005)
12. Horák, A., Pala, K., Rambousek, A.: The Global WordNet Grid Software Design. In: Proceedings of the Fourth Global WordNet Conference, University of Szegéd, Szegéd (2008)
13. Horák, A., Smrž, P.: VisDic – wordnet browsing and editing tool. In: Proceedings of the Second International WordNet Conference – GWC 2004, Brno, Czech Republic, pp. 136–141 (2003), http://nlp.fi.muni.cz/projekty/visdic/
14. Horák, A., Vossen, P., Rambousek, A.: A Distributed Database System for Developing Ontological and Lexical Resources in Harmony. In: Gelbukh, A. (ed.) CICLing 2008. LNCS, vol. 4919, pp. 1–15. Springer, Heidelberg (2008)
15. ICNC: Czech National Corpus - SYN2000 (2000), http://www.korpus.cz
16. Joffe, D., de Schryver, G.M.: TshwaneLex – professional off-the-shelf lexicography software. In: Third International Workshop on Dictionary Writing Systems: Program and List of Accepted Abstracts. Faculty of Informatics. Masaryk University, Brno (2004), http://tshwanedje.com/tshwanelex/
17. Kilgarriff, A., Rychlý, P., Smrž, P., Tugwell, D.: The Sketch Engine. In: Proceedings of the Eleventh EURALEX International Congress, pp. 105–116. Universite de Bretagne-Sud, Lorient (2004)
18. McNamara, M.: Dictionaries for all: XML to final product. In: XML Conference 2003, Philadelphia, USA (2003)
19. Rangelova, A., Králík, J.: Wider Framework of the Research Plan Creation of a Lexical Database of the Czech Language of the Beginning of the 21st Century. In: Proceedings of the Computer Treatment of Slavic and East European Languages 2007, Bratislava, Slovakia, pp. 209–217 (2007)
20. Reaney, P., Wilson, R.: A Dictionary of English Surnames. Oxford University Press, Oxford Oxfordshire (1997)
21. Sedláček, R.: Morphemic Analyser for Czech. PhD thesis, Masaryk University, Brno, Czech Republic (2005)
22. Vossen, P.: KYOTO Project (ICT-211423), Knowledge Yielding Ontologies for Transition-based Organization (2008), http://www.kyoto-project.eu/

Multidimensional and Multimodal Information in EcoLexicon

Pilar León-Araúz, Arianne Reimerink, and Pamela Faber

Abstract. EcoLexicon is a multilingual terminological knowledge base (TKB) on the environment that targets different user groups who wish to expand their knowledge of the environment for the purpose of text comprehension and/or generation. Users can freely access EcoLexicon, and are able to find the information needed, thanks to a user-friendly visual interface with different modules for conceptual, linguistic, and graphical data. The main goal of this TKB is user knowledge acquisition. This paper briefly explains the theoretical premises and methodology applied in EcoLexicon for knowledge extraction and representation. It also shows how environmental concepts are represented, interrelated, and contextualized. EcoLexicon combines the advantages of a relational database, allowing for a quick deployment and feeding of the platform, and an ontology, enhancing user queries. The internal coherence at all levels of a dynamic knowledge representation shows that even complex domains can be represented in a user-friendly way.

1 Introduction

EcoLexicon[1] is a multilingual terminological knowledge base (TKB) on the environment. The knowledge base was initially implemented in Spanish, English and German. Currently, three more languages are being added: Modern Greek, Russian and Dutch. So far it has 3,341 concepts, 17,390 terms and 6,390 conceptual propositions. It targets different user groups, such as translators, technical writers, environmental experts, etc., who wish to expand their knowledge of the environment for the purpose of text comprehension or generation. These users can freely access EcoLexicon, and are able to find the information needed, thanks to a user-friendly

Pilar León-Araúz · Arianne Reimerink · Pamela Faber
Department of Translation and Interpreting, University of Granada,
Buensuceso, 11 18002 Granada, Spain
e-mail: {pleon,arianne,pfaber}@ugr.es

[1] http://ecolexicon.ugr.es

A. Przepiórkowski et al. (Eds.): *Computational Linguistics*, SCI 458, pp. 143–161, 2013.
DOI: 10.1007/978-3-642-34399-5_8

visual interface with different modules for conceptual, linguistic, and multimodal information. The main and ultimate goal of EcoLexicon is user knowledge acquisition, which can only be achieved if TKBs account for the natural dynamism of knowledge mainly caused by context and multidimensionality.

2 Theoretical Underpinnings of EcoLexicon

EcoLexicon is primarily based on theoretical and methodological premises derived from cognitive linguistics and corpus linguistics. Context and situated cognition are the semantic foundations of our knowledge representation framework, whereas corpus analysis guides our knowledge extraction procedures.

2.1 Knowledge Extraction

According to corpus-based studies, when a term is studied in its linguistic context, information about its meaning and its use can be extracted [31, 9, 36]. For EcoLexicon, two corpora were created: a textual corpus and a visual corpus. The visual corpus consists of images selected according to the following criteria (iconicity, abstraction, and dynamism) as ways of referring to and representing specific attributes of specialized concepts. Iconicity refers to the representation of real-world objects through the abstraction of conceptual attributes. Abstraction, a matter of degree, refers to the cognitive effort required for the recognition and representation of the concept. Dynamism implies the representation of movement; either through explicit movement (e.g. video and animation) or through the illustration of the steps that make up a process (e.g. images showing different stages of a process). Iconic images are especially relevant for the representation of generic-specific and meronymic relations. The resemblance of the illustration to the real-world entity allows the user to identify the object in question by inferring its basic characteristics and linking them to previously stored knowledge structures. Abstraction facilitates the understanding of other conceptual relations, such as *location_of*, since schematic settings allow for the understanding of places and positions. Dynamism is conferred by the use of different symbols, such as arrows (representing movement), and textual information (linking the pictures to the real world) [17]. Images were also classified in terms of the morphological features described by Marsh and White regarding the functional relationship between images and texts [28].

The English textual corpus (5 million words) consists of specialized texts (e.g., scientific journal articles, PhD theses, etc.), semi-specialized texts (textbooks, manuals, etc.), and texts for the general public, all belonging to the multidisciplinary domain of the environment. The extraction of conceptual knowledge from the textual corpus combines manual direct term searches and knowledge pattern analysis. According to many research studies, knowlede patterns (KPs) are considered to be one of the most reliable methods for knowledge extraction [5, 6, 11, 27, 29]. This involves several complementary steps. Normally, the most recurrent knowledge patterns for each conceptual relation identified in previous research are used to find related term pairs [2, 13]. Afterwards, these terms become seed words that are used for direct term searches to find new KPs and relations. The methodology consists of the cyclic repetition of both procedures. Although previous studies propose a semi-automatized annotation-based approach, first of all certain selection criteria must be defined by manually identifying what information is useful, why it is useful, and how it is structured.

For instance, *formed by* is a very reliable pattern in our domain. However, certain pattern-based constraints must be defined in order to avoid polysemy, since it may convey different relations. Concordances in Figure 1 show the way formed by works in the three different facets it can express, although *result* (*formed by* the deposition of sediments) prevails over *part* (*formed by* a nucleus) and *material* (*formed by* alluvial sediments).

The disambiguation of this polysemic KP requires different steps. If the KP is followed by a verb, it is definitely related to *result*, as in *formed by* shearing along. Instead, if the KP is followed by a noun, it can be linked to any of the three facets. Then the difference lies in two factors: if the noun is a process concept type, the concept still falls into *result* (*formed by* the rocking of water; *formed by* glacial scouring; *formed by* evaporation; *formed by* marine action); if the noun is an object concept type, it can be either *part* or *material*, but if the noun is uncountable, it will always refer to *material* (*formed by* breccia and conglomerates), whereas countable nouns will always link wholes with *parts* (*formed by* cold and hyper-saline cascades, *formed by* alveolar septal structures).

Conversely, when analyzing some of these concepts as seed words, conceptual concordances show how different KPs convey different relations with other specialized concepts (Figure 2). The main relations reflected in EROSION concordances are *caused_by*, *affects*, *has_location*, and *has_result*, which highlight the procedural nature of the concept and the important role played by non-hierarchical relations in knowledge representations.

```
RESULT

 is the disturbing force of a seiche. Seiche is a wave formed by the rocking of water in an enclosed water are
s valley is a topographic low about 15 km wide and was formed by shearing along the transform that separates t
e lower berm is the natural or\par  normal berm and is formed by the uprush of normal wave action during the o
urce: Geomophology From Space   Delta coasts are those formed by the deposition of sediment at the mouth of a
effects). Also horizontal heterogenity such as burrows formed by biological activity, will affect the erosion
ep-walled ~80-m deep gorge in rock avalanche   debris, formed by headward erosion of seepage-fed streams emerg
s lie adjacent to ledge outcrops. Boulder beaches were formed by glacial scouring   and deposition and bedrock
level fluctuations. The higher berm, or storm berm, is formed by wave action during storm\par  conditions. Dur
he rest of the year. Consequently,   coastal sediments formed by evaporation of seawater are common, particula
 a historic phenomenon; the world's great deserts were formed by natural processes interacting over long inter
 Auldyn alluvial fans (Fig. 1). The alluvial fans were formed by the deposition of vast quantities of sediment
Deep-water waves: Sea and swell   Wind waves initially formed by the action of wind blowing over the sea surfa
 a sharp jolt or shaking. P-waves or primary waves are formed by the alternate expansion and contraction of be
non-marine processes, and secondary coasts, which were formed by marine action. Primary coasts happen because

PART

1989) and form ice. Barents Sea bottom water (BSBW) is formed by cold, hyper-saline cascades, which slide down
ohesive to cohesive, and (ii) the network structure is formed by another sediment fraction. Appropriate parame
tic laminated coatings. Some of the laminae are solely formed by alveolar septal structures. The size of the g
1989) and form ice. Barents Sea bottom water (BSBW) is formed by cold, hyper-saline cascades, which slide down
s are rounded and about 0.75 mm in diameter. They were formed by a nucleus, made of a detrital grain or root d

MATERIAL

 et al., 1999]. Both units correspond to alluvial fans formed by breccia and conglomerates, which distally pas
r  the Roman community.      The   cuspated   delta   was   formed   by   alluvial   sedi- ments   carried   by   the   ri
s are usually dark and regular with different laminae, formed by dark micrite with few detrital grains and by
```

Fig. 1 Conceptual facets expressed by KP *formed by*

EROSION is related to various kinds of agents, such as STORM SURGE (1, 7), WAVE ACTION (2, 13), RAIN (3), WIND (4), JETTY (5), CONSTRUCTION PROJECTS (6), MANGROVE REMOVAL (8), SURFACE RUNOFF (9), FLOOD (10), HUMAN-INDUCED FACTORS (11), STORM (12) and MEANDERING CHANNELS (14). They can be retrieved thanks to all KPs expressing the relation *caused_by*, such as *resultant* (1), *agent for* (2, 3), *due to* (6, 7), *responsible for* (11) and *lead to* (13). This relation can also be conveyed through compound adjective phrases, such as *flood-induced* (10) or *storm-caused* (12) and any expression containing *cause* as a verb or noun: *one of the causes of* (9), *cause* (4, 5, 8) and *caused by* (14).

EROSION is also linked to the patients it *affects*, such as WATER (15), SEDIMENTS (16), COASTLINES (16), BEACHES (17), BUILDINGS (18), DELTAS (19) and CLIFFS (20). However, the affected entities, or patients, are often equivalent to locations (eg. if EROSION *affects* BEACHES it actually *takes place* at the BEACH). The difference lies in the kind of KPs linking the propositions. The *affects* relation is often reflected by the preposition *of* (10) or by verbs like *threatens* (18), *damaged by* (17) or *provides* (19). In contrast, the *has_location* relation is conveyed through directional prepositions (*around*, 21; *along*, 22; *downdrift*, 23) or spatial expressions, such as *takes place* (24). In this way, EROSION is linked to the following locations: LITTORAL BARRIERS (21), COASTS (22) and STRUCTURES (23). *Result* is an essential dimension in the description of any process since it is not only initiated by an agent affecting a patient in a particular location, but also has certain effects, which can be the creation of a new entity (SEDIMENTS, 25; PRIMARY COASTS, 26; BEACH MATERIAL, 27; SHORELINES, 28; MARSHES, 29; BAYS, 31) or the beginning of another process (SEAWATER INTRUSION, 31; PROFILE STEEPENING, 32).

```
Caused_by

1    , Alabama. Significant storm surge and resultant beach erosion were  associated with Ivan's landfall. However,
2    nd climate  On the Castellón coast, the main agent for erosion is wave action, and this is therefore  responsi
3    f a stream. The first   factor, rain, is the agent for erosion, but the degree of erosion is governed by   oth
4    rts (BW) and semiarid steppe (BS). Wind can also cause erosion and deposition in environments where sediments
5    etty. Reflection of waves from a jetty may  also cause erosion of adjacent shorelines. However, erosion furthe
6    oastal zone management. However, in some cases coastal erosion\par  can be due to construction projects that a
7    tude of about 0,3 M m3 per year.  Acute erosion  Acute erosion due to storm surges  (waves and water levels at
8    er. Mangrove removal is also reported to cause coastal erosion and change sedimentation patterns and shoreline
9    [edit] Erosion  Surface runoff is one of the causes of erosion of the earth's surface. Reduced   crop producti
10   pes. Local disturbances, for instance by flood-induced erosion, redistribution of sediment or accumulation of
11   ors and human-induced factors  responsible for coastal erosion and highlight the time and space patterns withi
12   ocess is typical of a cyclical process of storm-caused erosion in winter, followed by progradation\par  owing
13   can cause excessive wave action that can lead to beach erosion. Trash dumped from boats can be washed up onto
14    that have reached base level develop broad valleys by erosion caused by meandering channels. The stream chann

Affects

15   ing these sensitive creatures.  In some cases, coastal erosion can have adverse effects on water quality and h
16   ine Depositional Coasts  The erosion of coastlines and erosion of sediments being carried to the   shoreline b
17    use of dredged material to restore beaches damaged by erosion. EPA works with the U.S. Coast Guard to regulat
18   reasonable points, though when push comes to shove and erosion threatens buildings, traditional beach maintena
19   ks and arches found on irregular rocky coastlines; and erosion provides the material which forms deltas and ba
20    near the base of the cliff. Constant undercutting and erosion causes the cliffs to retreat landward.

Has_location

21   ed by the position of sand accumulation\par  and beach erosion around littoral barriers. A coastal structure i
22   hes. Kuenen (1950) estimates\par  that beach and cliff erosion along all coasts of the world totals about 0.12g
23   ce and divergence of wave energy over an offshore bar, erosion downdrift of a structure such as a groin, sudde
24   proportional to the longshore transport rate, and\par  erosion takes place downdrift at about the same rate. T

Has_result

25    Excessive loads of silt and other sediments caused by erosion can suffocate bottom-dwelling plants and animal
26   islands or coral reefs.  Primary coasts are created by erosion (the wearing away of soil or rock), deposition
27   \par  transported. Beach material is also derived from erosion of the coastal formations caused by\par  waves
28   ed to the   passage of the ice. Shorelines produced by erosion of glacial till deposits differ   markedly from
29   beaches   and marshes, are being formed as a result of erosion   and transportation of unconsolidated material
30   ion of the seashore and a rise in SLR. The results  of erosion could lead to further seawater intrusion that c
31   fs are developed  in landslide debris. In this cliffs, erosion of softer material has created bays. The  expec
32   s of steep systems, a sea-level rise may cause coastal erosion resulting in profile steepening, and therefore
```

Fig. 2 Non-hierarchical relations associated with EROSION

As can be seen, all these related concepts are quite heterogeneous. They belong to different paradigms in terms of category membership and/or hierarchical range. For instance, some of the agents of EROSION are natural (WIND, WAVE ACTION) or artificial (JETTY, MANGROVE REMOVAL) and others are general concepts (STORM) or very specific ones (MEANDERING CHANNEL). This explains why knowledge extraction must still be performed manually. Nevertheless, it also illustrates one of the major problems in knowledge representation: multidimensionality [37]. This is better exemplified in the following concordances (Figure 3), since multidimensionality is most often codified in the *is_a* relation.

In the scientific discourse community, concepts are not always described in the same way because they depend on perspective and subject-fields. For instance, EROSION is described as a natural process of REMOVAL (33), a GEOMORPHOLOGICAL PROCESS (34), a COASTAL PROCESS (35) or a STORMWATER IMPACT (36). The first two cases can be regarded as traditional ontological hyperonyms. The choice of one or the other depends on the upper-level structure of the representational system and its level of abstraction. However, COASTAL PROCESS and STORMWATER IMPACT frame the concept in more concrete subject-fields and referential settings.

The same applies to subtypes, where the multidimensional nature of EROSION is clearly shown. EROSION can thus be classified according to the dimensions of *result* (SHEET, RILL, GULLY, 37; DIFFERENTIAL EROSION, 38), *direction* (LATERAL, 39; HEADWARD EROSION, 49), *agent* (WAVE, 41; FLUVIAL, 42; WIND, 43, 46; WATER, 44; GLACIAL EROSION; 45) and *patient* (SEDIMENT, 47; DUNE, 48; SHORELINE EROSION, 49). In section 3, the consequences of multidimensionality for knowledge representation are shown.

2.2 Knowledge Representation

According to Meyer et al. [30], TKBs should reflect conceptual structures similarly to how concepts are related in the mind. The organization of semantic information in the brain should thus underlie any theoretical assumption concerning the retrieval and acquisition of specialized knowledge concepts as well as the design of specialized knowledge resources [16]. In this sense, since categorization itself is a

```
Is_a

33    vided by the  area (A) of the drainage basin      (1) Erosion is the natural process of removal of soil by wa
34    in the Netherlands, geomorphological processes such as erosion, transport and sedimentation of sandy materials
35    BURY AND DUXBURY, 1996).     Coastal processes such as erosion and accretion are site-specific, season-specifi
36    these  catchments include:  Stormwater impacts such as erosion, channelisation, sediment deposition and sedime

Type_of

37    eroded by shallow overland flow (sheet, rill and gully erosion) and delivered to the drainage network. Channel
38    m the great  local relief, the result of differential erosion by glacier ice. Figure 9-20   includes two sche
39    ing flood events, the dikes are subject to the lateral erosion of the river   trying to reoccupy its former co4
40    d enlarges these small channels and generates headward erosion directed towards the aggrading active channel (
41    out five percent of the material on most beaches. Wave erosion of rocky coasts is usually slow, even where the
42     of the Earth's land surface is   dominated by fluvial erosion, lakes that do occur are threatened with either
43    ind  climate, topography and surface roughness.  Wind erosion risk applies only when soils are dry and not co
44    oportional to the steepness of the land surface. Water erosion is in proportion to the shear stress exerted by
45    ley to become both wider and deeper over time. Glacial erosion also results in a change in the valley's cross-
46    dominate in periglacial environments: nivation; eolian erosion and deposition; and fluvial erosion and deposit
47     erosion processes. 215 CHAPTER 13 EQUATIONS: SEDIMENT Erosion caused by rainfall and runoff is computed with
48    gineers to simulate  cross-shore beach, berm, and dune erosion produced by storm waves and water levels. The l
49    uctures constructed to date have resulted in shoreline erosion   in their lee. Furthermore, the key environmen
```

Fig. 3 Hierarchical relations associated with EROSION

dynamic context-dependent process, the representation and acquisition of specialized knowledge should certainly focus on contextual variation. From a neurological perspective, Barsalou [8] states that a concept produces a wide variety of situated conceptualizations in specific contexts, which clearly determines the type and number of concepts to be related to. Context has been explored in some depth by disciplines such as psychology, linguistics, and artificial intelligence. Even though all of these approaches have provided valuable insights, there seems to be no consensus on the definition of context since it is invariably conceived for different purposes, depending on the field.

In linguistics, context is especially mentioned in relation to pragmatic and cognitive notions, such as *speech acts* [3, 39], *conventions* [19], *maxims* [22], *Relevance Theory* [40], *framing* [21], and *common ground* [12].

From a computational perspective, contexts may be useful to put together a set of related axioms. In this way, contexts are a means for referring to a group of related assertions about which something can be said [23]. Since context, knowledge, and reasoning are closely intertwined [10], artificial intelligence formalizes context to perform automatic inferences and reasoning [24, 23]; to identify relational constraints for context-aware applications [14]; to improve automatic information retrieval; to resolve ambiguities in natural language processing, *inter alia*. Nevertheless, whatever the approach, context is defined as a dynamic construct. It is thus surprising that term bases are often restricted to generic-specific and part-whole relations, when conceptual dynamism can only be fully reflected through non-hierarchical relations. These are mostly related to the notions of movement, action, and change, which are directly linked to human experience, perceptually salient conceptual features and processes. In our approach, we consider that a given term does not have a meaning, but rather a meaning potential that will always be exploited in different ways that are dependent upon the discourse context [15]. In this sense, we believe that the formalization of context should account for the relational constraints shown by specialized concepts according to their situational nature.

Accordingly, dynamism in the environmental domain comes from the effects of context and multidimensionality. In EcoLexicon, this is reflected through: (1) the elaboration of category membership templates; (2) the inclusion of multimodal information associated with each entry; (3) the representation of multidimensionality and the situated nature of concepts through an inventory of both hierarchical and non-hierarchical relations (*is_a, part_of, delimited_by, causes, located_in, effected_by, made_of, has_function, result_of, takes_place_in, affects, phase_of, attribute_of*).

At a macrostructural level, all knowledge extracted from the corpus has been organized in a frame-like structure or prototypical domain event, namely, the Environmental Event (EE) (Figure 4; [17, 25, 34]). This prototypical domain event or action-environment interface [7] provides a template applicable to all levels of information structuring from a process-oriented perspective [18].

The EE is conceptualised as a dynamic process that is initiated by an agent (either natural or human), affects a specific kind of patient (an environmental entity) and produces a result in a geographical area. These macro-categories (AGENT,

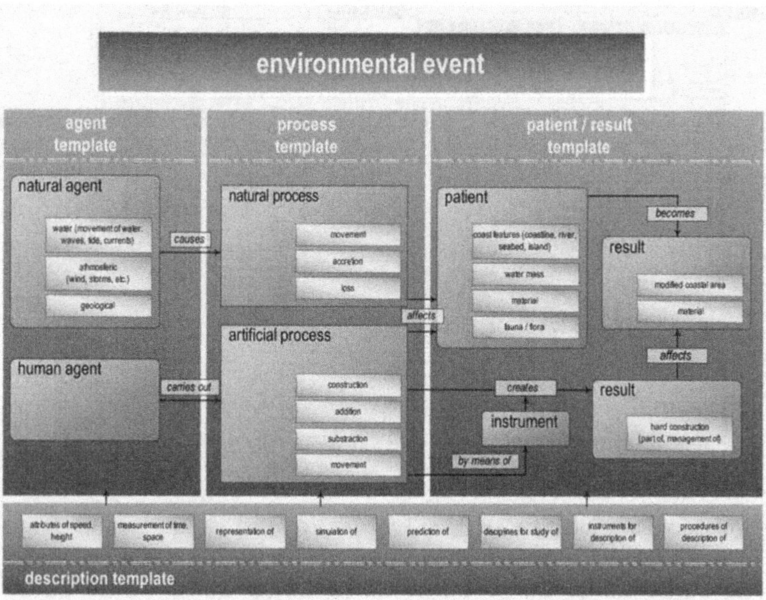

Fig. 4 The Environmental Event

PROCESS, PATIENT/RESULT) are the semantic roles inherent to this specialized domain, and the EE provides a model to represent their interrelationships at different levels. This means that all environmental concepts have been assigned to different semantic categories according to the prototypical semantic roles of this specialized domain.

3 EcoLexicon

Each entry in EcoLexicon provides a wide range of interrelated information. Figure 5 shows the entry for EROSION. Users are not obliged to view all this information at the same time, but can browse through the interface depending on their needs.

Under the heading Domains, an ontological structure shows the exact position of the concept in the class hierarchy. EROSION, for example, *is_a* natural process of loss (bottom-left corner of the window). All concepts are defined both in English and Spanish. Definitions are shown when the cursor is placed on the concept. Contexts and concordances appear when clicking on the terms, and inform different users about both conceptual and linguistic aspects, such as the KPs reflecting conceptual relations or the types of terminological variants that may designate the same concept. Conceptual relations are displayed in a dynamic network of related concepts (right-hand side of the window). Users are free to click on any of these concepts and thus further expand their knowledge of this sector of the specialized

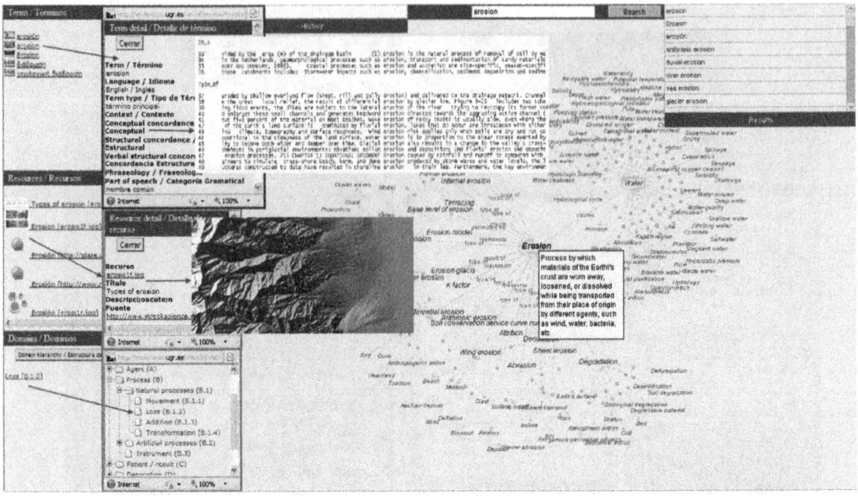

Fig. 5 EcoLexicon user interface

domain. The terminological units, under the heading Terms provide linguistic information, and show the designations of the concept in English, Spanish, German, and Modern Greek. Finally, knowledge representation would be partial if TKBs were to use only linguistic information as a means to communicate knowledge. In EcoLexicon, visual information has thus been included since it facilitates a dynamic access to specialized knowledge. Graphical and web resources are displayed when clicking on the links in the box Resources (in the left-hand margin towards the middle).

In the next sections we will focus step by step on the main resources of EcoLexicon and explain how they are interrelated (relational database and ontology, semantic networks and definitions and images).

3.1 Relational Database and Ontology

Data in EcoLexicon are primarily hosted in a relational database (RDB). Nevertheless, relational modelling has some limitations, such as its limited capability to represent real-world entities since natural human implicit knowledge cannot be inferred. Relational models are suited to organize data structure and integrity, whereas ontologies try to specify the meaning of their underlying conceptualization [4]. In EcoLexicon, semantic information is stored in the ontology, while the rest is stored in the relational database. The classes in our incipient ontology correspond to the basic semantic roles identified for the environmental domain (AGENT-PROCESS-PATIENT-RESULT-LOCATION) and concepts are their instances. Conceptual relations in EcoLexicon are enhanced by an additional degree of OWL semantic expressiveness provided by property characteristics. In fact, one of the main advantages of ontologies is that they make reasoning and inferences possible. For example, *part_of* relations can benefit from transitivity, as shown in Figure 6.

In Figure 6, a SPARQL query is made in order to retrieve which concepts are *part_of* Concept 3262, which refers to the concept SEWER. On the right side, DRAINAGE SYSTEM is retrieved as a direct *part_of* relation, whereas SEWAGE COLLECTION AND DISPOSAL SYSTEM and SEWAGE DISPOSAL SYSTEM are implicitly inferred through the Jena reasoner.

However, meronymy cannot always be a transitive relation. This is why six different meronymic relations have been defined. For example, if *located_at* were considered as a *part_of* relation, that would cause fallacious transitivity [32]. If a GABION is *part_of* a GROYNE and a GROYNE *part_of* the SEA, the ontology would infer that GABIONS are *part_of* the SEA, which is false. However, it is true that if a HARD DEFENCE STRUCTURE is *located_at* the BEACH and the BEACH is *part_of* the COAST, then the DEFENCE STRUCTURE is *located_at* the COAST [26]. In this sense, we are planning to include "property chain inclusions" in EcoLexicon, as defined in W3C recommendations.

3.2 Semantic Networks: Context and Dynamism

According to corpus-based information, concepts in EcoLexicon appear related to others in the form of multidimensional semantic networks. Multidimensionality is commonly regarded as a way of enriching traditional static representations by enhancing knowledge acquisition through different points of view in the same semantic network [26]. However, multidimensionality in the environmental domain has caused a great deal of information overload, which ends up jeopardizing knowledge acquisition.

This is mainly caused by versatile concepts, such as WATER (Figure 7), which are usually top-level general concepts involved in a myriad of events. For instance, in its conceptual network, WATER is linked to the same extent to diverse natural and artificial processes, such as EROSION or DESALINATION. Corpus data has provided 72 conceptual relations for the first hierarchical level of WATER.

However, WATER rarely, if ever, activates those relations at the same time, as they evoke completely different situations. Our claim is that any specialized

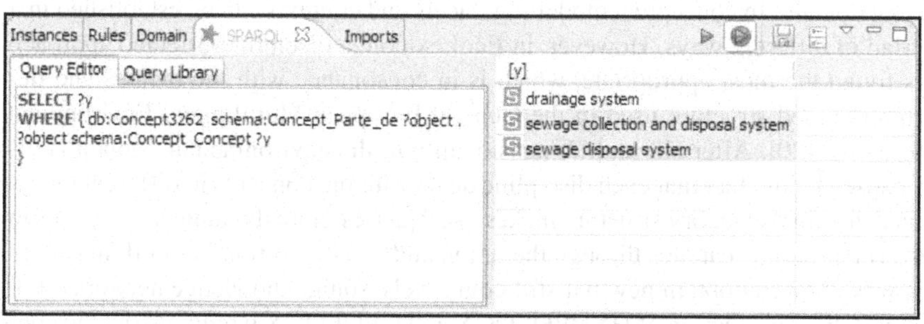

Fig. 6 Concept SEWER in the ontology and inferred transitivity

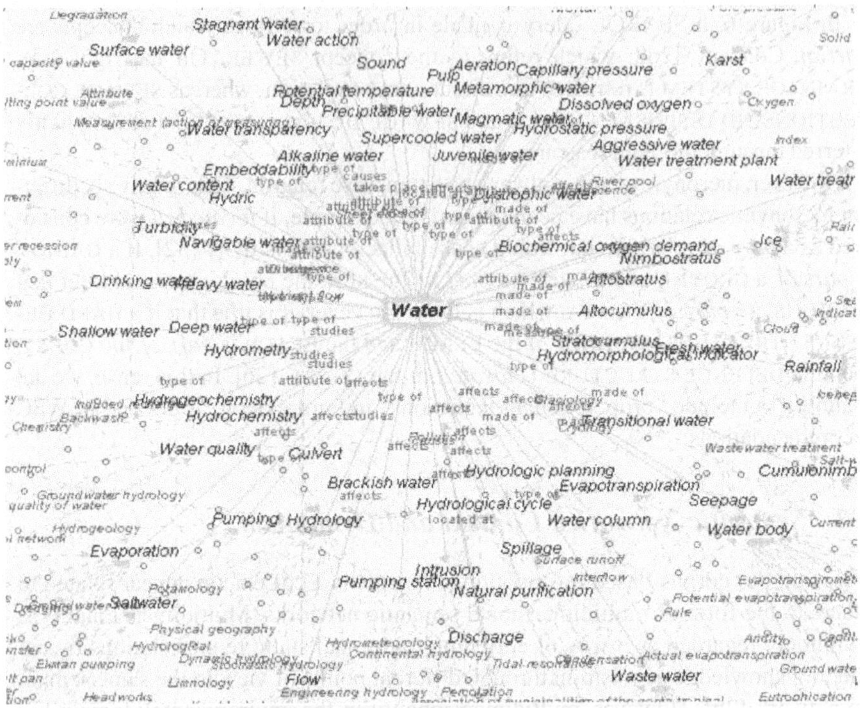

Fig. 7 Information overload in the network of WATER

domain contains sub-domains in which conceptual dimensions become more or less salient, depending on the activation of specific contexts. As a result, a more believable representational system should account for reconceptualization according to the situated nature of concepts. In EcoLexicon, this is done by dividing the global specialized environmental field in different discipline-oriented contextual domains (Figure 8). Reconceptualization is thus based on prototypes and context. Prototype theory [38] has been mainly applied to category member salience. Nevertheless, in our approach we focus on the salience of conceptual propositions within each contextual domain. In conceptual modelling, facets and contexts can be established in a myriad of different ways. However, in EcoLexicon, a discipline-oriented approach was found the most appropriate, which is in consonance with the themes and descriptors based structure used in the General European Multilingual Environment Thesaurus [20]. After all, multidimensionality in the environmental domain is often caused by the fact that each discipline deals with the concepts in different terms. This contextual category structure makes user queries more dynamic since they can perform different searches through the union and/or intersection of our domains. In this way, they can obtain new but still cognitively-sound knowledge networks. For instance, the intersection of HYDROLOGY and GEOLOGY would restrict the conceptual structure to only HYDROGEOLOGICAL propositions, whereas the union

EcoLexicon terminological knowledge base

Contextual Domains

1. ENVIRONMENTAL PROTECTION

1.1. Environmental Law
1.2. Environmental Education
  1.2.1. Sustainable Tourism
1.3. Natural resources management

2. SCIENCE

2.1. Geography
2.2. Biology
  2.2.1. Biological Oceanography
  2.2.2. Botany
  2.2.3. Zoology
  2.2.4. Microbiology
  2.2.5. Molecular Biology
  2.2.6. Biochemistry
2.3. Physics
  2.3.1. Geophysics
  2.3.2. Physical Oceanography
2.4. Geology
  2.4.1. Hydrogeology
  2.4.2. Geophysics
  2.4.3. Geochemistry
  2.4.4. Geological Oceanography
  2.4.5. Geomorphology
2.5. Hydrology
  2.5.1. Hydrogeology
  2.5.2. Hydrometeorology
2.6. Chemistry
  2.6.1. Geochemistry
  2.6.2. Biochemistry
  2.6.3. Chemical Oceanography
2.7. Atmospheric Sciences
  2.7.1. Meteorology
    2.7.1.1. Meteorological Oceanography
    2.7.1.2. Hydrometeorology
  2.7.2. Climatology
2.8. Ecology
  2.8.1. Human Ecology
2.9. Soil Sciences
2.10 Oceanography
  2.10.1. Biological Oceanography
  2.10.2. Physical Oceanography
  2.10.3. Meteorological Oceanography
  2.10.4. Geological Oceanography
  2.10.5. Chemical Oceanography

3. ENGINEERING

3.1. Marine Engineering
3.2. Civil Engineering
  3.2.1. Transport and Infrastructure Engineering
  3.2.2. Hydraulic Engineering
  3.2.3. Coastal Engineering
  3.2.4. Mining Engineering
  3.2.5. Environmental Engineering
    3.2.5.1. Waste Management
    3.2.5.2. Water Treatment and Supply
    3.2.5.3. Air Quality Management
    3.2.5.4. Soil Quality Management
3.3. Agronomy Engineering
3.4. Chemical Engineering
3.5. Energy Engineering
  3.5.1. Renewable Energy

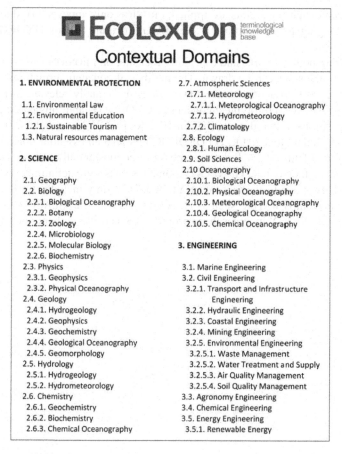

Fig. 8 Contextual domains

of ENVIRONMENTAL, MINING, TRANSPORT, HYDRAULIC and COASTAL ENGINEERING would make up the whole domain of CIVIL ENGINEERING.

Contextual constraints are neither applied to individual concepts nor to individual relations. Instead, they are applied to each conceptual proposition. For instance, CONCRETE is linked to WATER through a *made_of* relation, but this proposition is not relevant if users only want to know how WATER naturally interacts with landscape. Consequently, that proposition will only appear in an ENGINEERING context [35]. Nevertheless, not only versatile concepts, such as WATER, are constrained, since information overload can also affect any other concept that is linked to versatile ones. For instance, EROSION takes the following shape in a context-free network (Figure 9), which appears overloaded mainly because it is closely linked to WATER as one of its most important agents.

When contextual constraints are applied, EROSION only appears linked to propositions belonging to the context of GEOLOGY (Figure 10) or HYDROLOGY (Figure 11).

Comparing both networks and especially focusing on EROSION and WATER, the following conclusions can be drawn. The number of conceptual relations changes from one network to another since EROSION is not equally relevant in both domains. EROSION is a prototypical concept of GEOLOGY, and thus participates in more propositions in that domain than in HYDROLOGY. Nevertheless, since it is also strongly linked to WATER, HYDROLOGY is also an essential domain in the representation of EROSION. Relation types do not substantially change from one network to the other, but the GEOLOGY domain shows a greater number of *type_of* relations. This is due to the fact that HYDROLOGY only includes types of EROSION whose agent is WATER, such as FLUVIAL EROSION and GLACIER EROSION. In contrast, GEOLOGY includes those propositions as well as others, such as WIND

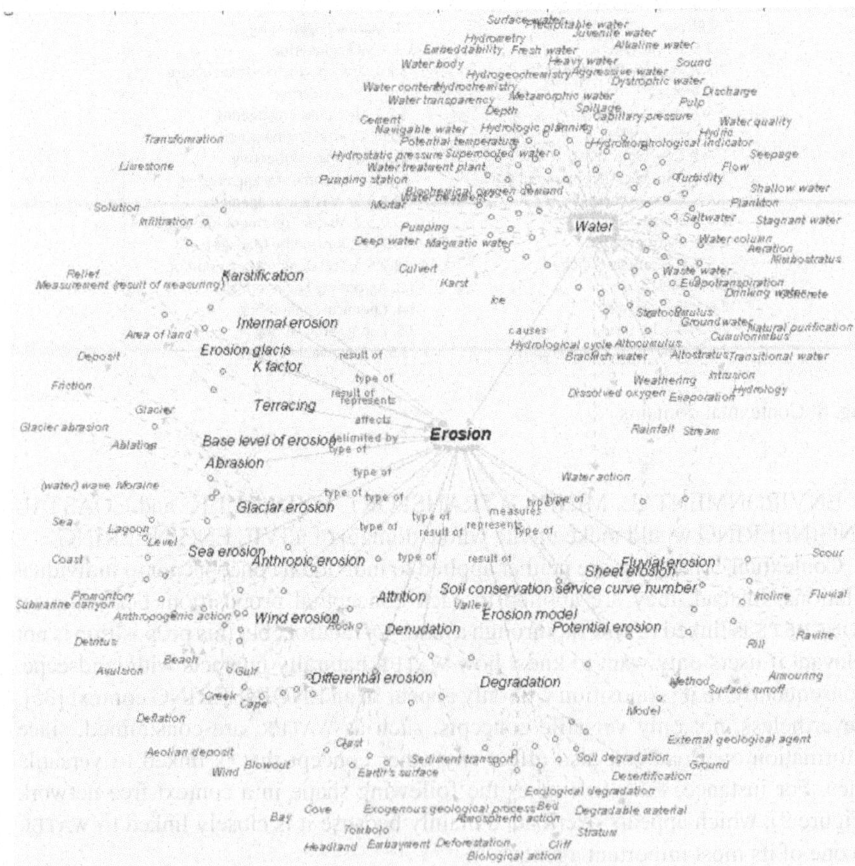

Fig. 9 EROSION context-free network

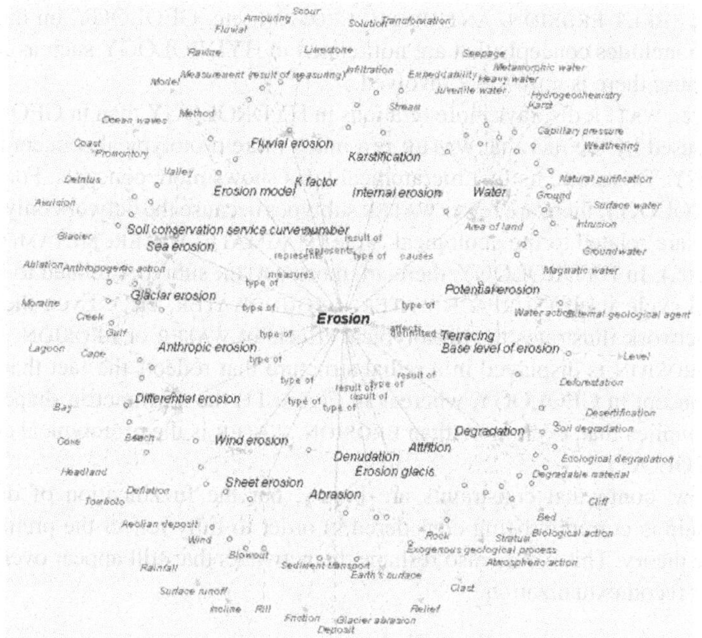

Fig. 10 EROSION in the GEOLOGY contextual domain

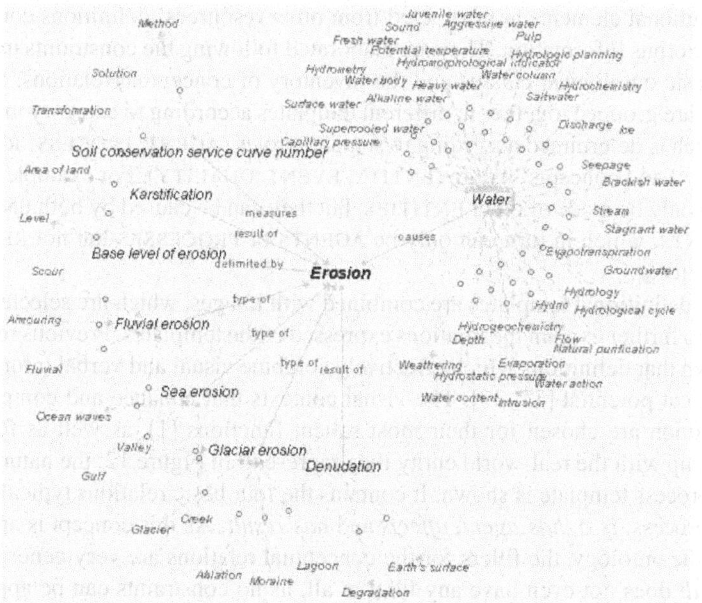

Fig. 11 EROSION in the HYDROLOGY contextual domain

EROSION, SHEET EROSION, ANTHROPIC EROSION, etc. GEOLOGY, on the other hand, also includes concepts that are not related to HYDROLOGY such as ATTRITION because there is no WATER involved.

However, WATER displays more relations in HYDROLOGY than in GEOLOGY. This is caused by the fact that WATER is a much more prototypical concept in HYDROLOGY. Therefore, its first hierarchical level shows more concepts. For example, in GEOLOGY, there are fewer WATER subtypes because the network only shows those that are related to the geological cycle (MAGMATIC WATER, METAMORPHIC WATER, etc.). In HYDROLOGY, there are more WATER subtypes related to the hydrological cycle itself (SURFACE WATER, GROUNDWATER, etc.). Even the shape of each network illustrates the prototypical effects of WATER or EROSION. In Figure 10, EROSION is displayed in a radial structure that reflects the fact that it is a central concept in GEOLOGY, whereas in Figure 11, the asymmetric shape of the network implies that, even more than EROSION, WATER is the prototypical concept of HYDROLOGY.

For now, contextual constraints are binary, but the fuzzification of domain-membership is currently being considered in order to fully reflect the premises of prototype theory. This would also reshape the networks that still appear overloaded even after recontextualization.

3.3 Definitional Templates: Text and Images

Definition construction follows a top-down and a bottom-up approach. This means that definitional elements are extracted from other resources' definitions combined with our corpus information. They are elaborated following the constraints imposed by the basic ontological classes and the inventory of conceptual relations. Similar concepts are grouped together in different templates according to category membership, which is determined according to semantic roles (AGENT, PROCESS, RESULT, LOCATION) and concepts' nature (ENTITY, EVENT, QUALITY). For example, ENTITIES can only be made of other ENTITIES, but they can be caused by both ENTITIES and EVENTS, which in turn can only be AGENTS or PROCESSES but not RESULTS or LOCATIONS.

These definitional templates are combined with images, which are selected from the web to further explain the relations expressed in the templates. Previous research has shown that definitions which effectively combine visual and verbal information have a great potential [17, 33]. The visual contexts that enhance and complement the definition are chosen for their most salient functions [1], as well as for their relationship with the real-world entity they represent. In Figure 12, the natural geological process template is shown. It contains the four basic relations typical of any natural process: *is_a*, *has_agent*, *affects* and *has_result*. As this concept is at a high level in the ontology, the fillers for the conceptual relations are very general ones. *Has_result* does not even have any filler at all, as no constraints can be applied at this general level. Only the *agent* dimension is constrained to a geological entity. As this level of the hierarchy is very general and therefore rather abstract, the image

NATURAL GEOLOGICAL PROCESS		
[IS_A]	Natural process	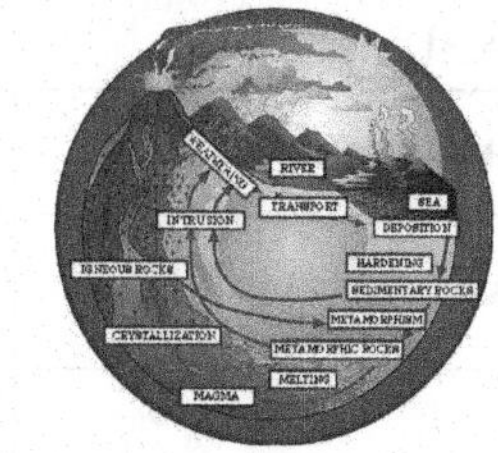
[HAS_AGENT]	Natural geological agent	
[AFFECTS]	(Entity, Process)	
[HAS_RESULT]	(Entity, Process)	

Fig. 12 NATURAL GEOLOGICAL PROCESS template

EROSION		
[IS_A]	Natural geological process	
[HAS_AGENT]	Gravity Water (river, stream, rain) Ice (glacier) Wind Animals	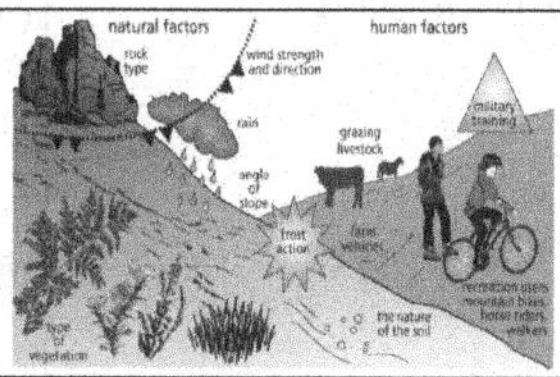
[AFFECTS]	Earth's surface (beaches, mountains, soil...)	
[HAS_RESULT]	Landslide Rill Gully Sheet ...	
[HAS_PHASES]	Weathering Transport Deposition	

Fig. 13 EROSION template

chosen is general and abstract as well. The geological cycle describes how all natural geological processes, such as EROSION, the WATER CYCLE, ROCK FORMATION, etc. interact and are all interdependent.

EROSION is the next level in the hierarchy and constrains the natural geological agent of the process to GRAVITY, WIND, WATER, ICE, and ANIMALS, with all their subtypes. The filler of the *affects* relation is the Earth's surface and all its subparts. The *result* dimension has even more fillers. Therefore, only a few of them have been included in Figure 13. A new dimension is added at this level: *has_phases*. The images chosen at this level of abstraction combine all the agents of EROSION and another that shows all types of possible landscapes resulting from EROSION (Figure 13).

WATER EROSION		
[IS_A]	Erosion	
[HAS_AGENT]	Water (river, stream, rain, wave, current…)	
[AFFECTS]	Earth's surface (beaches, mountains, soil...)	
[HAS_RESULT]	Rill	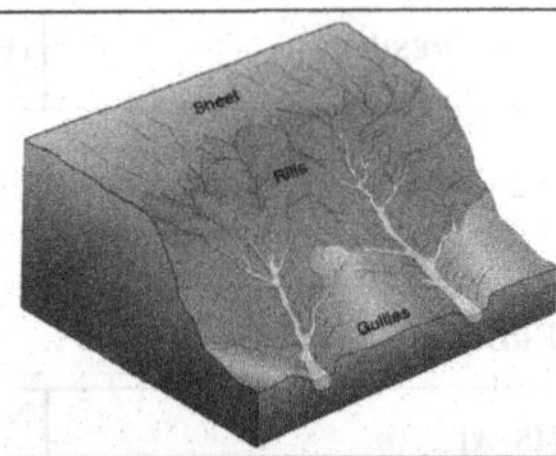
	Gully	
	Sheet	
	Cliff	
	Beach	
[HAS_PHASES]	Weathering	
	Transport	
	Deposition	

Fig. 14 WATER EROSION template

These images cannot give detailed information given that EROSION contains many subtypes, depending on the agent involved and the result obtained. For example, if we keep going down in the hierarchy, the level of specificity increases and the patients and results are closer to their real world referents. This is in consonance with Rosch's basic level and prototype theory [38]. According to prototype theory, basic level concepts belong to the first level of abstraction for which we can develop a concrete mental image. Although EROSION seems to be at the basic level (according to general language dictionaries, for example), when dealing with specialized knowledge, the basic level moves downwards in the hierarchy. This is why WATER EROSION can be better illustrated (Figure 14).

The template of WATER EROSION constrains the agent dimension further to WATER. However, even though *agents* are constrained to only one concept, it is hard to find a prototypical image of the concept WATER within the WATER EROSION event, since WATER can take the shape of rain, a wave, a river, etc. This does not mean that it is impossible to find prototypical images of general concepts such as water. For example, water also participates in other events such as the coriolis effect. Since this is a very specific event, the concept water is more constrained and therefore is easily depicted by a water spiral.

The *patient* and *phases* dimensions are the same as the superordinate EROSION, but the *result* dimension shows clear examples at this level of description. Here three images have been added showing the three phases of WATER EROSION. Moreover, two images show the *result* dimension. Going even further down in the hierarchy, other dimensions become more specific. For example, COASTAL EROSION and SURFACE EROSION are types of WATER EROSION that constrain the *location* dimension. The linguistic description of the concepts in EcoLexicon follows these templates insofar as type, quantity, and configuration of information are concerned. In this way, definitions show a uniform structure that complement the information encoded in conceptual networks, and directly refer to and evoke the underlying conceptual structure of the domain. These templates can be considered a conceptual grammar which thus ensures a high degree of systematisation.

4 Conclusions

In this paper we have briefly explained the methodology applied in EcoLexicon for knowledge extraction and representation. Corpus analysis, combining direct term searches and knowledge pattern analysis as well as images, has fed the EcoLexicon knowledge base with reliable information. However, this information has to be represented coherently and systematically. EcoLexicon combines the advantages of a relational database, allowing for a quick deployment and feeding of the platform, and an ontology, enhancing user queries. The internal coherence at all levels of a dynamic knowledge representation shows that even complex domains can be represented in a user-friendly way. This methodology solves two challenges derived from multidimensionality: (1) it offers a qualitative criterion to represent specialized concepts according to recent research on situated cognition [8] both in dynamic networks and multimodal definitions; (2) it is a quantitative and efficient solution to the problem of information overload. Further steps in EcoLexicon will be the automatization of some of its extraction procedures, as well as the evaluation of the resource through usability tests.

Acknowledgement. This research has been funded by project FFI2008-06080-C03-01/FILO, from the Spanish Ministry of Science and Innovation.

References

1. Anglin, G., Vaez, H., Cunningham, K.: Visual representations and learning: the role of static and animated graphic. Visualization and Learning (33), 865–917 (2004)
2. Auger, A., Barrière, C.: Pattern-based approaches to semantic relation extraction: A state-of-the-art. Special Issue on Pattern-Based Approaches to Semantic Relation Extraction, Terminology 14(1), 1–19 (2008)
3. Austin, J.: How to do things with words. Clarendon, Oxford (1962)
4. Barrasa, J.: Modelo para la definición automática de correspondencias semánticas entre ontologías y modelos relacionales. Ph.D. thesis, UPM, Madrid (2007)
5. Barrière, C.: Knowledge-rich contexts discovery. In: Proceedings of the 17th Canadian Conference on Artificial Intelligence (AI 2004), Ontario, Canada (2004)
6. Barrière, C., Agbago, A.: Terminoweb: A software environment for term study in rich contexts. In: Proceedings of the International Conference on Terminology, Standardisation and Technology Transfer (TSTT 2006), Beijing, China (2006)
7. Barsalou, L.: Linking images and words: the description of specialized concepts. Language and Cognitive Processes (18), 513–562 (2003)
8. Barsalou, L.: Simulation, situated conceptualization and prediction. Philosophical Transactions of the Royal Society of London: Biological Sciences 364, 1281–1289 (2009)
9. Bowker, L., Pearson, J.: Working with Specialized Language. In: A Practical Guide to Using Corpora. Routledge, London (2002)
10. Brézillon, P.: Task-Realization Models in Contextual Graphs. In: Dey, A., Kokinov, B., Leake, D.B., Turner, R. (eds.) CONTEXT 2005. LNCS (LNAI), vol. 3554, pp. 55–68. Springer, Heidelberg (2005)
11. Cimiano, P., Staab, S.: Learning concept hierarchies from text with a guided agglomerative clustering algorithm. In: Proceedings of the Workshop on Learning and Extending Lexical Ontologies with Machine Learning Methods, Bonn, Germany (2005)
12. Clark, H.H.: Using Language. Cambridge University Press, Cambridge (1996)
13. Condamines, A., Rebeyrolle, J.: Searching for and identifying conceptual relationships via a corpus-based approach to a terminological knowledge base (ctkb). Special Issue on Recent Advances in Computational Terminology, Terminology (127-148) (2001)
14. Dey, A.: Understanding and using context. Personal Ubiquitous Computing 5(1), 4–7 (2001)
15. Evans, V.: Cognitive linguistics. In: Cummings, L. (ed.) Encyclopedia of Pragmatics (2009)
16. Faber, P.: Conceptual modelling in specialized knowledge resources. In: Proceedings of the XII International Conference Cognitive Modelling in Linguistics, Dubrovnik, Croatia (2010)
17. Faber, P., León-Araúz, P., Prieto Velasco, J., Reimerink, A.: Linking images and words: the description of specialized concepts. International Journal of Lexicography 1(20), 39–65 (2007)

18. Faber, P., Montero Martínez, S., Castro Prieto, M., Senso Ruiz, J., Prieto Velasco, J., León-Araúz, P., Márquez Linares, C., Vega Expósito, M.: Process-oriented terminology management in the domain of coastal engineering. Terminology 12(2), 189–213 (2006)
19. Gadamer, H.: Truth and Method. Continuum, New York (1995)
20. GEMET: About GEMET. General Multilingual Environmental Thesaurus (2004), `http://www.eionet.europa.eu/gemet/about`
21. Goffman, E.: Frame Analysis. Harper and Row, New York (1974)
22. Grice, H.P.: Logic and conversation. In: Syntax and Semantics: Speech Acts, vol. 3, pp. 41–58. Academic Press, New York (1975)
23. Guha, R.: Contexts: A formalization and some applications. Ph.D. thesis, Stanford University (1991)
24. Lenat, D.: Cyc: a large-scale investment in knowledge infrastructure. Communications of the ACM 38(11), 33–38 (1995)
25. León-Araúz, P.: Representación multidimensional del conocimiento especializado: el uso de marcos desde la macroestructura hasta la microestructura. Ph.D. thesis, University of Granada (2009)
26. León-Araúz, P., Faber, P.: Natural and contextual constraints for domain-specific relations. In: Proceedings of Semantic relations. Theory and Applications, Valetta, Malta (2010)
27. León-Araúz, P., Reimerink, A.: Knowledge extraction and multidimensionality in the environmental domain. In: Proceedings of the TKE Conference, Dublin, Ireland (2010)
28. Marsh, E., White, M.: A taxonomy of relationships between images and text. Journal of Documentation 59(6), 647–672 (2003)
29. Marshman, E., L'Homme, M.: Disambiguation of Lexical Markers of Cause and Effect. In: Picht, H. (ed.) Modern Approaches to Terminological Theories and Applications, Linguistic Insights, pp. 261–285. Peter Lang, Bern (2006)
30. Meyer, I., Bowker, L., Eck, K.: Cogniterm: An experiment in building a knowledge-based term bank. In: Proceedings of Euralex 1992, pp. 159–172 (1992)
31. Meyer, I., Mackintosh, K.: The corpus from a terminographer's viewpoint. International Journal of Corpus Linguistics 1(2), 257–285 (1996)
32. Murphy, M.: Semantic relations and the lexicon: Antonym, synonymy, and other paradigms. Cambridge University Press, Cambridge (2003)
33. Prieto Velasco, J.: Información gráfica y grados de especialidad en el discurso científico-técnico: un estudio de corpus. Ph.D. thesis, University of Granada (2008)
34. Reimerink, A., Faber, P.: A frame-based knowledge base for the environment. In: Proceedings of Towards e-Environment, Prague, Czech Republic (2009)
35. Reimerink, A., León-Araúz, P., Magaña, P.: Ecolexicon: an environmental tkb. In: Proceedings of the International Conference on Language Resources and Evaluation (LREC 2010), Valetta, Malta (2010)
36. Reimerink, A., García de Quesada, M., Montero Martínez, S.: Contextual information in terminological bases: a multimodal approach. Journal of Pragmatics 42(7), 1928–1950 (2010)
37. Rogers, M.: Multidimensionality in concepts systems: a bilingual textual perspective. Terminology 10(2), 215–240 (2004)
38. Rosch, E.: Principles of categorization. In: Cognition and Categorization, pp. 27–48. Erlbaum, Hillsdale (1978)
39. Searle, J.: Speech acts: An essay in the philosophy of language. Cambridge University Press, Cambridge (1969)
40. Sperber, D., Wilson, D.: Relevance: Communication and Cognition, 2nd edn. Blackwell, Oxford (1995)

Techniques for Multilingual Security-Related Event Extraction from Online News

Martin Atkinson, Mian Du, Jakub Piskorski, Hristo Tanev,
Roman Yangarber, and Vanni Zavarella

Abstract. This chapter presents a number of techniques for multilingual event extraction, the main task is to accurately and efficiently detect key information about security-related events from electronic news media and summarize it in the form of database-like structures. Gathering such information over time is an important task for developing global news surveillance systems, particularly in the context of security threats and mass emergencies. In particular, this chapter describes novel techniques for dealing with specific extraction tasks, including: an event type classification method based on domain-specific inference rules, an approach to event geo-tagging based on utilisation of lexico-semantic patterns, a simple method for cross-lingual event information fusion, and techniques for scoring the relevance rank of automatically extracted facts.

1 Introduction

On-line news surveillance has been considered worldwide by various security authorities and law-enforcement organisations as a task of paramount importance for supporting early detection of certain threats and situation monitoring during crisis and mass emergencies. In particular, an ever-growing amount of information transmitted through the Internet led to an emergence of advanced tools that combine

Martin Atkinson · Hristo Tanev · Vanni Zavarella
JRC, Via Fermi 2749, Ispra, Italy
e-mail: {Martin.Atkinson, Hristo.Tanev, Vanni.Zavarella}@jrc.ec.europa.eu

Jakub Piskorski
Frontex, Rondo ONZ 1, Warsaw, Poland
e-mail: Jakub.Piskorski@frontex.europa.eu

Mian Du · Roman Yangarber
University of Helsinki, Department of Computer Science, PO Box 68, 00014 Finland
e-mail: {Mian.Du, Roman.Yangarber}@cs.helsinki.fi

A. Przepiórkowski et al. (Eds.): *Computational Linguistics*, SCI 458, pp. 163–186, 2013.
DOI: 10.1007/978-3-642-34399-5_9

techniques from text mining, machine learning, statistical analysis and computational linguistics to help intelligence experts to manage the overflow of information, filter out the relevant from the irrelevant, and to extract valuable, structured and actionable knowledge from on-line news sources.

A significant number of approaches to news mining and news exploration systems have been reported recently. The most prevalent way to organize news by such systems is to classify the incoming news into predefined or automatically discovered categories and to group topically similar news articles into clusters. However, as has been emphasized in [15], in order to facilitate an in-depth analysis of the news it is essential to extract structured information on the events from the news, i.e., to derive detailed information about them, ideally identifying *who did what to whom, through what methods (instruments), when, where and why* [1]. The current capabilities of news event extraction technology deployed in the security domain are exemplified in [29] (epidemiology) and [12] (armed conflicts), whereas [2] reports on general trends in the field of event extraction.

We have developed a Multilingual Event Extraction Framework that is capable of extracting structured information on security-related events from online news in 8 languages in near real time [4, 21]. The framework has been deployed in an operational environment and its use revealed that from an end-user practical perspective the three most important information extraction tasks in the context of security-related event extraction are: (a) to be as precise as possible in the classification of the event type, (b) to extract the actual location of the event, (c) to automatically rank the relevance of the automatically extracted event information. Furthermore, the ability to fuse event information from news in different languages was deemed as important too since news in different languages might provide complementary information. Therefore, we have carried out some explorations on new techniques to tackle the aforementioned problems, whose outcome is described in this chapter.

The remaining part of the chapter is structured as follows. In Section 2, the Multilingual Event Extraction Framework is briefly presented. An event type classification method that uses domain-specific inference rules is described in Section 3. Section 4 elaborates on an approach to event geo-tagging that uses lexico-semantic patterns. The outcome of an experiment on deploying a simple cross-lingual event information fusion technique for fine-tuning and enriching event extraction results is given in Section 5. A technique for scoring the relevance of extracted facts is described in Section 6. Finally, conclusions and future work outlook is given in Section 7.

2 Security-Related Event Extraction Framework

First, news articles are gathered by a large-scale news aggregation engine, the Europe Media Monitor (*EMM*)[1] developed at the Joint Research Centre of the European Commission (JRC) [3]. *EMM* retrieves more than 130,000 news articles per day from more than 3500 news feeds in 42 languages. These news articles are

[1] `http://press.jrc.it`

geo-located, tagged with meta-data and further filtered (classified) using standard keyword-based techniques in order to select those articles, which potentially refer to security-related events. In addition, the news articles harvested within a 4-hour window are grouped into clusters in every language individually according to content similarity (using hierarchical agglomerative clustering). The filtering and clustering process is performed every 10 minutes. Next, the stream of filtered news articles and clusters[2] is passed to the event extraction engine (also in 10-minute cycle), which consists of two core event extraction systems, namely, *NEXUS* [25, 21], developed at JRC, and *PULS* [9, 29] developed by the University of Helsinki. *NEXUS* follows a shallow cluster-centric approach (i.e. it analyzes only the initial part of news articles), which makes it more suitable for extracting information from the entire cluster of topically-related articles[3], whereas *PULS* follows a non-cluster centric approach and performs a more thorough analysis of the full text of each news article. *NEXUS* and *PULS* cover the extraction of a wide spectrum of security-related events, including: crisis events (e.g., man-made and natural disasters, armed conflicts, humanitarian crisis), border events (irregular migration incidents, human trafficking and cross-border crime activities), and medical hazards (e.g., outbreaks of infectious diseases). Currently, the Framework covers 8 languages including: English, Italian, Spanish, French, Portuguese, Russian, Arabic and Turkish. For all event types of interest there is a harmonised event template structure, which includes the following slots: `TYPE`, `SUBTYPE`, `DESCRIPTOR` (free text), `SNIPPET` (text fragment triggering the event), `PUB_DATE`, `DATE`, `LOCATION`, `RELEVANCE` (for the user), `SOURCE`, `PERPETRATOR`, `VICTIM`, `ITEM`, `MEANS`, `AFFECTED-COUNT`, `INJURED-COUNT`, `DEAD-COUNT`, `ARRESTED-COUNT`, and a boolean-valued slot `WOMEN/MINORS_INVOLVED`. The output produced by the core event extraction engine, i.e., a stream of instantiated event templates, is made accessible to the earth browser for visualization of the extracted events on a map. The KML[4] file at this location [5] can be used to see the automatically extracted event descriptions. In particular, dedicated layers show crisis, medical and border events all divided by language and period (i.e., last 24 hours, 7 days and 30 days). By clicking on an event icon structured information is displayed. The concrete results of deploying some (but not all) of the techniques outlined in this chapter are visible too. For further details regarding the Event Extraction Framework and the particular subcomponents thereof please refer to [4, 25, 21, 29].

[2] Please note that many security-related events are reported only in local news, e.g., in the domain of cross-border crimes. Therefore clusters often consist of single news articles. On the other hand, mass emergencies and severe crisis situations are reported not only by various news, but also by news in many different languages.

[3] But it can also be applied on single articles.

[4] KML - Keyhole Markup Language,
`http://www.opengeospatial.org/standards/kml`

[5] It can be opened using a KML-aware earth browser, e.g., Google Earth.
`http://emm-labs.jrc.it/events.kml5`

3 Event Type Classification, Using Domain-Specific Inference Rules

Often, events described in the news are complicated not only to be detected, but also to be classified. Some systems detect only one type of event, e.g., disease outbreaks [29]; clearly, in this case there is no need for classification. Others detect event types like terrorist attacks, shootings, armed conflicts, disasters, etc. [25]. Such event types are related directly to the nature of the event and can be directly inferred from the text that describes the event. For example, news about terrorist attacks use words and phrases like "*terrorist*", "*car bomb*", "*suicide bomber*", etc. From this point of view, event classification into such categories is similar to the text categorization task [27].

On the other hand, if we want to go behind the dynamics of the events and to find the motivation of the people who participate in them and the reasons, which lead to these events, it is necessary to make more elaborated processing of the information, presented in the news.

For example, in the context of border security, it is relevant to find news about arrests and deportations, related to irregular migration. Systems, which use action-detection patterns and rules, e.g., [25], [29] can detect news in which arrests or deportations are mentioned, as well as the descriptions of the people who have been arrested or deported. However, in order to find that such events are related to irregular migration, human trafficking, smuggling, or other cross-border activities, the event extraction system has to analyse the context, in which these events are described. We present an approach which searches domain-specific terms and then combines their semantics through domain-specific inference rules. For example, the following news text describes an arrest event, related to irregular entry attempt:

> *Police and Coast Guard has stopped 18 young Egyptians at dawn, as they were trying to reach a boat that would have taken them illegally to Italy (translation from Italian)*

There are two sub-events, described here: (i) young Egyptians try to reach a boat; (ii) the youngsters get arrested by the Coast Guard and Police. A classical event extraction system can detect that there are two events: an arrest and a movement of people. However, we have to combine information from both sub-events to infer that irregular entry attempt is described. A domain-specific inference system can conclude from the presence of the phrase "*Coast Guard*" that an event has to do with the borders, moreover the presence of the word "*boat*" means that the text implies some cross-border movements. On the other hand, "*Police*" and "*Coast Guard*" may imply that some illegal activity takes place. Finally, the text talks about "*Egyptians*" and "*Italy*", moreover the original language of the news is Italian; using a knowledge about the irregular migration domain, a system can use the fact that Italy is a preferred destination for irregular migrants from Egypt, therefore it can hypothesize that the event is related to the domain of irregular migration. Combining all these clues: cross-border movements, illegal activities and possible relation to irregular migration, a system can conclude that the news article is most probably about an irregular entry attempt.

An inference mechanism for such domain-specific reasoning can be implemented at different levels of complexity. However, in the multilingual context in which our Event Extraction Framework works, it is not possible to deploy elaborated language-processing tools, such as full syntactic parsers, because similar tools are not largely available for languages, other than English. On the other hand, using our weakly-supervised multilingual system for dictionary learning, Ontopopulis [26], we can relatively easily acquire domain-specific lexicon for each language, where each word or phrase is labeled with the relevant domain-specific semantic label, such as "irregular-migration-related", "cross-border movements", etc. Moreover, in the context of our Multilingual Event Extraction Framework, we have deployed local grammars for several languages in the domain of security. Then, inference rules can be defined in the set of semantic labels, considering also syntactic structures, captured by the local grammars. We have implemented such a reasoning approach in the domain of border security, therefore we will use this domain to illustrate our method, however the methodology described here does not depend on the domain.

3.1 Detection of Cross-Border Crimes

We built our inference mechanism inside the border-security version of the *Nexus* event extraction system [25]; we will call this version of the system *NEXUS Borders*. The purpose of the inference mechanism in *NEXUS Borders* is to infer the type of the cross-border criminal activity to which a news article is related. There are three big groups of cross-border crimes, which we consider: irregular migration, human trafficking and smuggling. Further, *NEXUS Borders* can infer several subtypes of these classes: irregular-entry attempt and irregular-migration related document forgery as subtypes of irregular migration, human-organs trafficking as a subtype of human trafficking, and smuggling of drugs, goods and CBRN (chemical, biological, radiological and nuclear) materials as subtypes of smuggling. *NEXUS Borders* considers these types of cross-border crimes as event types. Clearly, some event types like human trafficking and irregular migration can be divided into many more subtypes, however we modelled only the more important ones; the other subtypes are not currently identified and *NEXUS Borders* detects just the main type, e.g. irregular migration.

The type classification and the supporting inference mechanism work in the set of the so called *domain-specific semantic labels*, which describe domain-related concepts of different types at different levels of abstraction. They can be objects, actions, people, situations or their properties. The most important semantic labels are: "irregular entry attempt", "irregular migration", "trafficking", "potential irregular migrant", "vehicle", "cross-border movements", "operative context", "potential victim of human trafficking", "criminal organization", "prostitution", "smuggling" and others. Currently, *NEXUS Borders* exploits around 40 such semantic labels.

These semantic labels were manually defined by studying news about cross-border crimes. Each semantic label encodes a semantic concept from the domain. For example, the semantic label “potential irregular migrant” will be triggered in texts which directly or indirectly refer to people, who potentially can be irregular migrants. The semantic label “vehicle” is more concrete—it is triggered, when vehicles are mentioned. The label “operative context” is triggered, when the text is likely to describe some security-related operations or actions, related to cross-border criminal activities.

NEXUS Borders is supposed to detect articles, describing operative events, related to the cross-border activities mentioned so far. These operative events are arrests, interceptions of vehicles, deportations, detections of irregular-entry attempts and security operations. However, the priority of the event extraction system is to detect the cross-border activity, to which the event is related, rather than the operative event, reported in the article. Accordingly, the inference mechanism finds the type of cross-border criminal activity, to which an article refers, e.g., smuggling or human trafficking, but not the type of the operative event, reported in the article, which can be arrest, interception, confiscation, etc.

On the other hand, *NEXUS Borders* extracts information about the operative events, i.e., arrested people, intercepted vehicles, confiscated items, etc., as well as information about people and entities, more directly related to the cross-border crimes: people who cross the border irregularly, items, which are being smuggled, etc. The extracted information is used by our type inference approach together with the lexical content of the article to infer the type (and subtype) of the related cross-border criminal activity.

3.2 *Inference Algorithm*

Our algorithm works in two steps: First, a *primary set* of semantic labels is generated from the lexical content of the article and the domain-specific syntactic constructions, such as arrest or deportation descriptions, e.g., "*two people were arrested*". Secondly, a forward-chaining inference algorithm is used to generate all the possible additional semantic labels from the primary set and our reasoning rules.

3.2.1 Generating the Primary Set of Semantic Labels

This is the language-specific part of the algorithm. For each language which *NEXUS Borders* is able to process (currently Italian, Spanish, French and Arabic), a semantic dictionary was obtained semi-automatically, using the Ontopopulis lexicon learning algorithm [26]. In this dictionary each lexical entry (word or phrase) is related to one or more of the semantic labels. Each pair lexical entry - semantic label has assigned one of the two weights - “low” or “high”. For example, in Italian, the word “*traffico*” (“*trafficking*”) has the label “trafficking”, assigned to it with high weight and also the label “potential irregular migrant” with low weight. Semantic labels, assigned with high weight to a lexical entry, are always triggered when the corresponding word or phrase appears in the text. For example, the word “*traffico*” will trigger the label "trafficking", independently of the context. On the other hand, the low weight pairs lexical entry - semantic label trigger the label only when the lexical entry appears in a domain-related syntactic construction. For example, the word "*traffico*" will trigger "potential irregular migrant" in the phrase "*traffico di persone*" ("*trafficking of people*"), since this phrase can be parsed by the domain-specific syntactic pattern *traffico di [PEOPLE]*.

These domain specific syntactic constructions have one of the two forms:
LEFT-CONTEXT (PEOPLE|VEHICLE|SMUGGLED-ITEM|DOCUMENT) or
(PEOPLE|VEHICLE|SMUGGLED-ITEM|DOCUMENT) RIGHT-CONTEXT.
We use domain specific grammar and a list of predefined linear patterns to model the left and right contexts. The actions and situations, modeled through these constructions can be one of the following: intercepted vehicle (e.g., "*a boat was intercepted*"), death or injury, border crossing ("*two people passed the border*"), people in a vehicle ("*men in a rubber boat*"), arrests, rescuing of people, landing of people on the shore, exploitation and trafficking of people, smuggling, transportation and selling of smuggled items, and forged documents. The syntactic rules used to detect such constructions are also used to extract the actors (people), the means (vehicles), the smuggled items and the forged documents; these entities are then used to fill the corresponding slots of the event template. As an example, let's consider the following text fragment (translated from Spanish):

> *22 people without documents were arrested yesterday by a unit of United States Coast Guard, while trying to reach Puerto Rico.*

The following semantic label triggering will take place (on the left we show the triggering word and on the right the label) :

- "*people without documents*" → potential-irregular-migrant
- "*Coast Guard*" → border
- "*Coast Guard*" → operative-context
- "*reach*" → moving

3.2.2 Inference Rules

After the primary set of semantic labels is generated, the inference algorithm combines each pair of labels to check if a new label can be inferred. The newly inferred labels are added to the set of labels and a new attempt to infer labels takes place; this continues, until no more new labels can be obtained. *NEXUS Borders* uses about 30 inference rules, where each rule is of the type `A&B -> C`, where `A`, `B`, and `C` are semantic labels. The rule means that if labels `A` and `B` are present in the primary set or in the set of already inferred labels, then label `C` will also be added. Similar to the generation of the primary set, the rules are labeled as "low weight" and "high weight" ones. The high-weight rules are allowed to combine each pair of labels, generated from the processed article text. On the other hand, the low-weight rules are less reliable and can be triggered only inside domain-specific syntactic constructions. When more complex rules with three or more terms on the left side are used, e.g. `A&B&C -> D`, they are decomposed into simple rules , e.g. `A&B -> F` and

F&C -> D. This was done to make the inference mechanism more simple from the implementation point of view. In our example, the following "high-weight" rules will be triggered in the shown order:

1. border & moving → cross-border-movements
2. cross-border-movements & potential-irregular-migrant → irregular-entry-attempt
3. irregular-entry-attempt & operative-context → irregular-entry-attempt-operative

The semantic label "irregular-entry-attempt-operative", which is inferred at the last step, is a sufficient condition to assign the text to the type "irregular migration". Moreover, there are arrested people in this event, therefore *NEXUS Borders* uses an heuristic which says that if the label "irregular-entry-attempt-operative" is triggered and there are arrested people, then the event is of subtype "irregular entry attempt". This heuristic is necessary for more precise event classification. It ensures that the more specific subtypes are triggered only when sufficient evidence is presented, otherwise only the more generic type "irregular migration" will be assigned to the article. This inference mechanism is similar to the inference in the expert systems [28]. Although it is simple, this approach proved to be useful in our specific domain.

3.3 Experiments

We collected 60 events, detected by *NEXUS Borders* in Italian-language news, 20 for each event type - irregular migration, human trafficking and smuggling. We found out that the type classification precision was 75% for irregular migration events, 60% for human trafficking, and 95% for smuggling.

Most of the errors for the human trafficking events were due to non precise assignments of semantic labels to lexical items. The errors in irregular migration were mostly due to the fact that there were news which were related to irregular migration or cross-border crimes, but they were not considered relevant to the scope of the system, i.e., not related to particular incidents. In any case, all the news, for which the system detected an irregular migration event, were related to issues concerning cross-border criminal activities. The only error in detection of smuggling events was in case of a news article that talked about a movie whose plot was related to drug trafficking.

We carried another experiment with Turkish news, i.e., we downloaded 250 news articles from the Web and evaluated the performance of our system, considering both precision and recall. For irregular migration, precision was found to be 66%, recall 67%, and F1 0.66; for human trafficking precision was 63%, recall 51%, and F1 0.56; for smuggling results were - precision 60%, recall 71%, F1 0.65. The problem in this type of evaluation was that our system can return only one event per article and sometimes more than one event was present and annotated. To take into account this, we carried an overall evaluation, in which an article was considered correctly labeled, when the label of the system for this article coincided with any of the human annotator labels. In this way, we obtained precision 65%, recall 69%, and F1 0.67.

4 Event Geo-tagging

Accurate and fine-grained event geo-tagging is an important subtask of event extraction, particularly in the security domain, where information on the exact event location may be crucial for early warning purposes. A common way to tackle this task is to apply text geo-tagging techniques, which assign to an event-related article or a cluster of articles the "most relevant" geographical location. Geo-tagging typically consist of two steps [22]: in a "geo-parsing" step, place-referring expressions in text are detected; in a subsequent "geo-coding" step, text is actually mapped to the geographical coordinates that it is referring to[6].

In the standard approach, the only type of linguistic units which are relevant for the geo-tagging process are proper names referring to places. These are more effectively detected by performing lexical lookup on a large external resource such as a gazetteer, rather than by designing extraction patterns based on contextual features [17]. In our application, where a large, multi-lingual gazetteer of place, province, region and country names is used [21], extracting as a geo-entity any word which is listed in the lexicon is known to introduce a high level of ambiguity, for example because of the overlapping between place names and person/organization names, homography of place names among themselves and with other common words in the language. [21] lists a number of heuristics currently deployed to filter and/or disambiguate some candidate locations matched in text.

Nonetheless, there are contexts where these heuristics may not be sufficient to reduce the noise. This is particularly true in the border-security news domain, where target events typically involve some spatial motion of event participants and consequently they mention on average a significantly higher number of place names, referring for example to the source country of the migrants, intermediate path on their migration route, place of interception, country of origin of authorities involved, and so on. Clearly, most of such place names are not the actual location of the event. Take for example the following extract from a Spanish language news article about an irregular entry event[7]:

*TITLE: Encalla una barca con más de 300 inmigrantes a bordo frente a **Lampedusa***
SUBTITLE: Durante la noche llegaron unos 1.100 indocumentados
*BODY: Una barca procedente de **Libia** con más de 300 inmigrantes a bordo encalló esta madrugada en unos escollos en las proximidades del puerto de la isla italiana de **Lampedusa**, adonde durante la noche llegaron unos 1.100 indocumentados, informaron hoy fuentes de la Capitanía de Puerto.*

[6] Typically this involves applying a number of heuristics for filtering the expressions which are not geographical references in the target context, and disambiguating between locations sharing the same name.

[7] [TITLE: A boat with more than 300 immigrants on board agrounds in front of **Lampedusa**. SUBTITLE: About 1,100 undocumented arrived during the night. BODY: A boat from **Libya** with more than 300 immigrants on board ran aground early this morning in a reef near the port of the Italian island of **Lampedusa**, where some 1,100 undocumented arrived during the night, said sources from the Port Authority.]

The text contains three unambiguous references (in bold) to two geo-entities (Lampedusa, Libia), only one of which should be extracted as the actual geo-location of the target event (Lampedusa). Depending on additional geographical references available in the subsequent text in the article, it may well be that disambiguation heuristics may not filter out the wrong hit. Notice though, that this would be easy to correct by just looking at the local linguistic context around the place names, namely by considering the semantics of the Locative Prepositional Phrases (LPP) in which place names are nested (underlined in text). In this case, for example, for toponym "*Libia*", while part of the general spatial reference of the text, should be discarded as it fulfills an irrelevant semantic role, such as the ORIGIN or SOURCE of the agent involved in the irregular border crossing event.

By taking into account the semantics of the LPP related to a target event, we aim at specifying geo-tagging as the task of extracting the fine-grained event geo-location, rather than the general geographical "aboutness" of an article. In order to experiment on this, we apply a Cognitive Linguistics model of the semantics of LPPs, as adapted from [31].

4.1 A Semantic Decomposition Model of LPP

LPPs can be decomposed into 3 main semantic elements:

- Reference Entity (RE): this is an entity whose location in the space is taken as known. In our application context this will typically be an entity with specified geographical coordinates (a city, a mountain, an airport, etc.). Linguistically, this is usually realized as the object Noun Phrase of the PP, which could be just a place name (e.g., "*Cadiz*"), or a noun phrase containing a place name and possible modifiers (e.g. "*las costas de la surena provincia de Cadiz*" ["*the coast of southern Cadiz province*"]);
- a Place Value (PV) with respect to the Reference Entity, that is a definite point or surface or volume which takes the Reference Entity as its origin, for example its top surface, its interior, space below, etc. Reference Entity and Place Value can be considered as a single, complex element, referred to as Place.
- Locative Relation (LR) between the predicate X to which the locative phrase is attached and the Place: for example a Place can be the Position of X, or the Source or Goal of X, etc.

As an example, 3. and 4. below represent the semantic structures of 1. and 2., respectively:

1. $[_X$ *There is a book* $][_{PP}$ *on the table* $]$.
2. *He* $[_X$ *put a book* $][_{PP}$ *on the table* $]$.
3. $X[_{LR}POSITION[_{PLACE}[_{PV}SURFACE][_{RE}\,\text{table}]]]$
4. $X[_{LR}GOAL[_{PLACE}[_{PV}SURFACE][_{RE}\,\text{table}]]]$

Here the PPs in brackets in 1. and 2. share the same RE (a spatially located entity, the table) and describe the same PV with respect to it (its superior surface, as opposed to its interior space), but they differ as for the LR they express: in 1., the locative relation between X (the predicate expressing the existence of a book) and the table is that "*on the table*" (the Place element) is a POSITION, while in 2. the relation between X (the transfer of a book) and "*on the table*" is of type GOAL (i.e., the surface of the table is the target location of a motion event).

Linguists usually list a finite number of possible LR: POSITION, SOURCE, GOAL, ROUTE[8]. The LR is unique and mandatory: each LPP expresses one and only one locative relation[9]. The PV and LR are two independent semantic dimensions. PV are unique but optional, in the sense that a LPP can be unspecified with respect to PV, like in:

5. *John is at the supermarket*
6. *John comes from the supermarket*

which do not specify anything more than, resp., a POSITION and SOURCE LR between John and the supermarket.

Although PV can convey some highly relevant information for an accurate geo-location of events, we have not experimented on modeling their semantics in this work. The focus of our experiment instead is on LR: we want to filter event location candidates based on the LR of their corresponding LPP.

In many languages, LRs are expressed by prepositions: e.g., "*at*" usually conveys a POSITION meaning. Unfortunately, the semantics of prepositions is not sufficient to predict the LR of a LPP, as the examples 1. and 2. above show, where the same preposition ("*on*") conveys a POSITION meaning in 1., and a GOAL meaning in 2., respectively. The reason is that the LR of a LPP is actually given by some form of composition of the semantics of the preposition, the aspect of the predicate X the LPP is attached to, and the so called "eventuality" type expressed by the predicate itself.

Eventuality is a general term in linguistics referring to both event processes and states. For our purposes, we can distinguish just two types of eventualities, bounded and unbounded, where the former include various processes with an intrinsic temporal ending, while the latter cover states and activities (which do not have intrinsic ending by definition). To give an example, in:

7. *John is working in the office.*

the predicate represents an unbounded eventuality (a process with no intrinsic ending), consequently the LPP represents a LOCATION. Instead in:

8. *In the office building, he put a letter in the mailbox.*

---

[8] They can defined in the following ways. In:

$X[_{LR}Z[_{PLACE}[_{PV}][_{RE}]]]$

If Z=POSITION, PLACE is the position of X;

If Z=GOAL, PLACE is the terminus of X;

If Z=SOURCE, PLACE is the starting point of X;

If Z=ROUTE, it shows that PLACE is the space intervening from SOURCE to GOAL taken by X.

[9] To see this, notice that e.g., in English combinations of prepositions conveying two different LR are semantically unacceptable (e.g "*at from the supermarket*").

the predicate "*put a letter*" expresses a bounded eventuality and the LPP ("*in the mailbox*") conveys a GOAL meaning, despite the fact that the preposition is again "*in*".

This means that, in order to model the Locative Relation of the LPPs, we would need to mark up the aspect and verb class of the verb phrases the LPPs are attached to, together with some default semantics of the prepositions.

The analysis sketched above is mostly valid across languages. Nonetheless, in order to test the feasibility and usefulness of such semantical approach for event geo-location, we ran an experiment for the Spanish language, in the domain of border security.

4.2 Experiments

First, we asked a native speaker to annotate a short corpus of 80 Spanish news articles on border security related events (irregular border crossing at entry, smuggling, human trafficking). We show here a sample from the corpus[10]:

> ⟨**SENTENCE ID=1**⟩Al menos tres inmigrantes de la costa de Marruecos han muerto esta madrugada cuando la patera en la que ⟨**PREDICATE ID=1**⟩trataban de arribar⟨**/PREDICATE**⟩ ⟨**PP MODIFIES=1, LR=GOAL**⟩a las costas de Cadiz⟨**/PP**⟩ se ha destrozado tras chocar contra unas rocas, informa la Cadena SER⟨**/SENTENCE**⟩. ⟨**SENTENCE ID=2**⟩Los equipos de rescate ⟨**PREDICATE ID=2**⟩se dirigen⟨**/PREDICATE**⟩ al lugar de los hechos, ⟨**PP MODIFIES=2, LR=GOAL**⟩frente al Cabo de Trafalgar⟨**/PP**⟩, [...] ⟨**/SENTENCE**⟩.

For each sentence, we marked up event predicates relevant to the target domain. Then, for each predicate, we annotated any LPP containing place names and attached to it, with an attribute ("MODIFIES") encoding the syntactic link and an attribute LR whose value is the Locative Relation encoded by the LPP.[11] From this corpus, we collected counts on the correlations between, on the one hand, prepositional complexes (e.g."*en dirección a*", "*cerca de*", "*hasta*")[13] and verb or nominalized phrases, and the LPP's Locative Relations, on the other. We started by setting the default LR of each preposition complex as the one with occurrence counts largely predominant and we encoded this default value in a lexicon of LR patterns. The alternative LR values were also encoded in the LR pattern lexicon, and we manually modelled such correlations by generalizing on some features of the predicate phrases, such as Aspect, Verb class and Eventuality class. This resulted in a simple

---

[10] [At least three immigrants from the Moroccan coast have died this morning when the small boat on which they were trying to reach the coast of Cadiz is shattered after hitting a rock, Cadena SER reported. Rescue teams are directed to the scene of the events, off Cape Trafalgar,...]

[11] In order to increase the recall, we also included unnamed geo-references such as LPPs containing demonym modifiers (e.g., "*a las costas andaluzas*" [12]). We carried out semi-automatically the mappings with corresponding geo-names by using string distance metrics, and then added these modifiers to our gazetteer.

[13] [towards], [near], [to], respectively.

LR "calculus", which was implemented by a cascade of finite-state grammar rules[14], similar to the following:

```
rule :> ( ( predicate & [POS:"verb",TYPE:"BS", ASPECT:"perfective"]
            |
            predicate & [POS:"verb",TYPE:"BO", ASPECT:"perfective"]
          )
          ( gazetteer & [GTYPE:"LR_pattern", LOC_REL: "GOAL"]
            |
            gazetteer & [GTYPE:"LR_pattern", LOC_REL: "POSITION"]
          )
          location & [PLACE_VALUE:#pv]):phrase
  -> phrase: locative & [LOC_REL:"POSITION", PLACE_VALUE:#pv].
```

This rule combines a verb phrase, headed by a motion achievement verb with "perfective" ASPECT value, with a *location* entity nested in a LPP which is introduced by a LR pattern whose default LR values are either "POSITION" or "GOAL". It results in a LPP with "POSITION" LR value, modeling the fact that in: "*dos barcazas de inmigrantes llegan a la costa de Tenerife*"[15] *Tenerife* is interpreted as the actual location of the event agent. The *location* type is the LPP's object phrase containing the place reference, as recognized by rules applied at lower levels of the grammar cascade, and can cover expressions as complex as *"el puerto de Valencia"*, *" la zona sur de las provincia de Cadiz"*[16], etc. Moreover, verb classes were learned semi-automatically from a short set of seeds from the annotated corpus, using the Machine Learning tool Ontopopulis [26].

In order to evaluate on the role of LR values as a semantic filter to discard irrelevant event geo-location candidates, we place the semantic analysis in a pipeline with the output of our baseline geo-coding module, which consists of disambiguated matches of a place name multi-lingual gazetteer[17]. Place names are fed to a grammar cascade of around 50 rules, which parses the LPP that they are nested in and extracts their LR, together with information on the main event of the article.

We then apply two simple filters: (a) LR filter: that removes all candidate geo-locations which have LR values which are different than the POSITION, and (b) Event filter: removes all candidate geo-locations which are beyond a proximity threshold from the event mention in the article. We tested on one hand, how much this twofold filter could help improving the precision of event geo-location task, and on the other, how much relevant geographical information from the article would be lost.

A small evaluation was carried out on a corpus of randomly selected 40 Spanish articles gathered by the EMM engine during 10 consecutive days, from which *NEXUS Borders* detected border events. The corpus was then automatically annotated with information (name, type and triggering pattern) on the first occurring event in the article; a human annotator marked up the most relevant geo-location

[14] We used the Express finite state grammar engine [20].

[15] [two barges of immigrants come to the coast of Tenerife...].

[16] [the port of Valencia],[the south of the province of Cadiz].

[17] Consisting of ca. 1200000 entries, among place, province, region and country names, and considering language variants. This resource has grown from a number of open source information repositories like GAUL and Geonames.

(or set of geo-locations) of the extracted event in the article text. For each article, we compared this gold standard with: a baseline system (BL) consisting of the standard geo-coding approach and the target system (TG) consisting of the standard approach plus a combination of LR and Event filters[18]. In both cases, we counted correctly extracted geo-locations, false positives and false negatives. The Precision, Recall and F1 figures are shown in Table 1.

Table 1 Precision, Recall and F1-measure figures: baseline system vs. LR-based approach

	Precision	Recall	F1
BL	19.1	27.5	20.8
TG	28.5	38.4	30.2

The limited size of the test set did not allow us to carry out a detailed analysis. While modest in absolute terms, the results do suggest some hypotheses. Event geo-location is a much harder task, compared to the more "fuzzy" text geo-tagging task, where place inclusion relations (e.g., city-region-country) can be exploited in order to provide the global geo-reference of the article [19]. The significant gain in precision suggests that the proposed approach helps to focus on the specific event location, discarding irrelevant information from the article content. The gain in recall can be explained by the fact that typically the exact geo-location of an event consists of a small sized populated place, which our Baseline geo-coding algorithm tends to discard in favour of more prominent locations available in the article. Finally, we noticed that low recall figures are somehow correlated with the low recall of our Event Extraction engine in detecting multiple (including anaphoric) mentions of the same event, and consequent poor performance of the trivial text proximity heuristics that we adopted for mapping geo-entities to events.

5 Cross-Lingual Information Fusion

A vast bulk of research in the past focused on development of mono-lingual event extraction systems that operate on single documents. Several recent IE studies stressed the benefits of going beyond the classical single-document extraction and using information redundancy to estimate the correctness of the extracted facts [6, 10, 16, 19, 30]. Since security-related events are reported in the news multiple times by different media (cf. [18]) and in different languages, the ability to aggregate and fuse event information across news articles in several languages might potentially improve the quality of the extraction and is becoming increasingly important, as was

[18] In order to have enough information available, in the TG system we also used the candidate geo-locations which were discarded by the BL system.

[19] See [22] for some figures from a method similar to our Baseline. Notice that in our evaluation, we did not allow any approximation of event geo-location, so e.g., if the event occurred in Tenerife, neither Canarias nor Spain are accepted as correct hits.

acknowledged in [11]. Experiments on cross-lingual IE have been recently reported by various authors, e.g., [5, 24] describe attempts of cross-lingual bootstrapping of machine learning-based event extraction systems, while [23] presents an approach for mining facts from comparable corpora in different languages for answer validation in an entity profiling task. We have conducted similar in nature experiments on the usefulness of cross-lingual information fusion for refining the results returned by our Event Extraction engine. In principle, there are two polar ways of approaching cross-lingual information fusion in the context of multilingual news event extraction:

- run mono-lingual event extraction on the native language news articles, then translate extracted facts into one common language, and then perform information fusion
- translate all news articles into one common language for which a high-performance event extraction system exists, and run that system on the translated news articles, including cross-article information fusion

Clearly, intermediate approaches could be considered, e.g., first source text is summarized so that it includes only event-relevant information, then the summary is translated, and finally the target-language extraction system is run on the translated summary. Since it has been observed that approaches involving text translation perform notably worse, as reported for instance in [24], we limited our explorations to the first major approach based on utilisation of mono-lingual extraction systems.

5.1 Fusion Technique

First, for a given event all related news article clusters created by EMM [21] are processed by corresponding language-specific event extraction system, where the following slots were considered: (a) `TYPE`, (b) `LOCATION`, (c) other non-numerical slots: `PERPETRATOR`, `DEAD`, `INJURED`, `KIDNAPPED`, `ARRESTED`, `WEAPONS`, and (d) (integer-valued) numerical slots: `DEAD-COUNT`, `INJURED-COUNT`, and `KIDNAPPED-COUNT`. Next, event descriptions extracted by mono-lingual systems are normalized, i.e., all non-numerical slot fillers are translated (converted) into English, whereas geographical names are mapped to their canonical forms using the multilingual *GeoNames*[20] gazetteer. Subsequently, for each event the corresponding normalized event descriptions are merged into one via the application of a simple slot-value fusion method, based on the following general assumption: *'If a candidate slot value (returned by at least one of the mono-lingual systems) occurs frequently (more than once) as a filler of a given slot in a collection of event descriptions referring to a certain real-world event, and if this value was 'on average' extracted with high system confidence*[21]*, and if it refers to a more specific concept than the other values in the candidate slot filler set, that increases the likelihood that this slot value is correct'*.

[20] `http://www.geonames.org/`

[21] 'on average' meaning that the average system confidence was high.

More formally, let E denote an event and $E_D = \{e_1, \ldots, e_k\}$, where e_i is a list of slot-value pairs, to be the set of automatically extracted event descriptions that refer to E. Let $e(x)$ denote the value of slot x in the event description e. We extend this notion to a set of values for slot x in an event description collection $E_D(x) = \{v | \exists e \in E_D \wedge e(x) = v\}$. Next, let $E_D^{x=v} = \{e | e \in E_D \wedge e(x) = v\}$ be the set of event descriptions with certain value v for the slot x. Furthermore, we denote systems' confidence of extracting v as the value of $e(x)$ as $conf_e(x,v)$. The systems' confidence is based on a combination of factors: (a) the reliability of the pattern(s) used to extract a particular value (the likelihood that pattern extract the slot value correctly), (b) the number of articles in which the pattern was triggered (frequency), and (c) the overall reliability of the language-specific instance of the event extraction system. The main drive behind inclusion of the last factor into the confidence score was due to the fact that our empirical observations and past evaluations of the mono-lingual event extraction systems revealed that there are significant differences in their performance in terms of the quality of the results returned.

Let e^* denote the event description resulting from merging the event descriptions in E_D using fusion method M, which is defined as follows:

$$e^*(x) = \underset{v \in E_D(x)}{\operatorname{argmax}} \; Score_M(x,v)$$

where $Score_M(x,v)$ denotes a scoring function specific to method M. For filling non-numerical slots we used the following scoring function:

$$Score_M(x,v) = \sum_{e \in E_D^{x=v}} conf_e(x,v) \cdot \frac{1}{|E_D^{x=v}|} + \alpha \cdot |E_D^{x=v}| + \beta \cdot |\{v' \in E_D(x) | \wedge v' \supset v\}|$$

where $\alpha \geq 0$ is a factor determining the importance of the number of occurrences of v as a slot filler for x, and $\beta \geq 0$ is a factor which specifies the degree of boosting slot values, which happen to represent concepts that stand either in 'is-subsumed-by' or 'is-part-of' relation (denoted as '$\supset$') with other concepts in the same slot value set.[22] The rationale of using the latter factor is that, intuitively, a 'more-specific' value co-occurring with a related 'more-generic' concept is more likely to be the correct slot filler among those two. For instance, in $E_D(location) = \{Italy, Calabria, Naples\}$, *Naples* would be boosted by $\beta \cdot 2$ since *Naples* is a part of *Calabria* and *Italy*, whereas *Calabria* would be boosted by β since it is only part of *Italy*. Hence, *Naples* gets a higher chance of being selected as the location of the event. α and β were set differently for different slot types based on empirical observations, e.g., in case of locations β was significantly higher than α since boosting most specific candidate locations that are contained by other candidate locations appeared to be a sure-fire heuristic, even in cases when such locations appear only once in systems' responses.

[22] A small in-house ontology was used for that purpose. It has been created based on slot values returned by the event extraction system that was applied on a collection of a several thousands of news articles. In case of locations the inclusion relation was computed using the *GeoNames* gazetteer mentioned earlier.

As for the numerical slots, the definition of $Score_M(x,v)$ was simplified since the last part (β) does not apply to numbers. Furthermore, in case of candidate values, which are significantly distant one from another we selected a maximum[23], based on a simple assumption that the event is most likely evolving and numbers change continuously, the highest being the more up-to-date one. It is not necessarily the case that the last news article on a certain event reports the most up-to-date figures since there is certain latency between reporting on a given event in different countries. Therefore, the 'maximum' heuristic was chosen.

5.2 Experiments

For carrying out the experiments we have created a corpus consisting of 618 event descriptions automatically extracted by *NEXUS Crisis*[24] (3,49 slots extracted per event on average) from 1482 news sources on 22 randomly selected (non-continuing) days in 2010 from news in English, Spanish, Italian, French, Portuguese and Russian. The 618 event descriptions extracted refer to 523 real-world events, out of which 51 were reported in more then one language (27 events reported in 2 languages, 13 events in 3 languages, 4 events in 4 languages, 5 events in 5 languages, and 2 events in 6 languages). This accounts for circa 9,8% of all extracted events that could be potentially refined through cross-lingual information fusion. The 51 events reported in more than one language include: 33 violent events, 7 natural disasters, 9 man-made disasters and 2 other crisis-related events. For the aforementioned 51 events we manually created the gold-standard annotations (event descriptions) based on any information which could be found in the news articles in all 6 languages. In total, there were 4252 news articles that refer to the 51 events. The linking of event descriptions across languages for the purpose of the experiment was done manually using the Event Moderation Tool described in [4].

We have applied fusion technique described in Section 5.1 on the corpus. We measured extraction precision, recall and F-measure for each language-specific system instance of *NEXUS Crisis* and for the extraction based on cross-lingual information fusion. It is important to note that we assigned basically three scores (for non-numerical slots) for filling each slot: 0 (incorrect), 1 (correct), and 0.5 (partially correct), where 'partially correct' is assigned in cases where the slot fill represents a more generic concept than the one in the gold-standard, or in case of locations, if the slot fill refers to a administrative unit, which encompasses the specific place of an event, e.g., if an event happened in *Naples*, we assign the slot fill *Italy* the score 'partially correct'.

The results of the evaluation are shown in Table 2. Compared to the performance of the best mono-lingual system a gain of 6,4% and 4,8% in the overall recall and precision respectively could be obtained through fusing the responses of the

[23] Only candidate values extracted with confidence higher than a pre-specified threshold are considered.

[24] *NEXUS Crisis* extracts crisis events such as: natural and man-made disasters, armed conflicts, terrorist attacks, etc.

Table 2 Precision, recall and F-measure figures for (a) mono-lingual event extraction systems, (b) fusion of the responses of the mono-lingual systems

	All slots			Event Type			Location			Non-numerical slots			Numerical slots		
	P	R	F	P	R	F	P	R	F	P	R	F	P	R	F
English	83.5	78.8	81.1	86.6	80.7	83.5	82.6	80.7	81.6	**92.2**	80.8	86.1	73.9	70.8	72.3
Spanish	75.8	67.3	71.3	80.4	61.7	69.8	85.5	85.5	85.5	84.6	69.1	76.1	62.9	51.5	56.6
French	71.6	63.3	67.2	83.3	70.0	76.1	68.8	66.0	67.4	91.6	71.7	80.4	54.1	46.4	50.0
Italian	75.5	65.5	70.2	67.8	63.3	65.5	86.7	86.7	86.7	85.0	51.5	64.1	53.8	41.1	46.6
Russian	65.8	55.4	60.2	81.3	28.3	42.0	61.4	58.7	60.0	91.6	55.0	68.7	75.0	53.5	62.5
Portuguese	73.7	59.2	65.7	83.3	55.6	66.7	**87.5**	77.8	82.4	83.3	63.1	71.8	50.0	40.0	44.4
Fusion	**88.3**	**85.2**	**86.7**	**91.3**	**84.3**	**87.6**	87.3	**87.3**	**87.3**	91.5	**83.5**	**87.3**	**82.6**	**79.6**	**81.1**

mono-lingual systems. The obtained gain in precision and recall for event type and numerical slots was 4-5% and 8% respectively. The precision for extracting locations and non-numerical slots for the best-scoring mono-lingual system is better than the result of fusing the results returned by the mono-lingual systems. However, the recall for the same slot types is 0.6% and 2.7% respectively higher in case of fusing the response of the monolingual systems.

As regards the results of the particular mono-lingual systems, in some cases their performance varies significantly. This is mainly due to the fact that the English system is the most elaborated and fine-tuned one, whereas the system for the other languages are either less resourced or texts in these languages are more difficult to process, e.g., locations in Russian are inflected, hence their extraction is more difficult, i.e., performing simply gazetteer look-up yields worse results than for English.

We have carried out a small error analysis of cross-lingual information fusion. In case of fusing event type information, for 5 out of 51 events in the corpus none of the mono-lingual systems was able to assign any type information. Consequently, the cross-lingual fusion did not result in any improvement in case of those events. In case of 2 other events, all of the mono-lingual systems returned incorrect event type information, which resulted in incorrect cross-lingual fusion. Furthermore, in case of 2 events, the cross-lingual fusion resulted in selection of an event type (extracted by at least one of the mono-lingual systems), which is related to the event type in the gold standard, but the latter was not detected by any mono-lingual system. Finally, for 1 event, the fusion resulted in the selection of a wrong event type, although the correct event type was detected by at least one of the mono-lingual systems. The analysis of fusing location information revealed that: (a) in case of 3 events an incorrect location was selected, although at least one of the mono-lingual systems returned the correct answer, (b) for 4 events partially correct location was selected, and (c) for 2 events none of the mono-lingual systems returned a correct answer, hence, the error was propagated in the cross-lingual fusion process.

6 Relevance Ranking of the Extracted Events

The objective of attaching a relevance score to extracted events is to provide to the end-user an indication of the relative importance of a particular event to his/her task and needs. For example, if the amount of extracted data is large but the user's time is limited, s/he may choose to focus on the events with higher relevance first.

In our experiments, we used a generic rating scale from 0 to 5, arranged as follows. The score of 5 is assigned only to new, highly relevant events. A score of 4 means that the event is highly relevant to the border-security domain, though possibly not news-breaking but relating to on-going developments. A score of 3 is applied to current events that are still of relevance to the domain, but in a more general, non-specific form, such as review articles, statistics, etc., rather than specific incidents. The score of 2 is applied to older, historical events, of marginal relevance to the surveillance task. A score of 1 means hypothetical or other statements very distantly related to the topic. Zero score is reserved for completely irrelevant, probably incorrectly extracted. Although the boundaries between these discrete scores are not absolute, they do reflect categories that were designed in collaboration with end-users who are domain experts, and that we expect the users to agree upon.[25]

To predict the relevance of events, we collected a training corpus of events labeled with users' judgments, and experimented with several machine-learning methods, to build classifiers that predict the relevance of events extracted from new, unseen documents.

In this section we describe the features that we use in *PULS* for predicting the relevance of an event. The features refer to particular characteristics of the documents from which the events was extracted, as well as those of the events themselves. Features that refer to events are characterized in terms of the event *trigger* and its *attributes*. *PULS* system operates by pattern matching.[26] The trigger is the piece of text (a part of a sentence) on which an event pattern matches, signaling a mention of an event at that point in the document. The attributes of the event correspond to the fills in the database record. Several events may appear within a news article, and may have different relevance scores.

6.1 Features for Relevance Prediction

We distinguish two kinds of features: *lexical features* and *discourse features*. Lexical features are simpler low-level features based on bags of words. Discourse features are based on properties of the article text and of the events extracted from it. These features were devised through a detailed analysis of the domain and user-evaluated events, and were chosen based on their potential for relevance prediction. In essence,

[25] We do not yet have inter-annotator agreement measures on this scale.

[26] PULS has a knowledge base of domain-specific linguistic patterns, which map from surface-syntactic representation of the facts in the sentence to the semantic representation in the database records.

lexical features capture local information, while discourse features capture longer-range relationships within the document.

Discourse features capture information about the number of events within a document, positioning of the event in the document, the compactness of the placement of the event's attributes and the recency of event occurrence, and others. For example, the *number of events* may be an indicator of relevance, if we assume that highly relevant articles are also highly focused—they refer to only one or two events, whereas if the number of events is high, the article is more likely about a more general review, over a broader time-span. Likewise, the *relative position* of the event within the article may be an indication of relevance—the reporter is more likely to place highly relevant events earlier in the article, to catch the reader's attention. In general, each feature is a *weak* indicator of higher (or lower) relevance score. For a detailed examination of different types of discourse features is found in [8].

Lexical features for an event consists of bags of words. Only the "principal" parts of speech were used in our experiments (viz. nouns, verbs and adjectives). The words were brought to a normalized form via lemmatization. The lexical features for a given event consist of the lemmas found within the trigger sentence, as well as in a window (of a fixed size) of sentences immediately preceding and following the trigger sentence. The surrounding sentences provide additional context for disambiguation to help classify relevance more accurately.

6.2 Experiments

Our training data is summarized in table 3. The experiments reported in this section focus specifically on cross-border security events. We evaluated separately the relevance classification for "first" vs. "all" events within a document. The reason behind evaluating relevance assignment to *first events* separately is twofold. On one hand, this allows us to approximate classification at the *document* level—assuming that the first event in a document is in some sense the most "representative" or important. On the other hand, this type of classification may be a simpler problem to tackle, i.e., the system may achieve higher performance on the scaled-down task, while still providing useful information to the user.

For each classifier, we show the performance using discourse features only, lexical features only, and the combined set of features. Prior to training the classifiers, we employ standard methods for feature selection. We experimented with feature selection by means of information gain, correlation-based feature subset evaluation [7], and wrapper for AUC optimization [13]. In the table, we report results using information gain, which performed well under most conditions.

Table 3 The amount of labeled training data used in the relevance-classification experiments

	High relevance	*Low relevance*	*Zero relevance*	*Total*
First events only (same as the number of documents)	196	428	94	718
All events	319	662	124	1105

Table 4 Relevance classification results on security domain. Accuracy (and F-measure, in parentheses) for lexical features, discourse features, and combined features.

		All events			*First events only*		
Relevance	**Classifier**	**Lexical**	**Discourse**	**Combined**	**Lexical**	**Discourse**	**Combined**
High-low	SVM	82.2 (.537)	**85.1** (.618)	84.2 (.613)	87.2 (.625)	88.5 (.664)	**89.6** (.71)
	Naive Bayes	79.7 (.64)	80.7 (.598)	84.6 (**.702**)	85.8 (.679)	85.0 (.639)	89.2 (**.728**)
	Bayes Net	80.6 (.558)	79.1 (.615)	79.5 (.64)	82.6 (.529)	82.0 (.612)	82.5 (.619)
Zero-rest	SVM	83.9 (.907)	84.8 (.913)	**85.9** (.917)	80.6 (.888)	81.6 (.895)	83.0 (.897)
	Naive Bayes	85.3 (.915)	84.1 (.908)	85.7 (**.918**)	82.7 (.898)	82.5 (.895)	**83.8** (**.902**)
	Bayes Net	82.4 (.903)	81.7 (.891)	82.1 (.893)	78.3 (.876)	78.8 (.868)	78.2 (.864)

The results in Table 4 are from random 10-fold cross-validation on the labeled data set. The bold score in each column indicates the best classifier score achieved for the corresponding conditions. Naive Bayes and SVM seem to perform about equally well under most conditions. The table also contrasts two kinds of classification experiments, each one being (for simplicity) *binary* classification. In the "high-vs-low" experiments, the classifier only determines whether the given event receives high relevance (i.e., 4 or 5) or low (i.e., 3 or less). In the "zero-vs-rest" experiments, the classifier only decides whether the event is completely useless (class 0). We expect that class zero events are of lowest utility, and should be presented to the user only after all others have been presented (if at all).

In general, using the given feature set, the binary classification currently yield better performance than regression, which is likely due to the fact that the boundary between any two adjacent scores (e.g., 3 vs. 4, or 0 vs. 1) is rather vague—it is difficult to capture precisely in data annotation guidelines. From the current results, for the present it appears that zero-vs-rest classification is, in general, an easier problem to tackle (at least in terms of F-measure). We should note, however, that performance could likely be improved by increasing the number of features and the amount of training data.

7 Conclusions

In this chapter we have presented some IE techniques to improve the performance of a Multilingual Event Extraction Framework deployed in real-world operational environment that is capable of extracting detailed structured information on a wide range of security-related events from online news in 8 languages.

First, we presented an inference mechanism for event type classification (already deployed in the online version of the Event Extraction Framework). It is based on production rules, domain-specific knowledge and semi-automatically retrieved lexicon. The results of the preliminary evaluation for Italian and Turkish language carried out in the domain of border security turned to be encouraging. We plan to carry out a more detailed evaluation for other languages and we also intend to compare the presented approach to the standard text classification techniques, such as SVM.

Secondly, we described an approach to event geo-tagging based on the utilisation of lexico-semantic patterns. The goal of the experiment was rather a proof of concept that certain semantic features of the Locative Prepositional Phrases could help filtering some of the candidate event locations. Therefore, we manually designed large parts of the system, aiming mainly at the correctness of the analysis. Nonetheless, we believe that most of the lexical and grammatical knowledge we encoded can be acquired semi-automatically via Machine Learning techniques, considering also that the prediction rules are largely valid across languages. Although the task itself turned out to be challenging, the results of a preliminary evaluation are relatively promising. We plan to perform an extended evaluation of the method for the Spanish language, and then to experiment on adapting the lexicons and grammars to other languages.

Next, we presented the results of deploying a simple a cross-lingual information fusion technique to improve the recall/precision of the system. Circa 10% of the events extracted by the mono-lingual systems could be refined, and a gain in overall recall and precision of circa 5-6% could be obtained through fusing the responses of the mono-lingual systems. Since the time window for grouping event descriptions referring to a given event was limited to 1 day only the aforementioned figure of 10% constitutes an approximation of a lower bound for the fraction of events, whose descriptions can be potentially refined. Experiments aiming to explore going beyond the 1-day time window are envisaged. Future work in this area will also focus on: carrying out a direct one-to-one comparison between the English system (the one with the highest impact) and each of the mono-lingual systems in order to get a better insight into the real contribution of exploiting news in each language, in-depth analysis of the alternative approach based on translation of all news articles into one common language for which a high-performance event extraction system exists, exploring more elaborated fusion techniques [10] and techniques for automatically grouping event descriptions across various languages.

Finally, we presented a technique for automatically scoring the relevance of extracted facts for the end-user. The classification experiments demonstrate that, with reasonable accuracy, the system can help focus the end-users attention on those events and documents which are more likely to be of higher value to the crisis-management task. In summary, the results in this chapter demonstrate how the presented techniques complement each other in the overall task of extracting events in the target domain, and indicate several promising directions for future research and further improvements.

References

1. Appelt, D.: Introduction to Information Extraction Technology. Tutorial Held at IJCAI 1999 (1999)
2. Ashish, N., Appelt, D., Freitag, D., Zelenko, D.: In: Proceedings of the Workshop on Event Extraction and Synthesis. Held in conjunction with the AAAI 2006 (2006)
3. Atkinson, M., van der Goot, E.: Near Real Time Information Mining in Multilingual News. In: Proceedings of WWW 2009 (2009)

4. Atkinson, M., Piskorski, J., Van der Goot, E., Yangarber, R.: Multilingual Real-Time Event Extraction for Border Security Intelligence Gathering. In: Counterterrorism and Open Source Intelligence Series. Lecture Notes in Social Networks, vol. 2 (2011)
5. Chen, Z., Ji, H.: Can one Language Bootstrap the Other: A Case Study on Event Extraction. In: Proceedings of the NAACL HLT 2009 Workshop on Semi-Supervised Learning for Natural Language Processing (2009)
6. Downey, D., Etzioni, O., Soderland, S.: A Probabilistic Model of Redundancy in Information Extraction. In: Proceedings of IJCAI 2005 (2005)
7. Hall, M.A.: Correlation-based Feature Selection for Discrete and Numeric Class Machine Learning. In: Proceedings of ICML (2000)
8. Huttunen, S., Vihavainen, A., von Etter, P., Yangarber, R.: Relevance Prediction in Information Extraction Using Discourse and Lexical Features. In: Proceedings of the 18th Nordic Conference on Computational Linguistics, NODALIDA (2011)
9. Grishman, R., Huttunen, S., Yangarber, R.: Real-time Event Extraction for Infectious Disease Outbreaks. In: Proceedings of HLT 2002 (2002)
10. Ji, H., Grishman, R.: Refining Event Extraction through Cross-Document Inference. In: Proceedings of ACL 2008, pp. 254–262 (2008)
11. Ji, H.: Challenges from Information Extraction to Information Fusion. In: Proceedings of ACL 2008, pp. 507–515 (2010)
12. King, G., Lowe, W.: An Automated Information Extraction Tool For International Conflict Data with Performance as Good as Human Coders. In: International Organization, vol. 57 (2003)
13. Kohavi, R., John, G.H.: Wrappers for Feature Subset Selection. In: Artificial Intelligence, vol. 57(1) (1997)
14. Lee, A., Passantino, M., Ji, H., Qi, G., Huang, T.: Enhancing Multi-lingual Information Extraction via Cross-Media Inference and Fusion. In: Proceedings of COLING 2010: Posters, pp. 630–638 (2010)
15. Li, J., Li, J., Tang, J.: A Flexible Topic-driven Framework for News Exploration. In: Proceedings of KDD 2007 (2007)
16. Liao, S., Grishman, R.: Using Document Level Cross-Event Inference to Improve Event Extraction. In: Proceedings of ACL 2010, pp. 789–797 (2010)
17. Mikheev, A., Moens, M., Grover, C.: Named Entity Recognition without Gazetteers. In: Proceedings of EACL 1999 (1999)
18. Naughton, M., Kushmerick, N., Carthy, J.: Event Extraction from Heterogeneous News Sources. In: Proceedings of the AAAI 2006 Workshop on Event Extraction and Synthesis (2006)
19. Patwardhan, S., Riloff, E.: Effective Information Extraction with Semantic Affinity Patterns and Relevant Regions. In: Proceedings of EMNLP-CONLL 2007 (2007)
20. Piskorski, J.: ExPRESS: Extraction Pattern Recognition Engine and Specification Suite. In: Proceedings of the International Workshop Finite-State Methods and Natural Language Processing (2007)
21. Piskorski, J., Tanev, H., Atkinson, M., van der Goot, E., Zavarella, V.: Online News Event Extraction for Global Crisis Surveillance. In: Nguyen, N.T. (ed.) Transactions on CCI V. LNCS, vol. 6910, pp. 182–212. Springer, Heidelberg (2011)
22. Pouliquen, B., Kimler, M., Steinberger, R., Ignat, C., Oellinger, T., Blackler, K., Fluart, F., Zaghouani, W., Widiger, A., Forslund, A.-C., Best, C.: Geocoding Multilingual Texts: Recognition, Disambiguation and Visualisation. In: Proceedings of LREC 2006, Genoa, Italy, pp. 24–26 (2006)

23. Snover, M., Li, X., Lin, W.-P., Chen, Z., Tamang, S., Ge, M., Lee, A., Li, Q., Li, H., Anzaroot, S., Ji, H.: Cross-lingual Slot Filling from Comparable Corpora. In: Proceedings of the 4th Workshop on Building and Using Comparable Corpora: Comparable Corpora and the Web, pp. 110–119 (2011)
24. Sudo, K., Sekine, S., Grishman, R.: Cross-lingual Information Extraction System Evaluation. In: Proceedings of COLING 2004 (2004)
25. Tanev, H., Piskorski, J., Atkinson, M.: Real-Time News Event Extraction for Global Crisis Monitoring. In: Proceedings of NLDB 2008 (2008)
26. Tanev, H., Zavarella, V., Linge, J., Kabadjov, M., Piskorski, J., Atkinson, M., Steinberger, R.: Exploiting Machine Learning Techniques to Build an Event Extraction System for Portuguese and Spanish. Linguamatica (NLP Journal for Iberian Languages) 2 (2009)
27. Thorsten, J.: Text Categorization with Support Vector Machines: Learning with Many Relevant Features. In: Nédellec, C., Rouveirol, C. (eds.) ECML 1998. LNCS, vol. 1398, Springer, Heidelberg (1998)
28. Tyler, A.R. (ed.): Expert Systems Research Trends. Nova Science Publishers, New York (2007)
29. Yangarber, R., Jokipii, L., Rauramo, A., Huttunen, S.: Extracting Information about Outbreaks of Infectious Epidemics. In: Proceedings of the HLT-EMNLP 2005 (2005)
30. Yangarber, R.: Verification of Facts across Document Boundaries. In: Proceedings of International Workshop on Intelligent Information Access (2006)
31. Zhang, N.N.: Movement within a Spatial Phrase. In: Cuyckens, H., Radden, G. (eds.) Perspectives on Prepositions. Linguistische Arbeiten. Band, vol. 454, pp. 47–63. Max Niemeyer, Tübingen (2002)

Automatic Metadata Generation in an Archaeological Digital Library: Semantic Annotation of Grey Literature

Andreas Vlachidis, Ceri Binding, Keith May, and Douglas Tudhope

Abstract. This paper discusses the automatic generation of rich metadata from excavation reports from the Archaeological Data Service library of grey literature (OASIS). The work is part of the STAR project, in collaboration with English Heritage. An extension of the CIDOC CRM ontology for the archaeological domain acts as a core ontology. Rich metadata is automatically extracted from grey literature, directed by the CRM, via a three phase process of semantic enrichment employing the GATE toolkit augmented with bespoke rules and knowledge resources. The paper demonstrates the potential of combining knowledge based resources (ontologies and thesauri) in information extraction, and techniques for delivering the automatically extracted metadata as XML annotations coupled with the grey literature reports and as RDF graphs decoupled from content. Examples from two consuming applications are discussed, the Andronikos web portal which serves the annotated XML files for visual inspection and the STAR project, research demonstrator which offers unified search across of archaeological excavation data and grey literature via the core ontology CRM-EH.

Keywords: Automatic Metadata Generation, CIDOC CRM, Digital Archaeology, Digital Library, GATE, Knowledge Organization Systems, Information Extraction, Semantic Annotation, Semantic Search, SKOS.

1 Introduction

In archaeology today, we see digital libraries of grey literature reports and of excavation datasets but they are not meaningfully connected. This paper discusses

Andreas Vlachidis · Ceri Binding · Douglas Tudhope
Hypermedia Research Unit, University of Glamorgan, UK
e-mail: avlachid@glam.ac.uk, cbinding@glam.ac.uk, dstudhope@glam.ac.uk

Keith May
English Heritage, UK
e-mail: keith.may@english-heritage.org.uk

A. Przepiórkowski et al. (Eds.): *Computational Linguistics*, SCI 458, pp. 187–202, 2013.
DOI: 10.1007/978-3-642-34399-5_10

the automatic generation of metadata that makes possible semantic search of grey literature connected with diverse archaeological datasets. Such metadata enjoy definitions that enable information retrieval and cross searching on the semantic level via ontological and terminological models and references. Thus they are described here as "rich metadata".

The work forms part of the broader Semantic Technologies for Archaeological Resources (STAR) project, which employs the CIDOC CRM (International Council of Museums Conceptual Reference Model) and its archaeological extension, CRM-EH, as a core ontology providing a contextual framework between different types of information sources and disparate datasets [15, 18]. In collaboration with English Heritage (EH), the STAR project has developed methods for linking digital archive databases, vocabularies and unpublished excavation reports for purposes of semantic cross search. The CRM-EH is necessary for expressing the semantics and complexities of relationships between the data elements and annotations, which underline semantically defined archaeological user queries. The project has also employed knowledge resources, such as EH domain glossaries and thesauri, expressed as SKOS vocabularies [12]. These knowledge resources assist semantically defined queries and NLP information extraction from excavation reports.

An extract from the OASIS (Online AccesS to the Index of archaeological investigations) grey literature library, provided by the Archaeological Data Service, forms the STAR free text corpus. The term *grey literature* is used to describe documents and source materials that cannot be found through the conventional means of publication - many excavation reports exist in this format. The OASIS project is a joint effort of UK archaeology research groups, institutions, and organisations, coordinated by the University of York [13]. It aims to improve the communication of fieldwork results to the wider archaeological community.

This paper reports on an investigation of automatic methods for generating rich metadata that connects concepts via the ontological arrangements of the CIDOC CRM and CRM-EH and via the terminological SKOS definition of adopted vocabulary (English Heritage thesauri).

The CIDOC CRM is an international standard (ISO21127:2006) semantic framework, aiming to promote shared understanding of cultural heritage information [8. 10]. The CRM is capable of mapping any type of cultural heritage information, as published by museums, libraries and archives [2]. Extensibility is an important aspect; the CRM can be specialized when it is required by a domain. A finer granularity of detail can be expressed for domain purposes while still retaining interoperability at the core CRM level.

A particular extension of CRM that addresses the needs for semantic interoperability in the archaeology domain is the English Heritage (EH) extension. Based on the archaeological notion of context, modeled as a subclass of CRM Place, the CRM-EH ontology describes entities and relationships relating to a series of archaeological events, both in the past and present (ie during excavation) [7].

Previously, the role of the CRM for rich semantic annotations of text documents has been explored via intellectual, non-automatic process, which has the potential for producing very fine grained annotation of specific, important documents, for example as part of the detailed Text Encoding Initiative markup [14]. Inevitably, this process will be resource intensive over a large corpus. Thus, automation of semantic annotation is highly desirable for advancing semantic interoperability over larger corpus.

The rest of this paper discusses the efforts of an automatic semantic annotation and rich metadata delivery of archaeological grey literature documents with respect to standard ontological and terminological definitions. The metadata extraction consists of a three phase process of semantic enrichment, employing the General Architecture for Text Engineering (GATE) toolkit [9]. The initial phase pre-processes the grey literature and vocabulary resources. The second phase identifies archaeological domain concepts in context. The final phase transforms GATE annotations to two different representations for consuming applications. The first representation is as XML coupled with the grey literature report – this permits colour coded inline viewing of the annotations and also semantic metadata ranked by frequency of occurrence. The second representation is as RDF triples for semantic applications. In both cases, the extracted metadata are expressed as CRM entities, qualified by SKOS archaeological vocabulary concepts. Two web applications use the resultant semantic annotations. The Andronikos web portal delivers the annotated XML files as hypertext documents for visual inspection of the information extraction results. The STAR research demonstrator offers a unified searching of both data and grey literature in terms of the core ontology. Examples of the semantically enriched grey literature from both applications are discussed in this paper, together with results from a preliminary evaluation exercise.

2 The Process of Semantic Enrichment

Information Extraction (IE) is a particular NLP technique relevant to the semantic enrichment of free text documents by extracting specific information snippets suitable for further manipulation [16, 6]. These semantic annotations in context enrich documents, enabling access on the basis of a conceptual structure, providing smooth traversal between unstructured text and conceptual models [5]. In addition, they can aid the integration of heterogeneous data sources by exploiting a conceptual structure and allowing users to search across resources for entities and relations, instead of words. Users can search for the term *'Paris'* and a semantic annotation mechanism can relate the term with the abstract concept of *'city'*, while providing a link to the term *'France'*, which relates to the abstract concept *'country'*. Employing a different conceptual model, the same term *'Paris'* can be related with the concept of *'mythical hero'* linked with the city of *'Troy'* from Homer's epic poem, the Iliad.

In this paper the discussion of the process of Semantic Enrichment is divided into the three broad phases of the IE pipeline, each subdivided into various sub-tasks (pipelines). The initial pre-processing phase prepares the OASIS corpus and knowledge resources. The section on the second main IE phase highlights some details of the pipeline used for the annotation of the textual resources with conceptual and terminological references. The last section discusses the techniques employed in the final phase that constructs RDF representations of the metadata for the semantically enriched documents. The discussion begins by introducing the underlying Language Engineering architecture and Knowledge Organization System resources that contribute to the process of semantic enrichment.

2.1 Underlying Knowledge Organization System Resources

Gazetteers are sets of lists, sometimes containing the names of entities such as cities, day of the week, etc. In GATE, gazetteer listings are used to find occurrences of terms in free text, and often support named entity recognition tasks. GATE gazetteers are not flat, which means that list entries can enjoy attributes which in turn can be invoked by Java Annotation Pattern Engine (JAPE) rules for the construction of sophisticated patterns. For example, a gazetteer containing the names of European cities can be enhanced with an attribute denoting the country of origin for each European city enlisted in the gazetteer resource. Therefore, a JAPE rule at later stages can exploit that particular attribute for targeting term matching only at UK cities or only at UK and Greek cities. In the semantic enrichment work described in this paper, gazetteers proved useful, being used to accommodate the wide range of Knowledge Organization System (KOS) resources made available to the STAR project by English Heritage.

One type of KOS used in the semantic enrichment process was the thesaurus, with its various semantic relationships. In STAR, the various EH thesauri and glossaries were represented in SKOS, allowing unique identifiers (URIs) for a concept and possible links between concepts (skos:exactMatch, skos:closerMatch) [12]. Four English Heritage thesauri and five glossary resources are incorporated in the process of semantic enrichment for identifying occurrences of various conceptual entities. The thesauri are the Archaeological Objects, Monument Types, Building Materials and the experimental Time-line Thesaurus [11]. The glossary resources are the Simple Names for Deposits and Cuts, Find Type Index, Material Index, Small Finds and the Bulk Find Material glossary. All the KOS resources had been previously expressed in SKOS format for the purposes of the STAR project [3, 4].

The glossary resources contain a small set of concepts which are highly relevant to the domain of archaeology. On the other hand, the thesauri resources contain a large set of concepts which relate to the general cultural and heritage domain. Initial entity recognition experiments exploited only the available glossary and Time-line Thesaurus vocabulary. This limited the semantic enrichment to a small set of highly relevant concepts in the context of archaeology. Therefore, new GATE techniques needed to be developed in order to allow the possibility of selectively exploiting the wider context of the EH thesauri. This makes possible richer terminology availability for the semantic enrichment process.

Exploiting the whole range of thesaurus concepts would expand the process of enrichment to concepts that are not very relevant to the archaeology domain. Therefore an optimum range is needed. A solution is to use concepts that come from those areas of thesaurus structures potentially useful to the enrichment process. Overlapping concepts between the (highly relevant) glossaries and thesauri can serve as appropriate entry points to the thesaurus structures. Thesaurus semantic relationships can then be exploited for expanding from the entry points across thesauri areas relevant to the task of archaeological semantic enrichment. In order to enable this semantic expansion within GATE, JAPE rules must be able to exploit thesaurus narrower and broader semantic relationships. Therefore, transformation of SKOS thesauri to GATE gazetteers allows the translation of thesauri properties to gazetteer attributes, enabling JAPE rules to exploit semantic relationships between gazetteer terms, as described in the following section.

2.2 *Transforming KOS to GATE Gazetteers*

The transformation of SKOS thesauri and glossary resources to GATE gazetteers is achieved via the use of XSLT templates. The templates exploit SKOS properties for adding attributes to gazetteer terms, which can be used by JAPE rules. It is important that the rules are capable of traversing through a thesaurus hierarchy, in order to produce matches that achieve a semantic expansion, which can expand beyond the limits of a single semantic level. Since JAPE is essentially a pattern matching rule engine, some work was required to enable the semantic expansion of SKOS concepts (with their unique identifiers) within GATE at the lookup stage of the information extraction pipeline.

Consider the following case; *Container (by function) >Food and Drink Serving Container > Drink Serving Container > Jug > Knight Jug*. A JAPE rule that used only the narrower concept property would be able to semantically expand only on immediately narrower concepts. Therefore, a gazetteer attribute was built during the transformation from SKOS to GATE gazetteer to reflect the path of unique SKOS identifiers from the concept to the top of its hierarchy. For example:

```
KnightJug@skosConcept=149773@path=/101601/101204/1013
40/101023
Jug@skosConcept=101601@path=/101204/101340/101023
Drink Serving
Container@skosConcept=101204@path=/101340/101023
Food and Drink Serving
Container@skosConcept=101340@path=/101023
Container@skosConcept=101023@path=/101023
```

A simple JAPE rule can then exploit the above gazetteer attributes by matching all terms that contain a skosConcept and a path attribute of a particular reference; for example 101023 matches all concepts in the gazetteer within the container hierarchy. The XSLT transformation also takes into account SKOS alternative concept labels (thesaurus non-preferred terms) and makes them available as gazetteer entries that have the same skosConcept and path attributes as their preferred label counterparts.

In addition during the transformation, particular glossary concepts are given an extra attribute (skos:exactMatch) to accommodate the previously defined mapping between a glossary and a thesaurus. For example the concept Hearth of the glossary Simple Names for Deposits and Cuts is mapped to the concept Hearth of the thesaurus Monument Types class Archaeological Feature:

```
hearth@skosConcept=ehg003.37@skosExactMatch=70374
```

This allows the potential for JAPE rules to optionally expand the glossary concept Hearth to associated concepts within the Monuments thesaurus, depending on the overall context.

2.3 Main Knowledge-Based Information Extraction Phase

The main IE process annotates grey literature documents with conceptual and terminological references. Thus process is carried out by the OPTIMA pipeline (figure 1) developed for the project (Object, Place, TIme and Material). These four concepts were considered key metadata elements for the purposes of the project's concern with archaeological excavations and are the focus of the pipeline. The IE process uses a large number of JAPE patterns and utilises the gazetteer resources created by the pre-processing phase. We term this particular approach, Knowledge Based Information Extraction, since the combination of core ontology and knowledge resources plays a major role in the process of semantic enrichment. This drives the IE task by using JAPE patterns which exploit and produce semantic relationships. This section gives an overview of the main functionality of the pipeline.

The first stage of the main IE pipeline invokes the gazetteer resources and generates the initial lookup annotations. The various knowledge resources that participate in the pipeline are capable of identifying lookups that fall within the

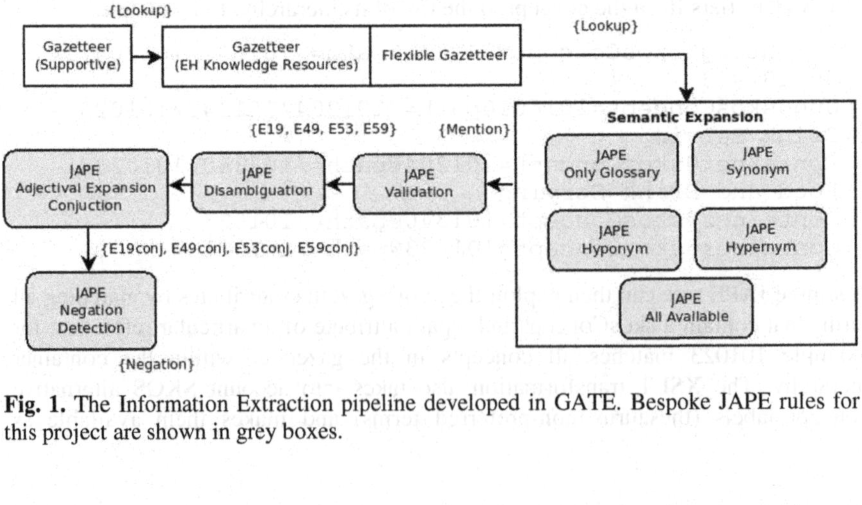

Fig. 1. The Information Extraction pipeline developed in GATE. Bespoke JAPE rules for this project are shown in grey boxes.

four concept categories mentioned above. For example, the Archaeological Objects thesaurus contains concepts of physical objects, the Building Materials thesaurus contains concepts of materials, the Timeline thesaurus contains time appellations, such as periods, while glossaries are a source of more specific archaeological terminology, such as the Simple Names for Deposits and Cuts, which covers archaeological contexts. There are cases where a term can be found within two different knowledge resources, thus potentially having two different conceptual references (ie multiple senses). For example, the term brick can refer to a material or to a physical object, as can terms such as, glass, stone, iron, gold, etc. in the particular domain practice of archaeology. Therefore the first stage of the pipeline generates two types of lookup annotation, single sense and multiple sense. Multiple sense lookup annotations are disambiguated at later stages.

The second stage of the pipeline validates the lookup annotations and aligns annotations to CRM entities. Annotations that are not part of noun phrases and annotations that are part of headings, table of contents and single worded phrases are suppressed for purposes of the current project. It is important to validate lookup annotations, especially those that are not part of noun phrases, because gazetteer matches are invoked via a morphological analyser and matches are created on the root of words. This technique allows matches within a broader orthographical context, including singular and plural forms of matches, but also generates matches for verb senses that have to be suppressed by the validation stage. In addition, during this stage the pipeline performs negation detection over lookup annotations. The next stage of the pipeline disambiguates multiple sense lookup annotations. The disambiguation technique is based on JAPE patterns that examine word pairs and Part-of-Speech input. Lookup annotations that cannot be disambiguated maintain all their possible senses, leaving the judgement to the end user.

The pipeline is capable of exploiting the semantic relationships of the knowledge resources that are accommodated as gazetteer resources. The pipeline can be configured in one of three different modes of semantic expansion developed for purposes of the research within GATE, *Synonym*, *Hyponym*, and *Hypernym* expansion. *Synonym* expansion utilises the glossary resources and expands on the synonyms of glossary terms available in the thesauri resources. *Hyponym* is similar to the *Synonym* expansion in utilising the glossary terms and their synonyms but also traverses over the hierarchy of the thesaurus structures to include (transitively) all narrower concepts available for the glossary. The *Hypernym* expansion mode includes the above two modes of expansion and also exploits the broader concept relationships within the thesauri structures. In terms of volume, the last mode of expansion will expand the semantic enrichment task to include the largest set of concepts, while the first mode of expansion will include the smallest and the most precise set of concepts. For example, for the glossary concept enclosure, *Synonym* expansion will include the concept, *garth*, in addition to enclosure. *Hyponym* expansion will also include the concepts *curvilinear enclosure, ditched enclosure, rectilinear enclosure,* etc. and their narrower concepts, such as *oval enclosure, double ditched enclosure, polygonal enclosure*, etc.; *Hypernym* expansion will also include all Monument by Form concepts, such as *arch, boundary, barrier, ditch* and their narrower concepts, such as *fence, hedge boundary ditch*, etc. Clearly, there are recall/precision trade-offs associated with the different expansion modes and these are a topic for investigation in the forthcoming evaluation work.

2.4 Transformation of Semantic Annotations to RDF Triples

The last phase of the semantic enrichment is the transformation of the semantic annotations produced in the previous phase to RDF triples. For this purpose, CRM-EH semantic annotations are exported initially from the GATE environment as XML documents. The GATE exporter produces annotations in the form of XML tags, which are coupled with the associated content. The exported XML tags enjoy properties produced during the IE process, such as skos:Concept and skos:exactMatch unique terminological identifiers, gateID unique annotation identifiers and a note for capturing the context surrounding a semantic annotation. In addition, each grey literature document has a unique name that constitutes a unique identification of the file within the OASIS corpus. This unique file name is used in conjunction with the unique gateID property of each annotation to create a corpus wide unique identifier for each individual annotation. In addition, the SKOS concepts assigned to the annotations are associated with underlying CRM Types, using the relationship (*is_represented_by*) modeling the association asserted for data items (mapped to CRM) and SKOS concepts [18]. For example a semantic annotation of the CRM-EH class, *Context*, can be associated with the SKOS concept, *pit*. This supports cross search between data and grey literature in terms of CRM and SKOS.

The transformation from XML files to RDF triples required the development of bespoke techniques based on the XML Document Object Model, using the scripting language PHP for building the transformation templates. The decision to use a server-side scripting language like PHP is supported by the two main requirements, a MySQL database for retrieving the unique file name of each document and a visual interface for parameterisation of the transformations, allowing easy selection of semantic annotations that participate in the transformation. PHP allowed rapid development and proved robust with a large set of documents. The final RDF documents are decoupled from the content. However, as explained above each annotation resource is tied to a corpus wide unique identifier.

The following example presents the details of the RDF transformation. Consider the following rich phrase *'pits were uniformly filled with large quantities of pottery'*. The phrase can be modelled by a CRM-EH ContextFindDepositionEvent, connecting the find, *pottery*, with the context, *pit*. The coupled XML output would have the following structure (the note property is omitted and URIs truncated for simplicity):

```
<EHE1004.ContextFindDepositionEvent
gate:gateId="281105">
 <EHE0007.Context gate:gateId="281155"
skos:Concept="#ehg003.55">pits</EHE0007.Context>
```

```
were uniformly filled with large quantities of
 <EHE0009.ContextFind gate:gateId="281158"
skos="#ehg027.2">pottery</EHE0009.ContextFind>
</EHE1004.ContextFindDepositionEvent>
```

The RDF transformation would have the following structure:

```
<crmeh:EHE1004.ContextFindDepositionEvent
rdf:about="http://base#suffl-6115.281105">
<dc:source rdf:resource="http://base#suffolkc1-6115"
/>
<dc:source rdf:resource="http://base#ehe0001.oasis"
/>
<crm:P2F.has_type rdf:resource="http://base#suffl-
6115.281156" />
 <crm:P3F.has_note>
  <crm:E62.String>
   <rdf:value>pits were uniformly filled with large
quantities of pottery</rdf:value>
  </crm:E62.String>
 </crm:P3F.has_note>
<crm:P26F.moved_to rdf:resource="http://base#suffl-
6115.281155" />
<crm:P25F.moved rdf:resource="http:// base#suffl-
6115.281158" />
</crmeh:EHE1004.ContextFindDepositionEvent>
```

3 Example Uses of Rich Metadata

The automatically produced metadata are utilised by two web applications, the STAR research demonstrator and the Andronikos portal [17, 1]. The STAR demonstrator uses the decoupled RDF files to support cross searching between grey literature documents and disparate datasets [3], in terms of the core CRM-EH conceptual model. A SPARQL engine supports the semantic search capabilities of the demonstrator, while an interactive interface hides the underlying model complexity and offers search (and browsing) for Samples, Finds, Contexts or interpretive Groups with their properties and relationships. On the other hand, the Andronikos portal uses the coupled XML files for constructing and delivering the semantic annotations in an easy to follow human readable format. While the portal was developed for project purposes to assist visual inspection of the information extraction outcomes, it is seen as indicative of potential digital library applications where access to the semantically enriched text is desired.

The metadata take account of lexical ambiguities such as polysemy (same word having multiple meanings). For example, find all archaeological Contexts of type *'cut'*, where the term *'cut'* is ambiguous. The semantic enrichment mechanism manages to disambiguate the verb from the noun form and to reveal phrases which make use of *'cut'* as relevant to archaeological use, eg *'levelling layers sealed the base of a brick wall cut into layer'*, or *'It measured 0.3m in diameter and 0.2m deep with a circular cut'* and avoids the annotation of non archaeological uses of *cut*, such as *'although the current 'cut-off' channel is now 500m'*. In addition, the metadata can take account of a form of polysemous ambiguity. For example, the word *'brick'* in an archaeology usage can refer to a material or to a physical object. In the phrase *'yellowish-brown sandy deposit containing frequent unbonded brick'*, the term refers to a physical object, whereas in the phrase *'A layer of small brick tiles forming the street paving'*, the term refers to a material.

The STAR Demonstrator makes use of the rich metadata for semantic search, building on CRM and SKOS unique identifers. For example, searches are possible of the form: Context of type X containing Find of type Y. The two different extracts of screendumps in Figure 2 show a Context of type *'hearth'* containing Context Find of type *'coin'*, together with a Context Find of type *'Animal Remains'* within a Context of type *'pit'*.

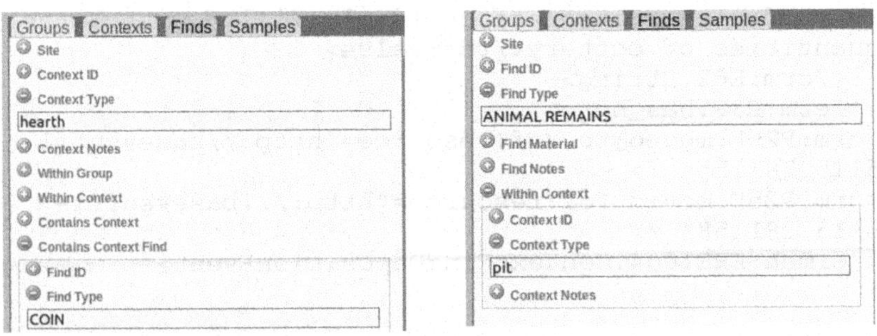

Fig. 2 The STAR demonstrator search of semantic metadata

Fig. 3 Search Results from the STAR demonstrator

The cross-search capability of the STAR demonstrator retrieves results from both datasets and grey literature reports (a variety of datasets for the *hearth* query). As seen above, the searches returned results for different annotation types (Contexts, Finds) and from different resources (grey literature resources commence with a hash bar in this demonstrator interface). For example (Figure 3), the first search retrieves from Grey Literature #archaeol8-6428.134861 a *Hearth* containing a *coin;* the original text was *'It differs from the other coin finds, however, in that it was associated with a hearth'*. Similarly the second search retrieves from Grey Literature an *animal bone* within a context of type *pit;* the original text was *'the test pit produced a range of artefactual material which included animal bone (medium/large ungulate)'*. The semantic enrichment makes it possible for the STAR demonstrator to overcome lexical boundaries and retrieve synonymous terms, as evident in the example of *'Animal Remains'* where the term *'Animal bone'* is retrieved.

The Andronikos web portal uses XML outputs of rich metadata for generating and linking HTML pages, which accommodate semantic annotations of grey literature documents. The annotations are divided into three abstractions: (i) preprocessing annotations, such as Headings, Table of Contents and Summary; (ii) single CRM annotations such as Physical Object, Place, Time and Material; (iii) CRM-EH archaeology specific annotations of rich phrases. Therefore, it is possible to optionally expose particular document abstractions according to different application strategies. Thus, in certain cases, the Summary sections (an example is given in Figure 4) might be targeted (or prioritised) for retrieval as being strongly representative of a grey literature report. Alternatively, the most frequently appearing CRM entities (see Figure 5) in a report might be considered a useful entry strategy. Yet again, a cross search might be interested in any occurrences within grey literature reports of highly specific, rich CRM annotations. Andronikos also makes available links to the XML and RDF versions of grey literature documents, which can be downloaded and further transformed or manipulated.

Annotated Document: archaeol1-19366_1.xml

Summary

SUMMARY An archaeological excavation was undertaken at land off Norwich Road, Caister-on-Sea, Norfolk in advance of the construction of a new foodstore. The site lies 200m to the southeast of a Roman shore fort that was established in the early 3rd century. The excavation followed earlier desk-based and trial trenching evaluations. These had demonstrated the high archaeological potential of the site, with ditches and pits of 2nd to 4th century Roman date present across the whole of the development area. The earliest archaeological evidence

Fig. 4 Summary Section of Grey literature, Andronikos web-portal

TERM	SKOS	Count	TERM	SKOS	Count
anglian	#136306	14	fired clay	#ehg027.5	40
medieval	#134745	19	plate	#96797	53
20th century	#134841	19	artefacts	#ehg020.7	55
prehistoric	#134718	46	tile	#ehg027.3	57
roman	#134738	216	pottery	#ehg027.2	81

Fig. 5 Frequent CRM entities (Time Appellation – left side, Physical Object right side) in a grey literature report, Andronikos web-portal

The summary section of the report (Figure 4) discusses evidence that relates to the Roman period, while the CRM overview (Figure 5) show the most frequently used SKOS concepts in the report for the CRM entities, Time Appellation and Physical Object, in this case *Roman* and *pottery*. Both CRM and SKOS annotations are produced by the IE pipeline.

4 Evaluation

This section discusses the results of a preliminary evaluation exercise which aims to inform an ongoing larger scale 'gold standard' evaluation. The main objective of the preliminary evaluation is to examine the effectiveness of the manual annotation instructions for supporting the annotation task, as well as to explore the capacity of GATE modules for supporting the evaluation and the inter-annotator agreement (IAA) analysis tasks. IAA analysis is performed in order to reveal differences between annotators and to refine the instructions of manual annotation before proceeding to the main evaluation phase.

The evaluation used a small corpus of 10 summary extracts. Considering the logistic constraints of the manual annotation task, summary extracts, originating from archaeological excavation and evaluation reports were considered to be the best available option. They are potentially useful targets of semantic enrichment work, as discussed in the previous section..

The manual annotation instructions were written to reflect the end-user aims of the evaluation, hiding complex and unnecessary ontological details. Annotators were instructed to annotate at the level of archaeological concepts rather than identifying more abstract ontological entities. The instruction directed the task of manual annotation at the concepts of archaeological place, archaeological finds, the material of archaeological finds and time appellations. Three annotators were employed by the pilot evaluation task from the target archaeology domain, a senior archaeologist (SA), a commercial archaeologist (CA) and a research student (RS) of archaeology. The evaluation corpus of the ten summary extract (5 excavation and 5 evaluation reports) containing in total 2898 words, was made available in MS Word format. The annotators were instructed to use particular highlight colours and underline tools in order to produce their annotations. Although, the annotation task could have applied the GATE Ontology Annotation Tool (OAT), the use of Microsoft Word was preferred, since manual annotators had no previous experience working with GATE but were fluent in Word. The resulting manual annotations sets were systematically transferred to GATE as CRM and CRM-EH oriented annotations using the Ontology Annotation Tool (OAT).

Upon completion of annotation transfer, the differences between the individual manual annotation sets were analysed using the IAA module of GATE. The module was configured to report the inter-annotator agreement score in terms of Precision, Recall and F-Measure metrics. The metrics were reported on both Average and Lenient mode. The Average mode treats partial matches as half matches whereas the Lenient mode treats partial matches as full matches. An example of a partial match might be *'flints'* instead of *'worked flints'* for the phrase *'worked flints found in several pits'*.

The overall IAA F-Measure score for the three annotation sets on the average mode of reporting was 58% while on the lenient mode the score increased to 68%. In the lenient mode of reporting, individual differences in terms of annotation borders are not encountered since partial matches are considered full matches. An increment of 10% from average to lenient mode is evidence of the disagreement between annotators on the scope of annotation boundaries. The following tables present the IAA annotator score for the three different annotation sets in terms of overall agreement including all entity types and also agreement on some individual entities.

Table 1 IAA scores for the three different annotation sets, reported on Average and Lenient mode. SA: Senior Archaeologist, CA: Commercial Archaeologist, RS: Research Student.

	Precision		**Recall**		**F-Measure**	
	Av.	**Le.**	**Av.**	**Le.**	**Av.**	**Le.**
SA-CA-RS	0.52	0.62	0.64	0.76	0.58	0.68
SA-CA	0.6	0.6	0.5	0.5	0.54	0.54
SA-RS	0.62	0.75	0.83	1	0.71	0.86
CA-RS	0.38	0.5	0.6	0.8	0.46	0.61

Table 2 IAA scores of individual entities reported on Average and Lenient mode. The scores are reported on the IAA score of the three Annotation Sets SA-CA-RS.

	Precision		**Recall**		**F-Measure**	
	Av.	*Le.*	*Av.*	*Le.*	*Av.*	*Le.*
E19.Physical_Object	0.67	0.89	0.57	0.76	0.61	0.82
E49.TimeAppellation	0.82	0.88	0.77	0.81	0.79	0.84
E53.Place	0.7	0.84	0.68	0.8	0.69	0.82
E57.Material	0.33	0.36	0.65	0.71	0.43	0.47

The data of table 1 reveal fairly low IAA scores, especially when results are reported in the Average mode, which is not untypical in some application domains. The Lenient mode provides improved scores as expected, since disagreement on the annotation boundaries is not taken into account. The best performing F-Measure is between Senior Archaeologist and Research Student, scoring 71% on the Average mode and 86% on the Lenient mode respectively. On the other hand, the IAA scores involving the commercial archaeologist (CA) are lower, due largely to different interpretation of occurrences of the Time Appellation and Place entities. For example, the CA tended to annotate the term *'phase'* as for example, *'phase 1, phase 2'* etc., as Time Appellation, whereas the other annotators considered it outside the intended scope. The CA also annotated the term *'trench'* as Place. Trenches are usually produced during excavation and so might not be considered places associated with archaeological events in the past. This distinction was not made clear in the Instructions.

Examining table 2 reveals that the agreement on *E57.Material* entity is significantly lower both in Average and Lenient modes of reporting. This suggests that *material* poses particular problems of ambiguity or recognition. Also the there is a 20% difference between Average and Lenient mode on the F-Measure score of the *E19.Physical_Object* entity, indicative of the different annotation boundaries used by annotators.

Performance of the pipeline was also evaluated against Recall and Precision using the three individual annotation sets from the exercise using the GATE corpus benchmark utility. The overall scores are shown in Table 3. Although these are only preliminary results, they tend to support the potential use of thesaurus expansion in IE to improve vocabulary matching. Taking the use of Synonyms (a concept will have various synonym terms) as a base point, we see that both Precision and Recall are generally slightly improved by Hyponym expansion, which brings in narrower concepts (with their terms) in the thesaurus to the starting concept (by a small margin, the best precision score (75%) is delivered by the Synonym mode against the input of the senior archaeologist). On the other hand, Hypernym expansion, which also brings in narrower and broader concepts, tends to decrease precision but improve the recall by a greater amount, The best recall score (79%) is delivered by the Hyponym mode against the senior archaeologist input. The results show a slight improvement in F-Measure rates over all annotators for both Hyponym and Hypernym expansion modes. In addition, results show that the system produces better F-Measure scores when performing on the Hypernym expansion mode due to the improvement in recall rates. However, these are very preliminary results. Subsequent and larger scale evaluation will examine the use of thesaurus expansion further and will consider whether the F-Measure scores continue to show a similar pattern.

Table 3 Precision, Recall and F-Measure scores

	Synonym	**Hyponym**	**Hypernym**
SA			
Precision	0.75	0.74	0.68
Recall	0.68	0.7	0.79
F-Measure	0.71	0.72	0.73
CA			
Precision	0.65	0.66	0.64
Recall	0.56	0.6	0.68
F-Measure	0.6	0.63	0.66
RS			
Precision	0.71	0.72	0.7
Recall	0.6	0.62	0.7
F-Measure	0.65	0.67	0.7

The preliminary evaluation also revealed a range of terms which were selected by annotators but were not matched by the IE pipeline. Example terms include *Bowl, Castle, Cemetery, Flake, Fort, Garden, Hollow, House, Inhumation, Knife, Nail, Plough, Pond, Playground, Ridge and Furrow, Ring, Settlement, Shed, Tool, Weapon.* In subsequent discussion with archaeological domain experts, it was judged that many of these terns were valid. Some of these terms were in fact present in the thesaurus resources but had no appropriate bridging entry point from the archaeology glossaries. Some way of widening the terminology resources to take account of these terms would increase the eventual recall of the IE process.

Based on the results of the pilot evaluation, the manual instruction should be refined to improve the manual annotation task. For example, the instructions for the annotation of material entities should make clear that the annotation task should be focused on materials that have an archaeological interest and are associated with physical objects, rather than materials in general. Similarly, the instructions for the annotation of place entities should be clarified particularly for rich phrases. Instruction should highlight that entity annotations spans can contain both conjunct and adjectival moderators when applicable. In addition, annotation examples should be included for each annotation type targeted by the task in order to ease comprehension. These should attempt to clarify the cases of phrasal annotation and how boundaries should be treated. It was also seen that annotators tended individually to miss different valid annotations in the rather laborious task. The experience suggests the need for a final resolution phase before reaching a 'gold standard' that resolves differences between annotators. An ongoing larger scale 'gold standard' evaluation is informed by the results of this evaluation exercise.

5 Conclusions

The discussion has revealed the viability of automatic generation of rich metadata for enabling semantic search of grey literature connected with archaeological datasets. The methods of Information Extraction, driven by the core ontology CIDOC CRM and its extension CRM-EH, in combination with SKOS resources, were central to the process of automatic metadata generation. A large scale evaluation excercise is in process to evaluate the information extraction performance in general and in dealing with lexical ambiguities and the accuracy of rich phrase annotation in particular.

Specific contributions of the work reported here include the potential of combining knowledge based resources (including ontologies and thesauri) in information extraction, techniques for using SKOS and CRM resources within GATE based information extraction, techniques for automatic rich metadata generation and expression as coupled XML and also as RDF triples for different consuming applications, including semantic cross search over excavation datasets and archaeological grey literature reports. In addition, the currernt evaluation results demostrate the ability of the system to peform in diffrent modes of semantic expansion that exploit vocabulary resources which may allow the possibility of tailoring the system to favour recall or precision.

In general, the current study demonstrates the capability for CRM based methods to drive automatic generation of rich metadata in domain specific digital libraries. Such metadata can be expressed in interoperable formats such as XML and RDF graphs, which can be exploited by digital library systems to enable cross-search functionality between disparate resources. Work is underway investigating generalisation of the methods to related areas in the cultural heritage domain.

Acknowledgements. The STAR project was supported by the Arts and Humanities Research Council [grant number AH/D001528/1]. Thanks are due to the Archaeology Data Service for provision of the OASIS corpus and to Phil Carlisle (English Heritage) for providing domain thesauri.

References

1. Andronikos web-portal of semantic indices of the OASIS corpus, http://andronikos.kyklos.co.uk
2. Babeu, A., Bamman, D., Crane, G., Kummer, R., Weaver, G.: Named Entity Identification and Cyberinfrastructure. In: Kovács, L., Fuhr, N., Meghini, C. (eds.) ECDL 2007. LNCS, vol. 4675, pp. 259–270. Springer, Heidelberg (2007)
3. Binding, C., May, K., Tudhope, D.: Semantic Interoperability in Archaeological Datasets: Data Mapping and Extraction Via the CIDOC CRM. In: Christensen-Dalsgaard, B., Castelli, D., Ammitzbøll Jurik, B., Lippincott, J. (eds.) ECDL 2008. LNCS, vol. 5173, pp. 280–290. Springer, Heidelberg (2008)
4. Binding, C.: Implementing Archaeological Time Periods Using CIDOC CRM and SKOS. In: Aroyo, L., Antoniou, G., Hyvönen, E., ten Teije, A., Stuckenschmidt, H., Cabral, L., Tudorache, T. (eds.) ESWC 2010, Part I. LNCS, vol. 6088, pp. 273–287. Springer, Heidelberg (2010)
5. Bontcheva, K., Duke, T., Glover, N., Kings, I.: Semantic Information Access. In: Semantic Web Semantic Web Technology: Trends and Research in Ontology Based Systems, Wiley, Sussex (2006)
6. Cowie, J., Lehnert, W.: Information extraction. Communications ACM 39(1), 80–91 (1996)
7. Cripps, P., Greenhalgh, A., Fellows, D., May, K., Robinson, D.: Ontological Modelling of the work of the Centre for Archaeology (2004), http://hypermedia.research.glam.ac.uk/resources/crm
8. Crofts, N., Doerr, M., Gill, T., Stead, S., Stiff, M.: Definition of the CIDOC Conceptual Reference Model, http://www.cidoc-crm.org/docs/cidoc_crm_version_5.0.2.pdf
9. Cunningham, H., Maynard, D., Bontcheva, K., Tablan, V.: GATE: A Framework and Graphical Development Environment for Robust NLP Tools and Applications. In: Proc. 40th Meeting of the Association for Computational Linguistics (ACL 2002), Philadelphia (2002)
10. Doerr, M.: The CIDOC Conceptual Reference Module: an Ontological Approach to Semantic Interoperability of Metadata. AI Magazine 2493, 75–92 (2003)
11. English Heritage Thesauri, http://thesaurus.english-heritage.org.uk/
12. Isaac, A., Summers, E.: SKOS Simple Knowledge Organization System Primer (2009), http://www.w3.org/TR/skos-primer
13. Online AccesS to the Index of archaeological investigationS (OASIS), http://www.oasis.ac.uk/
14. Ore, C.-E., Eide, Ø.: TEI and cultural heritage ontologies: Exchange of information? Literary and Linguist Computing 24(2), 161–172 (2009)
15. May, K., Binding, C., Tudhope, D.: A STAR is born: some emerging Semantic Technologies for Archaeological Resources. In: Proceedings Computer Applications and Quantitative Methods in Archaeology (CAA 2008), Budapest (2008)
16. Moens, M.: Information Extraction Algorithms and Prospects in a Retrieval Context. Springer, New York (2006)
17. Semantic Technologies for Archaeological Resources (STAR) demonstrator. University of Glamorgan, http://hypermedia.research.glam.ac.uk/resources/star-demonstrator/
18. Tudhope, D., Binding, C., May, K.: Semantic interoperability issues from a case study in archaeology. In: Kollias, S., Cousins, J. (eds.) Proc. First International Workshop SIEDL 2008, Semantic Interoperability in the European Digital Library, Associated with 5th European Semantic Web Conference, Tenerife, pp. 88–99 (2008)

Towards Automatic Detection of Various Types of Prominence in Read Aloud Russian Texts

Nina Volskaya, Daniil Kocharov, Pavel Skrelin, and Ekaterina Shumovskaya

Abstract. The paper describes an attempt of the automatic detection of prominence which was marked by experts in prosodic annotation of recorded read aloud Russian texts (duration of the analyzed recording is 2 hours). The material is part of the CORPRES, a large speech corpus of Russian professionally read speech. Prominence labels obtained in prosodic annotation of recorded material were compared with the essential acoustical features, which revealed that perceived prominence that is defined by listeners judgments, refers to syllables that stand out from the phonetic environment in several ways, some of them were not considered in previous studies. It can be manifested as the deviation from the declination line either in a form of an F0 upstep or F0 drop, their choice being defined by the type of the global melodic contour of the intonational unit. This information along with the acoustical measurements has been used in automatic detection of prominence in Russian read aloud texts.

The F0 declination observed in Russian, as in many other languages, conveys linguistic information. Being characteristic of terminal declaratives and not typical for a particular type of interrogatives (general questions), it may signal a sentence type. A change in the global F0 declination line conveys information about the syntactic structure of the sentence: a declination reset often occurs at the intonation unit boundaries, representing different types of clauses. Experimental data obtained in the analysis of the recorded Russian read speech show that phonetic manifestation of prominence may take the form of deviations from the F0 global trend: an upward twist from F0 declination line is perceptually relevant. It often corresponds to the perceived prominence and triggers the assignment of the emphatic accent to a particular word in the prosodic annotation.

Nina Volskaya · Daniil Kocharov · Pavel Skrelin · Ekaterina Shumovskaya
Saint-Petersburg State University, Universitetskaya emb., 11, 199034,
Saint-Petersburg, Russia
e-mail: {volni,kocharov,skrelin}@phonetics.pu.ru
katya-volskaya@yandex.ru

A. Przepiórkowski et al. (Eds.): *Computational Linguistics*, SCI 458, pp. 203–215, 2013.
DOI: 10.1007/978-3-642-34399-5_11

The paper contains the results of an experimental study aimed at establishing a correspondence between phonetically manifested and perceived prominence. It considers the possibility and efficiency evaluation of using a declination reset measure for the task of automatic prominence detection.

1 Introduction

In natural language prominence can be conveyed by a number of ways. A contrastive or emphatic accent normally involves using a set of prosodic parameters: pitch change, intensity and temporal variations. In Russian a pitch change usually takes the form of the intensified nuclear high fall. Observations over real speech behavior and analysis of the experimental data show that phonetic manifestation of prominence is more varied: thus, emphasis can also be achieved by an upstep in pitch in relation to the down-stepping adjacent pitch accents, thus breaking the global declination line and initiating it again after resetting. This type of prominence is characteristic of terminal declaratives. It can also be found in non-terminal intonational phrases, ending with a nuclear fall. On the other hand, in non-terminal intonational units, ending in a nuclear rise, a prominent accent is characterized by an abrupt drop of F0. Thus in the intonational phrase a prominent accent may coexist with the nucleus.

Prominence location is predictable from the context and the information status of the word, since it is not accidental: it normally falls on "lexical intensifiers" like qualifying adverbs (so, very, awfully, terribly etc.). At the same time, certain groups of words trigger the assignment of an emphatic accent to a neighboring word.

Pronounced upward and downward deviations from the F0 declination line as a feature of prominence can be used in ASR systems, and as one of the simple and effective ways for conveying emphasis in modeling intonation for speech synthesis.

The solution of the prominence detection task can be also used in the spoken term detection, video and audio data indexing. Intonation prominence is very important linguistic information for speech understanding, i.e. [1] showed that the use of word prominence degree helped to disambiguate the meaning of the utterances. The experiments were performed using data from the COrpus of Russian Professionally REad Speech (CORPRES) developed at Saint-Petersburg State University [2]. Overall corpus size is 60 hours and it contains more than 500 000 running words.

The corpus includes samples of different speaking styles recorded from 4 male and 4 female speakers. Six levels of annotation cover all phonetic (segmental) and prosodic information about the recorded speech data and include labels for pitch marks, phonetic events, narrow and wide phonetic transcription, orthographic and prosodic transcription. The intonation types and prominence were labeled manually by expert phoneticians and cross-verified.

The use of declination reset significantly improved the efficiency of automatic prominence detection system. The increase of the prosodically prominent word detection is up to 44% and the overall system improvement is up to 6% relative to the baseline system using other prosodic features.

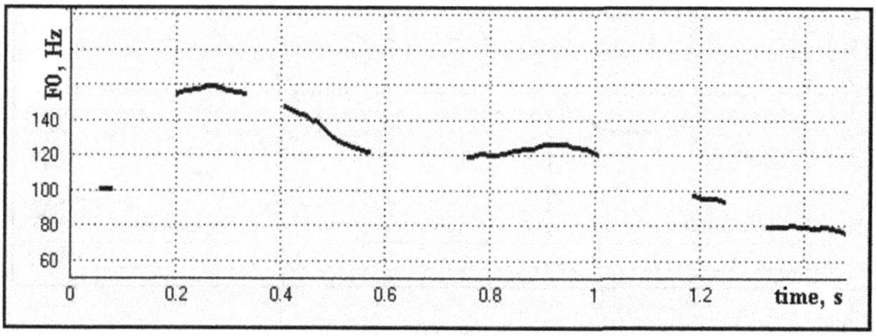

Fig. 1 F_0 declination in a neutral Russian declarative

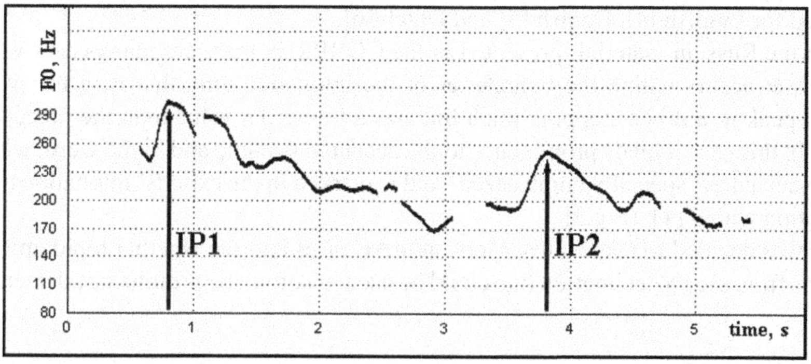

Fig. 2 F_0 declination course in a compound sentence: declination reset at the boundary between IP1 and IP2

2 Declination and Declination Reset

The "tendency of F_0 to decrease slowly from beginning to end of an utterance" known as the declination [3], has been observed for many languages. In neutral Russian terminal declaratives, which do not contain contrasted or otherwise emphasized elements, pitch accents often form a succession of downward steps or glides, where each accented peak is lowered relative to the preceding one (Fig. 1): the whole sentence glides gradually down.

A large experimental literature suggests that declination is perceptually relevant. It means, that a deviation from the descending scale of pitch peaks may signal some sort of information: for example, an intonation unit boundary – the end of one unit and the beginning of the next one, – or emphasis.

In Russian, in real speech situation an accented syllable which "begins" a new intonational phrase, often initiates a new declination line. Moreover, in a succession of two intonational phrases, representing, for example, coordinate clauses, the declination of the second intonational phrase is reset at a lower F_0 starting point (Fig. 2).

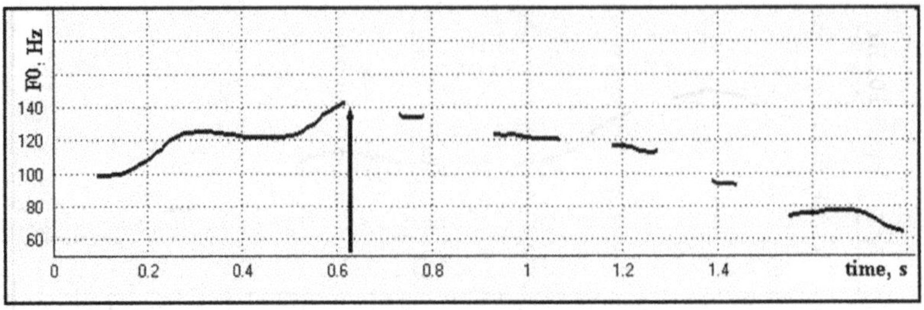

Fig. 3 The case of F_0 declination reset perceived as prominence

The occurrence of the declination reset at clause boundaries has been also reported for English [4], Danish [5] and Dutch [6].

In the Russian material, presented in the CORPRES there are many cases when the reset occurs within the boundaries of the intonation unit, that is, a following pitch peak in a down-stepping pitch line has a higher F_0 value than the preceding one. In this case it lends prominence to the accented syllable and to the word, which is perceived as "specially emphasized" and is marked in the experts' intonation transcription with a [+]. (Fig.3)

It was decided to take a closer look at these cases and see whether rapid upward jumps from the F_0 declination line could be used in automatic prominence detection.

3 Feature Extraction

The declination reset measure is a binary scalar that has value 1 if there is a declination reset within a word and 0, if a reset is absent.

The F_0 maximum is calculated for the stressed vowel in a word. If there are two stressed vowels within a given word then the F_0 maximum (F_{max_i}) is calculated for both of them. Each word is characterized by a maximum value of F_0. This feature is defined as zero in cases when there are no stressed vowels in a word, like prepositions.

The next stage is calculating a difference between the F_0 maximum of a given word and that of the preceding word (ΔF_{max_i}). If in the preceding word the F_0 maximum is equal to zero denoting absence of the stressed vowel then we take into account a previous word.

$$\Delta F_{max_i} = \begin{cases} F_{max_i} - F_{max_{i-1}} \ if \ F_{max_{i-1}} \neq 0, \\ F_{max_i} - F_{max_{i-2}} \ if \ F_{max_{i-1}} = 0. \end{cases} \quad (1)$$

There are two restrictions to this method:

1. it is impossible to calculate F_{max_1};
2. if the first word is unstressed then we cannot calculate F_{max_2} either.

In such cases F_{max_i} is defined as -1.

The final declination reset measure (DR_i) is calculated as following:

$$DR_i = \begin{cases} 0, & if\ \Delta F_{max_i} < 0, \\ 1, & if\ \Delta F_{max_i} \geq 0. \end{cases} \quad (2)$$

This calculation reflects the declination reset in the experiments described further.

4 Correspondence of Declination Reset and Prominence

The goal of this experiment was to estimate the correspondence between the acoustic manifestation of the declination reset in a tone unit and perception of the word as prominent by expert phoneticians.

The experimental material (a part of CORPRES – 5 hours of the speech data recorded from one male speaker) was limited to intonation units with falling intonation, where F_0 declination is expected. Among them terminal declaratives with different types of nuclear tones (level, low falling and high falling), declaratives with a contrastive nuclear fall, two types of exclamations and non-final declaratives with a non-low terminal fall. In non-final intonational phrases with rising melodic contours F_0 declination is much less evident, and in general questions it is practically absent [8].

Table 1 contains the data about presence or absence of declination reset in groups of words representing different types of prominence accent. The data is given as percent in relation to the total number of words within each group.

As it follows from Table 1, in the groups "No accent" and "Displaced sentence accent" declination reset in observed in about 31% and 39% of words accordingly, in the group "Sentence accent" the declination reset characterizes only 19% of the total number of words. The majority of cases where declination reset is present, correspond to words, carrying emphatic or contrastive accent.

The data confirms the assumption that the speakers use the declination reset as an effective mean for providing emphasis to a particular word in the intonation unit, since about 82% of cases of perceived prominence (emphatic and contrastive accent) in the experimental material correspond to an upward twist of F_0 from its due course.

The predictive power of the declination reset for detecting a prosodically prominent word in the intonation unit is illustrated in Table 2, which shows the distribution

Table 1 Presence or absence of declination reset in groups of words representing different types of prominence accent (% to the total number of words within each group

	No accent	Sentence accent	Emphatic accent	Displaced sentence accent
No decl. reset	69.39	80.96	18.11	60.84
Decl. reset	30.61	19.04	81.89	39.16
Total number of words	5227	2521	243	263

Table 2 Distribution of declination reset (%) over units with different types of accents

	No decl. reset (%)	Decl. reset (%)
No accent	61.77	67.17
Sentence accent	34.76	20.15
Emphatic accent	0.75	8.36
Displaced sentence accent	2.72	4.32

of words characterized by presence or absence of the declination reset over groups representing different accent types.

Analysis of the experimental data shows that on the one hand perceived prominence is greatly influenced by the presence of a boosted accent in the F_0 declination line; on the other hand, the declination reset cannot be used as the only feature in detecting prosodically prominent words in the intonation unit since 60–70% of words have no prominence regardless of its presence or absence.

The effectiveness of this cue as an additional one in the set of acoustic features was tested further in the experiment using the earlier developed system for automatic detection of prosodically prominent words [9].

The description of the system and the effectiveness of the applied method, using declination reset along with other acoustic cues is presented below.

5 Baseline System of Automatic Prominence Detection

The system used for efficiency evaluation was designed to detect prominent words in Russian speech and is presented in [9]. The system is shortly described in this section.

5.1 Acoustic Features

A speaker prosodically emphasizes a particular word in an utterance to make it stand out of the surrounding words. It is generally assumed that relative temporal, melodic and loudness variations are closely connected with prominence. The acoustic correlates of these prosodic cues are used in the baseline system.

The means were calculated for sound samples in the speech corpus.

Pitch change. The melodic features were extracted from a preprocessed and smoothed F_0 curve. The pitch contour is a result of automatic pitch detection system developed at the Department of Phonetics of Saint-Petersburg State University [10].

The goal of preprocessing is to eliminate microprosodic events and to get a smoothed melodic contour. This allows to get rid of calculation errors occurring due to microprosodic perturbations. The following melodic features were selected for prominence detection based on the analysis of other research and solutions as well as a series of experiments:

- maximum F_0 within a word,
- minimum F_0 within a word,
- mean of the F_0 within a word,
- standard deviation of the F_0 within a word.

The rate of F_0 change is also taken into account to model the extent of F_0 change. We assume that the change of F_0 is higher within a prominent word. Thus the following features are used as well:

- maximum ΔF_0 within a word,
- minimum ΔF_0 within a word,
- mean of the ΔF_0 within a word,
- standard deviation of the ΔF_0 within a word.

Intensity. Since perceived prominence also correlates with loudness, corresponding to the speech signal intensity (or, more precisely, with its spectrum intensity), the following features were used for the analysis:

- maximum speech signal intensity within a word,
- minimum speech signal intensity within a word,
- mean of the speech signal intensity within a word,
- standard deviation of the speech signal intensity within a word.

The signal intensity was measured each 25 ms as average speech signal amplitude within a window of 25 ms length.

Duration. Two temporal features were used:

- total word duration in milliseconds,
- relative allophone duration in a current word expressed by the ratio of allophone duration within a current word to the mean duration of these allophone in the corpus.

5.2 *Statistical Processing of Acoustic Features*

The choice of statistical data processing and acoustic modeling method that allows to achieve the best efficiency of automatic prominent word detection is no less crucial than the choice of acoustic features. We used the classification and regression tree (CART) as a classification method for detecting prominent words.

Normally the entropy is used as a splitting criterion in a CART framework. However, it has been decided to use probability of prominent words as a splitting criterion in the current system. There are two reasons for that. When entropy is calculated, all classes are supposed to be equally probable, but the number of prominent words is 4 times smaller than the number of non-prominent words. In case of using entropy this could lead to the situation when there are objects of different classes in all CART leaves: many non-prominent words and several prominent words. The other reason is that there are just two classes in this case: prominent and non-prominent words. Thus, uncertainty degree is unambiguously defined by probability of one class.

6 Experimental Results

For estimating the efficiency of the automatic prominence detection a different part of the CORPRES was used: about 40 minutes of recording from each of 8 subjects in the corpus. The training data consist of 4 hours of recorded speech: about half an hour per speaker. The test data is 1 hour of the recorded material: 7–8 minutes per speaker. This part of research material contained all types of tone units with different melodic contours: rising, falling and level, thus it was not limited to falling contours only as in the first experiment described above. It was done in order to estimate the efficiency of the system in real conditions when there is no information about the tone unit melodic type. We considered all accented words as prominent, that is, those with a sentence accent, a displaced sentence accent and an emphatic accent. The total number of running words in the test corpus was 1799; 772 of them were prominent and 1027 were not prominent.

Table 3, contains experimental results obtained by using declination reset measure in addition to the set of the acoustic features described in Section 5.1.

Table 3 Detection error rates obtained by combining baseline acoustic features with declination reset (DR) correlate (%)

Acoustic features	Prominent words	Non prominent words	General efficiency
Baseline	50	23	34
Baseline + DR	28	31	30

The data show that significant improvement in error rate has been obtained in the task of detecting prominent words. The increase of the prosodically prominent word detection is up to 44% and the overall system improvement is up to 6%.

7 Prominence Perception in Non-terminal Phrases: Basis for Research and Prior Studies

As it was stated earlier, phonetic manifestation of prominence is more varied: in terminal declaratives emphasis can be achieved by an upstep in pitch in relation to the down-stepping adjacent pitch accents. On the other hand, in non-terminal intonational units, ending in a nuclear rise, a prominent accent is characterized by an abrupt drop of F_0. Thus in the intonational phrase a prominent accent coexists with the nucleus. As it was shown by [11], this type of prominence is typical for TV news readers.

The basis for her research were 18.5 minutes recordings of the Russian TV news speakers. The research started with the auditory analysis of the recordings with the aim to create a generalized intonational transcription (identifying tonal movement

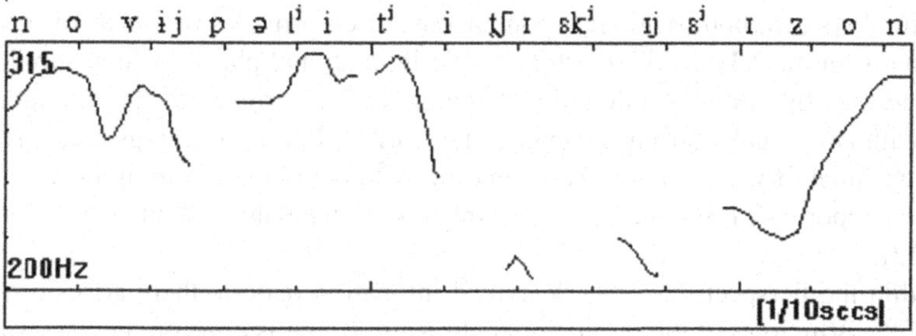

Fig. 4 F_0 course in the intonational phrase with two intonational centers: *politicheskij* ***sezon***

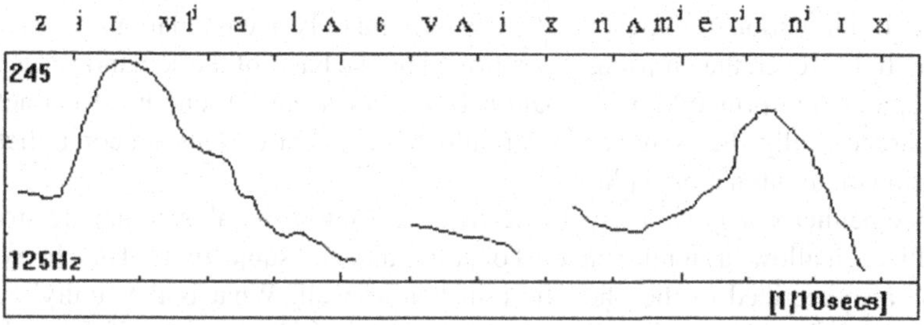

Fig. 5 F_0 course in the intonational phrase with two intonational centers: *zajavljal o svoih namerenijah*

on the perceptually prominent word and on the accented syllable of the intonational centre, the presence of the 'psychological' and physical pauses).

The choice over the phrases of two intonational centres was made on the basis of the prosodic transcription. Out of 700 phrases having been analyzed, the 'two-nuclei' phrases constituted 12.3 %.

'Two-nuclei' phrases are non-final phrases which have two prosodically prominent accents: the first one is characterized by the falling tone which provides prominence to a particular word in the intonational unit for reasons of contrast or emphasis; the second one, characterized by the phonologically rising tone, is considered as the nucleus proper.

In some of the examples given below the additional falling accent in the intonational phrase can be interpreted as a means for emphasizing a word : *"politicheskij" sezon – '**political** season'* – since prior to that details of the military campaign were provided; *zajav**ljal** o svoih namerenijah... – '**declared** his intentions'*- as it had **already** been done, etc. (see Figures 4, 5). It should be noted though that some examples do not demonstrate the necessity for such strong accents from the semantic perspective of the text (*... v etom dages**tan**skom gorode... – 'in this **Dagestan** city',... **prezidentu** Putinu... – 'to **President** Putin'*). Nevertheless, the melodic pattern used by the speakers was the same in both cases.

Natural speech, both read and spontaneous, often provides instances of the deviations from the 'classical' structure of the intonational phrase, which should be characterized by only one intonational center. The occurrence of the intonational units with two equally strong accents which carry different meanings and perform different functions, i.e. the emphatic and intonational phrase-forming or delimitative, was reported earlier and viewed as inherent to the natural spontaneous speech acts [12].

Within the perspective of the described intonation pattern, there arises a question of the principles of the intonational phrasing. If that is done on the basis of the phonetic criterion – that is, the intonational unity of a phrase, – then we have to consider the units with two centers as containing *two* intonational phrases. D.Hirst [13] suggested that a number of *phonological phrases* in an utterance is defined by the number of intonational centres, i.e. there is one and only one intonational centre in a phrase. If, however, the phrasing is performed on the basis of the semantic criterion, then, against the principle of intonational unity of a phrase, one has to conclude that, paradoxically, the existence of two intonational centres does not contradict the definition of an intonational phrase.

Since neither semantic nor syntactic criteria of intonational phrasing are strictly formalized, it allows a significant level of ambiguity and subjectivity. That, however, can be also ascribed to the phonetic criterion as well. What is the 'unity' of an intonational phrase from the point of view of the melodic contour? If the melodic contour changes sharply (like in the 'two-nuclei' phrases), would that be a sufficient reason to define the phrase boundary, or is a change of the melodic contour also possible *within* a phrase? Is it always the case that the existence of the hesitation pause define the phrase boundary? In other words: can there be a pause within a phrase? Answers to some of these questions were provided as a result of the auditory experiment.

8 Material and Methodology of the Auditory Experiment

21 phrases with two "intonational centres" were selected as a result of the auditory and instrumental analysis. Auditory analysis focused on 40 phrases: 21 'two-nuclei' and 19 'regular', at random choice, in order to prevent listeners from getting accustomed to the melodic contour. Each phrase was repeated twice with a 4 second interval. The purpose of the auditory experiment was to clarify how such phrases are perceived by the native speakers: whether a change of the melodic contour was a signal of prominence and whether it was of substantial significance for them to regard it as a sufficient phrase boundary signal. 38 listeners took part in this experiment (29 of them students of the philology of the first and second year and 9 people having no affiliation to philology). The task for the listeners was

1. to identify a word or words made prominent by the speaker,
2. to identify the location of a pause (if it is to occur).

As it can be seen from the Table 4, in 19 out of 21 cases of the 'two-nuclei' phrases, the majority of the listeners (71.1 % to 94.7 %) pointed out the additional accent

Table 4 Results of the auditory experiment

#	Intonational phrase	Two nuclei (%)	Pause position (|)	Contrastive accent (%)
1	... v etom dages**ta**nskom **go**rode	68.4	etom |	28.6
2	... vo vremja osetino-in**gu**shskogo kon**fli**kta...	94.7		71.4
3	... prezi**de**ntu Vladimiru **Pu**tinu...	50.0	prezidentu |	21.4
4	... sozdaj**ot** pravovuju os**no**vu...	65.8		35.7
5	... novij poli**ti**cheskij sezon...	76.3		35.7
6	... osenju devjanosto vtorogo **go**da	94.7		64.3
7	... s zhurnalistami vstretilis advokati lju**dej**...	89.5	vstretilis |	42.9
8	... spetzialnja lingvis**ti**cheskaja ko**mi**ssija...	78.9		50.0
9	... u rossijsko-gru**zi**nskoj gra**ni**tzi...	78.9		50.0
10	... chto i **mi**rnij dogo**vor**...	76.3	chto |	78.6
11	... chto **cha**st paketa **ak**tzij...	89.5	chast |, chto |	57.1
12	... sov**me**stnogo hozjastvennogo osvo**je**nija...	71.1		71.4
13	... ministr pe**cha**ti Michail **Le**sin	81.6	pechati |	71.4
14	... segodnja ruko**vo**dtsvo Chech**ni**...	73.7	segodnja |	28.6
15	... zaklu**chit** mirnij dogo**vor**...	71.1	zakluchit |	14.3
16	... zajav**ljal** o svoih na**me**renijah...	86.8		42.9
17	... chto so**ve**tnika prezi**de**nta...	73.7	chto |, sovetnika |	35.7
18	... provesti **rjad** terrorestichesk ih **ak**tov...	73.7		21.4
19	... o se**go**dnjashnih sobitijah v Rja**za**ni...	84.2	sobitijah |	71.4
20	... segodnja v Moskve s zhurna**li**stami...	86.8	Moskve |	78.6
21	... eto **me**stnij kon**fli**kt	78.9	eto |	78.6

(see Table 4, 'two nuclei'). Unlike it was initially assumed, the listeners did not identify pauses within the 'two-nuclei' phrases (in cases of the quick changes of the melodic contour) (see Table 4, 'pause position'). More so, in the phrase 'chto/mirnij dogovor', where the speaker did make a physical pause (see Figure 4), its existence was confirmed only by 11 out of 38 listeners.

In an additional experiment (14 participants of the key experiment) performed after the first one, the listeners were requested not only to mark the accented words but also to identify a specific type of prominence: emphasis or contrast. The following results were obtained: for the phrases # 2, 6, 10, 12, 13, 19, 20, 21 the number of positive responses was over 70 %, for all other phrases this number did not carry any statistical significance (see Table 4, 'contrastive accent').

The experiments demonstrated that the majority of the listeners do hear the additional accent in the majority of the 'two-nuclei' intonational phrases, but only in eight cases the accent was interpreted as contrastive by the majority of the listeners.

9 Conclusion

Experimental results showed that prominence perception is due to sudden changes of the F0: in terminal declaratives it is greatly influenced by the presence of the F0 declination reset within the boundaries of an intonational phrase. It has been effectively used in modeling intonation for Russian speech synthesis as a simple and effective way of providing emphasis to a particular word in the intonation unit. At the same time the data obtained in this study indicate that F0 declination reset is not always connected with prominence, therefore it cannot be used as the only guide to prominence detection. The effectiveness of this cue as an additional one in the set of prosodic features was tested. The set of features includes relative word and sound duration, maximum, minimum, mean and standard deviation of speech signal intensity, F0 and F0 derivative of a given word.

Significant improvement in the error rate was achieved while solving the task of detecting prominent words. The increase of the prosodically prominent word detection is up to 44% and the overall system improvement is up to 6%.

The results of the auditory experiments confirmed the assumption that the F0 drop is used by the speakers as a means for providing prominence in the non-final intonational phrases, which results in a compound melodic contour. Before automatic prominence detection procedure could be applied to non-final intonational phrases all theoretical issues regarding the phonetic realization and functional status of the accents as well as the integrity of the intonational unit should be settled.

Further research in this field should be focused on automatic prominence detection in Russian spontaneous speech and on the intonation unit boundary detection using declination reset data.

Acknowledgements. The authors acknowledge Saint-Petersburg State University for a research grant # 31.37.106.2011.

References

1. Beckman, M.E., Venditti, J.J.: Tagging prosody and discourse structure in elicited spontaneous speech. In: Proc. of Science and Technology Agency Priority Program Symposium on Spontaneous Speech, Tokyo, Japan, pp. 87–98 (2000)
2. Skrelin, P., Volskaya, N., Kocharov, D., Evgrafova, K., Glotova, O., Evdokimova, V.: A fully annotated corpus of Russian speech. In: Proc. of the Seventh Conference on International Language Resources and Evaluation (LREC 2010), Malta, pp. 109–112 (2010)
3. Collier, R., Gelfer, C.E.: Physiological explanations of F0 declination. In: Proc. of the Tenth International Congress of Phonetic Sciences, Dortrecht-Cinnaminsa, Foris, vol. 354–360, pp. 354–360 (1984)
4. Ladd, D.R.: Declination "reset" and the hierarchical organization of utterances. Journal of the Acoustical Society of America 84, 530–544 (1988)
5. Grønnum Thorsen, N.: Sentence intonation in textual context – supplementary data. Journal of the Acoustical Society of America 80, 1041–1047 (1986)
6. Collier, R.: F0 declination: the control of its setting, resetting and slope. In: Baer, Sasaki, C., Harris, K.S. (ed.) Laryngeal Function in Phonation and Respiration, pp. 403–421. Little and Brown, Boston (1987)
7. Volskaya, N.B., Skrelin, P.A.: Prosodic model for Russian. In: Proc. of Nordic Prosody X, pp. 249–260. Peter Lang (2009)
8. Volskaya, N.B.: Prosodic phrasing and intonation of the interrogative sentences of varying length. In: Materialy Mezhdunarodnoy Konferentsii 100 Let Experimental'noy Phonetike v Rossii, Saint Petersburg, pp. 54–57 (2001) (in Russian)
9. Kocharov, D.: Automatic detection of prominent words in Russian speech. In: Proc. of International Multiconference on Computer Science and Information Technology, Wisla, Poland, pp. 435–438 (2010)
10. Skrelin, P., Kocharov, D.: Automatic processing of the utterance prosody: prosodic features relevant for the automatic interpretation of the intonation contour. In: Trudy Tretiego Mezhdistciplinarnogo Seminara Analiz Russkoj Rechi, Saint Petersburg, pp. 41–46 (2009) (in Russian)
11. Volskaya, E.G.: On nonconventional types of sintagm intonation. In: Materialy Konferentsii 100 Let Eksperimentalnoj Fonetike v Rossii, Saint Petersburg, pp. 49–53 (2001) (in Russian)
12. Bondarko, L.V., Verbitskaya, L.A., Zinder, L.R., et al.: Phonetics of Spontaneous Speech, Moscow (1988) (in Russian)
13. Hirst, D.: Intonative Features: A Syntactic Approach to English Intonation. Mouton Publishers (1997)

Part III

Multilinguality

Translation Ambiguity Resolution Using Interactive Contextual Information

Farag Saad and Andreas Nürnberger

Abstract. One of the main problems that impact the performance of cross-lingual information retrieval (CLIR) systems is how the users express their information need in form of a query. Traditional CLIR systems are not fully effective when the user need is not expressed appropriately. The difficulty lies in dealing with the natural lexical ambiguity of the source and target language, which is not a trivial task that can be resolved fully automatic. Therefore, there is a need for systems that support users to overcome these shortcomings. In this chapter, we first discuss selected state-of-the art CLIR tools. The considered tools are web-based and support interactive translation processes. Based on the state-of-the-art review, we discuss identified issues and shortcomings and thus motivate requirements for CLIR tools that could improve retrieval performance. Finally, we propose an interactive approach that fulfils several of these requirements. It considers the user as an integral part of the CLIR process, in that the user can interact with provided contextual information in a way that allows to interactively improve the translation and thus improve the overall performance of the CLIR process.

1 Introduction

The increase of multilingual information on the World Wide Web has led to the necessity to develop methods and applications to make use of this multilingual information. However, language barriers are a serious issue to

Farag Saad
GESIS – Leibniz Institute for the Social Sciences, Unter Sachsenhausen 6-8, 50667 Cologne, Germany
e-mail: `farag.saad@gesis.org`

Andreas Nürnberger
Otto-von-Guericke-University Magdeburg, Universitätsplatz 2, 39106 Magdeburg, Germany
e-mail: `andreas.nuernberger@ovgu.de`

A. Przepiórkowski et al. (Eds.): *Computational Linguistics*, SCI 458, pp. 219–240, 2013.
DOI: 10.1007/978-3-642-34399-5_12

world communication and to economic and cultural exchange. In order to allow users, to overlap across languages, CLIR can be used. The main research obstacle that prevents CLIR from performing well is the lexical ambiguity of source and target languages. In every language, there are words which have multiple meanings, which will lead to the fact that the user query can have several possible translations. In addition to the classical information retrieval tasks, CLIR requires that the query (or the documents) be translated from one language into another. Query translation is widely used for CLIR tasks, as query translation requires fewer computational resources compared to translating a large set of retrieved documents [9]. Furthermore, users who are able to understand more than one language might not be able to effectively express their need in those languages. Those users with CLIR system support can cover more multilingual resources with a single query expressed in a language they are fluent in. Furthermore, CLIR based on query translation, can also be useful for users who can read a single language. Using query translation can narrow the examined documents, by the user, in the target language. This can reduce time and effort in comparison to translating all documents in the data set and then retrieving the relevant documents out of them. In some cases, CLIR can be useful for the monolingual user. For example, an industrial expert is looking for a specific pump, in a specific country and he/she would like to know if this pump is produced there. Using the CLIR system, the query will be translated and relevant documents will be provided. Based on examining these documents, the user might find images for the pump which meet the expectations about his/her information need. Furthermore, the user might then select one or two documents to automatically translate. Therefore, using query translation and then retrieval can be more beneficial than document translation and then retrieval [19].

In the past, most research has been focused on the retrieval effectiveness of CLIR systems through information retrieval test collection approaches [7], whereas few researchers have been focused on the user interface requirements with respect to the multilingual retrieval task [21]. Despite the clear effort which has been directed toward retrieval functionality and effectiveness, only little attention was paid to developing multilingual interaction tools, where users are really considered as an integral part of the retrieval process. One potential interpretation of this problem is that users of CLIR might not have sufficient knowledge of the target languages and therefore they are usually not involved in multilingual processes [22].

In the following we discuss in detail some of the state-of-the art CLIR interactive tools, categorizing them and identifying their limitations. Then, we present in more detail an interactive CLIR approach that we have developed to tackle several of these limitations. For the evaluation we are focusing on evaluating the automatic translation algorithm integrated in the proposed approach. The retrieval process itself is performed by use of search web services of a standard web search engine, which are called after the translation was performed. In other words, the main goal of the discussed approach is to support the user in performing the CLIR task by providing him/her with contextual information that describe the translation in the user's preferred language. Therefore, we had to use different measures that are appropriate to evaluate this specific task (See Section 4.1 and Section 5). In addition, we evaluated the approach from the user point of view in form of a user study.

2 Cross-Lingual Interactive Tools

CLIR system provides users with helpful information in the user's native language and based on this information, the user can provide the CLIR system with useful feedback that would likely help to improve the translation and thus improve the CLIR quality. Therefore, the accuracy of the CLIR system depends to a strong extent on the interaction between the user and the system [3, 5]. In this chapter, we selected CLIR tools for review on the basis of the general criteria that they should be web-based and provide methods that support to some extend interactive retrieval tasks, i.e., they are not limited to a pure ad-hoc query paradigm that merely provides a result list. These tools consider the user to some extend as an integral part of the retrieval process. We did not differentiate between different back-end technologies and did not consider, e.g., systems based on conceptual indexing [1], since they do not provide additional interaction metaphors for the interface itself. Since the translation is the most important part of any CLIR process, we further classified the selected CLIR tools into two categories, depending on the best match of their used features. These two categories are CLIR tools that provide automatic query translation (automatic disambiguation) and CLIR tools that provide a user based query translation (user-based disambiguation).

2.1 Tools Based on Automatic Translation Methods

MIRACLE: In order to support the interactive CLIR, the system uses the *user-assisted query translation* [20]. The user assisted-query translation feature supports the user to select the correct translation. However, there might be a case when the user might delete a correct translation. The system reacts, in that the searcher can see the effect of the choice and have possibilities to learn better control of the system. This is done by providing the following features, the meaning of the translation (loan word or proper name), using back translation, a list of possible synonyms are provided. Translation examples of usage are obtained from translated or topically-related text. In MIRACLE, there are two types of query translations, fully automatic query translation (using machine translation) and user-assisted query translation. In fully automatic translation the user can be involved only once. After the system translates the query and retrieves the search results, the user can refine the query if he/she is not satisfied after examining the search results. In the user-assisted query translation, four possible refinement steps give the user an opportunity to be involved in the translation process. First, based on evidence about the meanings of the proposed translations by the system, the user has an opportunity to deselect some of the proposed translations before the search can be performed. Second the user can reform the query based on evidence about the meanings of the proposed translations. Third, the user can reform the query based on examining the search results. Fourth, in case the search result doesn't satisfy the user's needs, the user has a possibility to deselect/reselect the translations. The rapid adaption to new languages was taken into account in the design of the MIRACLE system. The query language is always English, in MIRACLE. However, language resources that are available for English can be leveraged, regardless of the document language. Although MIRACLE overcomes some of the limitations of the previously mentioned CLIR interaction tools, it also has some limitations. For example, despite the use of automatic translation in MIRACLE, the user has no influence on refining the translation before the search can be conducted. In addition, to the previously mentioned limitation, in MIRACLE, single word translations are used, which forces the user to spend a lot of effort checking each single translation alternative with their meanings before he/she can select/deselect translations.

WORDS: Lopez-Ostenero et al. (2002) proposed a CLIR interactive approach which provides the user with the possibility of formulating and refining a query [18]. It includes a reference system (WORDS) that supports the user in selecting proper translations for each of the query terms. Furthermore, it includes the possibility of assisting the user in formulating his/her query, by providing him/her with a set of relevant phrases. The user has the possibility of selecting promising phrases, in the presented documents, in order to improve the search. The reference system (WORDS) includes a user query translation assistance and refinement. WORDS translates each query term (in Spanish) by providing all possible translations for each term in English. In order to give the user confidence in the translation, WORDS uses back translation (from English to Spanish). This allows the user to deselect any translation before the search can be performed. Once the user selects the suitable translation the search can be performed. In this case, English documents will be retrieved based on the English translation. Once the documents are retrieved (in English) the system provides the user with a summary of each retrieved document, in the user's own language. This summary includes translation of all noun phrases in the document, using the Systran machine translation system[1] and the document title is automatically translated. Based on this information, the user can mark the document as relevant, irrelevant or unsure. In order to refine the query to improve the search, the user can check the retrieved document translation (in Spanish) and point to any query term in the document. The system then points to the English query terms (one of the possible translations for the Spanish query term). The user then

[1] http://www.systran.de/

has the possibility to select or deselect any English term. This allows the user to keep only the appropriate translations for the Spanish query term. Despite lots of support for the user to perform the CLIR task, the user has to check all translation alternatives with their definitions in order to disambiguate translations. Furthermore, the query refinement depends on the automatic translation, which can not be accurate in all cases i.e., inaccurate translation leads to low retrieval performance.

Patent CLIR: Bian and Teng (2008) proposed a cross-lingual patent retrieval and classification system that makes use of the various free web translators to translate the user query [6]. The system was designed for Japanese/English cross-lingual patent retrieval. The proposed system provides monolingual and cross-lingual functionalities. The input to the system is the query or the selection of the topic file. The user then can use one of the different web translators to translate the query. The proposed system gives the user a possibility to modify the translation. The different system modules are described in the following. The *indexing module* processes and indexes the multi-lingual patent document sets. The system uses two types of indexing methods, a word-based method to index the English text collection and the bigram-based method to index the Japanese text collections. The *translation module* translates the query from the source language to the target language. The query is sent via the system to the selected online translator system by the user. The obtained translation is then obtained and displayed to the user. Since the user can use different translators at the same time, it is possible that the user can review and modify the translation based on the results from different translators. The *classification module* processes the retrieved patent documents in order to classify them based on the International Patent Classification (IPC)[2]. Therefore, the documents are retrieved based on the topic of the input patent (query). Then the first top ranked 3000 patent documents and their IPC code from the patent data collection are retrieved and the score of the IPC code is computed based on the similarity score between the query and the retrieved documents. Finally, the IPC codes are sorted by their score. Despite the possibility of refining the web translators translation integrated in the tool (selecting or removing translations from three different web translators integrated in the tool), users with low knowledge in the target language will have no possibility to select suitable translations from different translators i.e., no information in the user's own language to describe the translation so the user can interact with it effectively.

2.2 Tools Enabling Interactive User-Based Translation

Mulinex: Mulinex supports CLIR by giving the users possibilities to formulate, expand and disambiguate queries. Furthermore, the users are able to

[2] http://www.wipo.int/classifications/ipc/en/

filter the search results and read the retrieved documents by using only their native language [8]. Mulinex performs the multilingual functionality based on a dictionary-based query translation. In Mulinex, three languages are supported, French, German, and English. In Mulinex, the CLIR process is fully supported by the translation of the queries, documents and their summaries. Hereby, users do not need to have any knowledge about the target language. Mulinex provides a lot of functionality to support the retrieving of the documents in multilingual collections. Examples of these functionalities are translation of the user's query, interactive disambiguation of the query translation, interactive query expansion, on-demand translation of summaries and search results. The Mulinex interface is available in three languages English, German, and French. Since the search engine queries are usually between 2.4 and 2.7 in length [14] which typically does not provide enough context for automatic disambiguation, Mulinex using "query assistant" provides an opportunity for interactive query translation disambiguation. This task is performed by the "query assistant" by performing the back translation. The translated query terms are translated back into the original query language. However, this approach has some clear limitations. When no synonyms can be found in the dictionary, the technique is not helpful; and significant homonymy in the target language can result in confusing back translations [20].

Keizai: The goal of the Keizai project is to provide a Web-based CLIR system that accepts the query in English and searches Japanese and Korean web data [21]. Furthermore, the system displays English summaries of the top ranking retrieved documents. In Keizai the query terms are translated into Japanese or Korean languages along with their English definitions and thus this feature allows the user to disambiguate the translations. Based on the English definitions of the translated query terms, the user who does not understand Japanese or Korean language can select the appropriate translation out of several possible translations. Once the user selects those translations whose definitions are consistent with the information needed, the search can be performed. Only documents that are relevant to the selected translations will be retrieved. Keizai enhances the visualization by representing the retrieved documents together with small images, which they call "Document Thumbnail Visualizations". Using this document representation, the retrieved documents are retained with a familiar shape and format and thus the user can see how the query terms are distributed in the retrieved documents. Using this technique the authors investigated the potential advantage of the representation of the documents as one image within the context of different interactive text retrieval tasks. In Keizai, the authors could show that the visualization improved recall and efficiency.

MultiLexExplorer: The goal of the MultiLexExplorer tool is to support multilingual users in performing their web search. Furthermore, the tool supports the user in disambiguating word meanings by providing the user with information about the distribution of words in the web [11]. The tool allows users to explore combinations of query term translations by visualizing EuroWordNet[3] relations together with search results and search statistics obtained from web search engines. Based on the EuroWordNet, the tool supports the user with the following functionalities: methods to explore the context of a given word in the general hierarchy; support for searching in different languages, e.g., by translating word senses using the interlingual index of EuroWordNet; methods to disambiguate word sense for combinations of words; simple system interaction methods to change the search word, the number of retrieved documents, expanding the original query with additional terms; and methods to automatically categorize the retrieved web documents. Despite the good visual and functional design of MultiLexExplorer, it relies on the use of EuroWordNet, which only employs a limited number of languages.

3 Shortcomings of State-of-the Art CLIR Tools

In the following, we identify and discuss major shortcomings of the CLIR tools discussed above. Therefore, we focus on specific aspects that need to be considered in order to support a user appropriately in an interactive search process.

Translation Confidence: An important point, which has been studied in depth in the state-of-the art CLIR tools analysis, is the translation confidence. How we expect the user to rely on the translation provided by the CLIR tool when he/she is not able to understand or even read this translation. Based on the analysis of six CLIR tools, we found out that only two CLIR tools provide a possibility of giving the user some confidence in the translation. However, both CLIR tools used back translation, where the translation is translated back to the source language. If there is overlap between the query and the back translation then one might have some confidence in the translation. However, this approach suffers from a clear drawbacks, when no synonyms can be found in the dictionary, the technique is not helpful; and significant homonymy in the target language can result in confusing back translations [20]. Some state-of-the art cross lingual tools used the dictionary definitions to give the user some confidence in the translation. However, bilingual dictionaries, in which the definitions of source language are available for each translation for the target languages, are very rare and very laborious. Sometimes, in the existing of translation definition, it does not resolve the problem clearly because this definition is displayed for each translation term independent from other translated terms.

Automatic Translation: Most of the studied state-of-the-art CLIR tools provide no automatic translation and thus for automatic translation disambiguation. They are based on individual term translations, where the user is requested to perform the disambiguation process. This manual disambiguation process has

[3] `http://www.illc.uva.nl/EuroWordNet/`

to be done based on the translations definition, which in some cases are displayed with the translation itself. Despite the lack of translation definitions for many words in the usually very limited dictionary, the task of checking each translation alternative along with its definition in order to disambiguate is quite cumbersome for a user.

Translation Improvement: This was one of the important aspects that we carefully studied in the state-of-the art CLIR tools. We wanted to check whether the state-of-the art CLIR tools really consider the user as an integral part and whether the state-of-the art CLIR tools provides the user with significant information to perform the CLIR task. Only two out of the eight studied CLIR tools provided some kind of translation improvement. However, this support was deficient in various aspects. For example, some tool provide a translation improvement possibility by providing the user with the retrieved documents relevant to his/her information need. The user can initiate a new translation process, based on the examined retrieved documents (based on the search result the user can use different query terms). However, there is no possibility of improving the translation during the translation process, which leads the user to lose time and be frustrated. One tool provides a translation improvement possibility by using EuroWordnet[4] relations. However, EuroWordNet employs only a limited number of languages.

Adaptation to New Language: A very important feature to consider when designing CLIR tools is the ability of the CLIR tool to handle more languages. One of the researched state-of-the art CLIR tools provides this possibility. However, it was not described how and to what extent.

4 [Mult]iSearcher: An Interactive CLIR Tool

Based on the above discussed tools and issues (See Section 2 and 3), we discuss in the following results of the design process of [Mult]iSearcher. The goal was to obtain a smooth software design that is on the one hand supported by significant back-end components and on the other hand gives the user some control over the query translation. Therefore, the user is considered as an integral part of the CLIR process, in that he/she can interact with the tool in a way that allows him/her to improve the translation and thus improve the CLIR process.

In the following, we first discuss typical design issues discovered by performing a wide user study. The initial prototpye used for this study had been developed in order to identify common problems of CLIR systems [3] . Based on the outcome of this user study a second prototype had been developped that is discussed in detail in Section 4.2. A more general language specific evaluation of this approach is described afterwards in Section 5.

[4] `http://www.illc.uva.nl/EuroWordNet/`

4.1 User Study

In order to reveal design issues and to get more insights into user's expectations of CLIR systems in general we conducted a user study with 15 participants. The type of users we selected are students and researchers who have no, little or good knowledge in the target language. Ten of the users were male and five were female. Age ranged from 22 to 43. In the following, we outline the major results of the study. Therefore we discuss each identified problem and discuss how to tackle these problems in a revised interface design:

Translation Confidence: Addressed how useful and accurate was the contextual information that describes the translation in the source language. The translation confidence gained full rate by the users with simple request of improvement. All users found the contextual information which is displayed a long with the translation very helpful in giving them a confidence in the translation. For the user who has no or little knowledge about the target language, the contextual information was very helpful in term of giving them full confidence about the translation that they see but they cannot understand. The improvement request, was about decreasing the size of the contextual information. Currently, the tool displays 5 documents (sentences) as contextual information. Users mentioned one or two short sentences would be enough and will simplify the task of having a confidence in the translation. However, decreasing the contextual information size will lead to insufficiency in the interactive terms that can be used to improve the translation.

We tackled this issue in the new design and compensated for this insufficiency. In the new design, the tool provides the user with a list of interactive terms, regardless of the contextual information. The user can select any term/terms to improve the translation, if needed, saving the user time. There will be two benefits from this step: the contextual information will be used only for translation confidence and the list of suggested interactive terms will be used to improve the translation, if needed.

Lack of Information Flow Control: For example, users complained that intensive-information is displayed at the same time e.g., the user is disturbed by seeing all contextual information for the five ranked translations displayed at once. The user mentioned that it would be helpful to control which information could be seen and when.

In the new design, we tackled this issue in that we gave the user a possibility to hide information, which is not of current interest. This is done by displaying only the contextual information for the top ranked translation (first translation) and displays only a few words in the contextual information for other translations. If the user is interested in checking other translations with its contextual information, the user only needs to click on "mehr anzeigen – show more". The tool then displays the full contextual information for the selected translation. At the same time, the tool automatically hides the contextual information for the previously selected translation. The user is then able to see and focus only on the selected translation and its contextual information.

Lack of Information in the Target Language: Some users, who have a good knowledge in the target language, would like to see information for the translation in the target language "gloss".

Interactive Terms Usefulness / Identifying Related Terms in the Contextual Information: Addressed the usefulness of the interactive terms in the contextual information that can be used to improve the translation. The suggested interactive terms by the tool, based on highly frequent co-occurrence data, in the context of the query, are in bold blue and underlined [3]. In many cases these were helpful as the users mentioned, however, in some cases the users needed more terms than the ones suggested by the tool. Although these terms are found in the context of the user query, the users mentioned that these terms in many cases they don't lead to an improvement in the translation. Another deficit which has been reported by some users is the lack of identifying terms in the contextual information e.g., currently terms related to the user query terms are binary compared e.g., (universitätsabschluss "university degree") wouldn't be recognized as relevant to (universität "university").

In the new design the contextual information is only used for translation confidence and a list of suggested terms to improve the translation is provided independently from the contextual information. In the old design, the interactive terms are obtained from the contextual information, which is a few documents in size. In the new design, the corpora is used as a whole, to obtain these suggested terms. Only terms that have a significant co-occurrence score, with the query terms, will be suggested. In order to tackle the second deficit which is the identifying of terms in the contextual information, we highlight the terms which exist in the user query with bold black and the remaining character(s)/word, which forms a synonym for the given query term with light grey (selecting these highlighting mechanism can avoid blind color people issue).

Lack of Detailed Error Notification: Users mentioned that there is a lack of detailed error notification being displayed when some error occurred e.g., when there is no translation available from the dictionary for some term/terms, the tool wouldn't notify the user. Instead the tool would show a message that there were no translations available. This results in confusion as to whether there is no translation available for a term/terms or whether the translation algorithm couldn't find enough statistical co-occurrence data to perform the translation.

In the new design, we tackled this issue by notifying the user that there is no translation available for a term/terms from the dictionary. In order to simplify the user task, the tool will automatically translate the rest of the terms. However, the rest of the terms must be at least two terms so the translation can be performed.

Lack of Interaction Mechanism: Users mentioned that there is a lack of control when they interact with the tool e.g., when the user selects a term/terms to improve the translation, the user has no possibility of removing this term, if he/she discovers that the selected term/terms doesn't improve the translation. In the old design, the user tackled this issue by resubmitting the original query. However, this results in wasted time and effort. Another lack of interaction mechanism is in order to improve the translation, the user has to select an interactive term/terms from the contextual information. This term is automatically added to the user query. This query will be resubmitted and new translations, based on the selected term/terms, will be provided. This mechanism is not welcomed by the user as it will enlarge the query each time the user interacts with the tool and selects relevant terms to improve the translation. The majority of the users prefer not to revise the original query they submit and only wish to rank the initially obtained translations.

In the new design, we tackled this issue by providing the user with more control, in that any term/terms can be selected/deselected by a simple mouse click. The tool will then immediately respond to any action by the user e.g., selecting a term would result in performing the re-ranking process (with the contextual information and the gloss for each translation) and alternatively, deselecting a term would return back to a previous state. In order to deal with the lack of improving the translation, we offered the user to use the interactive term/terms only for ranking purposes and they will not be added to the original query. This suggestion was welcomed by the user, which we took into account in the new design.

4.2 System Architecture

As Figure 1 shows, the interface flow starts when the user submits his/her query. The entered user query will pass through several interface components before the CLIR results can be displayed to the user. These interface components are: query pre-processing, automatic translation, contextual information and gloss, query post-processing and Error notification. Figure 1, shows how these components are related and how the information flows between the different interface components. In the following, we describe these components in detail.

Query Pre-processing Component: Before the query can be translated, it will be pre-processed. The first important pre-processing step is to check whether the query is misspelled or since usually the dictionary doesn't include all word forms instead only the root form, morphological analyzer can be used to find the root of each query term. For example, if the query is misspelled, the misspelling query term/terms, using the MultiSpell approach [4], will be identified and corrected. MultiSpell is a language-independent spell-checker that is based on an enhancement of a pure n-gram based model.

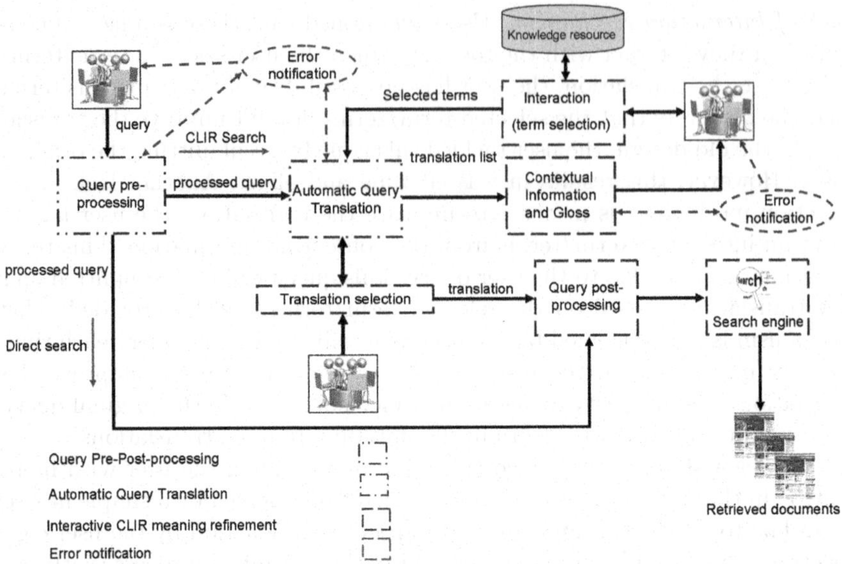

Fig. 1 The main components of the [Mult]iSearcher system architecture

Automatic Translation Component: Using the bilingual dictionary, each possible translations for each query term is obtained (translation set). Having a translation set for each query term, the translation combinations between terms in the translation set are generated. The result of this step is having all possible translations for the submitted user query (translation combinations). In order to select and rank the proper translations, statistical methods, based on target corpora (Mutual Information approach) [3] is used. The translations that maximize the statistical score measure are selected and ranked. In the following, we describe the Mutual Information Approach in more detail:

Mutual Information Approach. Giving a source of data, Mutual Information (MI) is a measure to calculate the correlation between terms in specific space (corpora or web). The MI approach has been frequently used in word sense disambiguation tasks e.g., [13]. The automatic translation process starts by translating each query term independently. This is done by obtaining a set of possible translations of each of the query terms from the dictionary. Based on the translation sets of each term, sets of all possible combinations between terms in the translation sets are generated. Using co-occurrence data extracted from monolingual corpora, the translations are then ranked based on a cohesion score computed using Mutual Information: Given a query $q = \{q_1, q_1, ..., q_n\}$, and its translation set $S_{qk} = \{q_k, t_i\}$, where $1 \leq k \leq n, 1 \leq i \leq m_k$ and m_k is the number of translations for query term k. The MI score of each translation combination can be computed as follows:

$$MI(q_{t_i}, q_{t_j}) = log_2 \frac{p(q_{t_i}, q_{t_j})}{p(q_{t_i})p(q_{t_j})} \quad (1)$$

The probability $p(q_{t_i}, q_{t_j})$ is estimated by counting how many time each two terms, in the translation combination, appear together in the corpora, e.g., how many time the term q_{t_1} and the term q_{t_2} co-occur together in the corpora. The probabilities $p(q_{t_i})$ and $p(q_{t_j})$ are estimated by counting the number of individual occurrences of each possible translated query term in the corpora. For a more detailed discussion about the MI based approach for disambiguating a translation and how it was improved to tackle the data sparseness issue, we refer the reader to our previous work [5].

Contextual Information and Gloss Component: Once the automatic translation is performed and displayed to the user, new issues as limited confidence of the user in the translation can arise. Therefore, a contextual information provider is integrated in the CLIR tool. The contextual information is displayed to the user – along with each proposed translation – in a language the user is familiar with [3]. The contextual information is not delivered to the user as raw text; instead a classified representation for each term, in the contextual information, is generated (see Figure 2). For example, an interesting point for the user, to rely on the translation, is to see the query terms in the contextual information. These terms are highlighted in bold black and are displayed with their context in different sentences. In order to improve

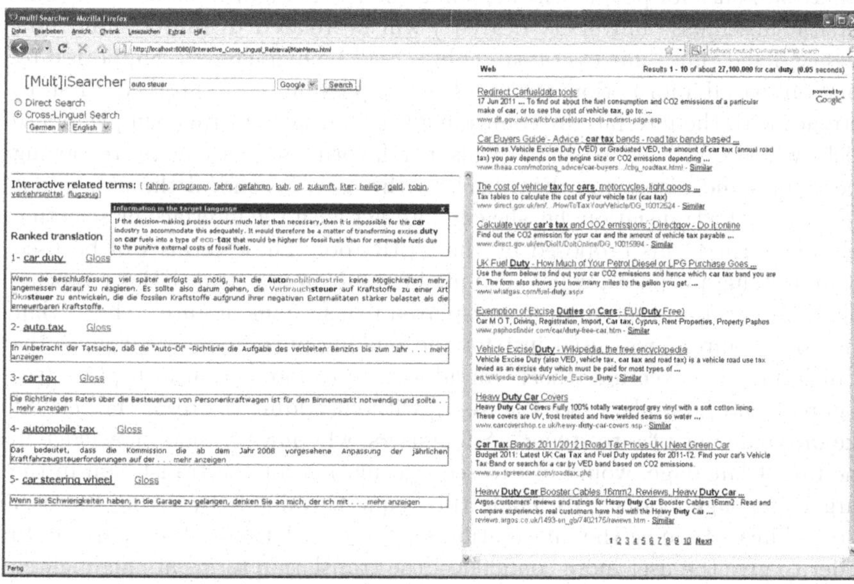

Fig. 2 The interface of the second prototype with the initial suggested translation (the retrieved documents are obtained by the Google-API)

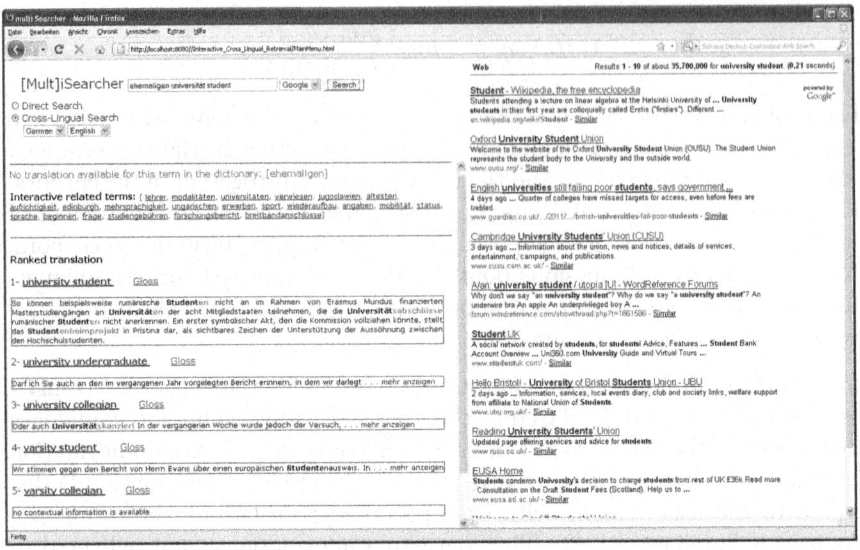

Fig. 3 The error notfication e.g, there is no translation for query term "ehemaligen (previous)"

this feature, synonyms for the given query terms, in the contextual information, are highlighted in light grey (selecting these highlighting mechanisms can avoid issues for people who are color blind). As Figure 3 shows that (universitätsabschluss "university degree") will be related to (universität "university") and thus "universitätsabschluss" will be highlighted. Supported by this contextual information, the user has two possibilities to proceed. First, to interact with the interface by confirming the translation with a simple mouse click, which will then be sent to his/her favorite search engine, retrieving the results and displaying them back to the user. Second, if the user is not sure about the translation, he/she can interact with the interface by selecting relevant term/terms proposed by the CLIR tool. These terms will be used for re-ranking purposes, which also might result in new translations appearing, different from the initial five displayed translations. Figure 4 shows that the term "fahren" was used just for the ranking purpose where it gives the translation "car steering wheel" an advantage to move from fifth place (see Figure 3) into first place (see Figure 4), without adding the term "fahren" to the original query "auto steuer". Some users, who have a good knowledge in the target language, would like to see information for the translation in the target language "gloss". We made this request available by giving the user a possibility of seeing the information for the translation in the interface. In order to give the user more confidence, the translation terms are highlighted in the gloss, in the same way as in the contextual information (see Figure 2).

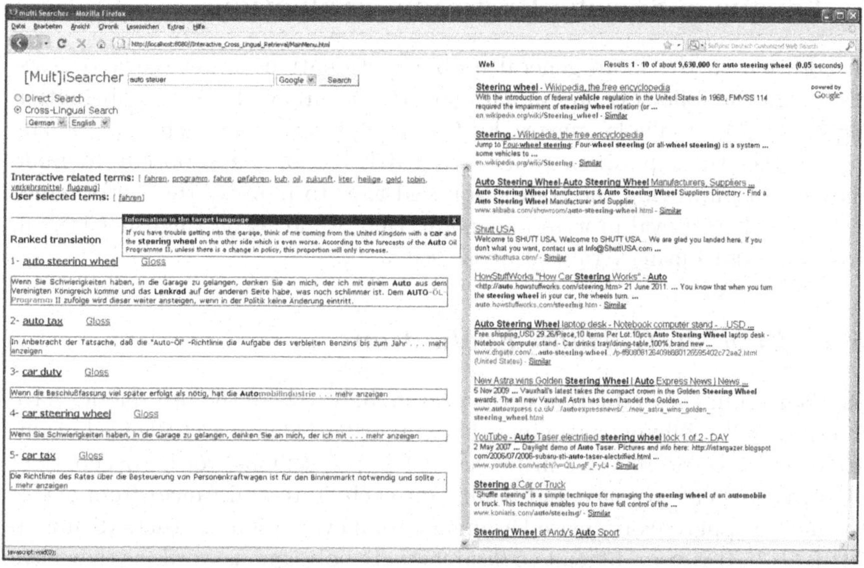

Fig. 4 The re-ranked translation based on the user interaction (the retrieved documents are obtained by the Google-API)

Query Post-processing Component: Once the translation is refined and acknowledged by the user, new issues may arise. The characteristics of highly inflectional languages, such as Arabic, very often result in poor information retrieval performance. As a result, current search engines suffer from serious performance with the direct query term-to-text-word matching for these languages. Thus, search engines need to be able to distinguish different variants of the same word. In order to tackle this issue, a language-independent conflation approach, based on enhancing the n-gram approach is integrated in the CLIR tool [2].

Error Notification Component: In order to support the user, in using the CLIR tool, an error notification component is integrated. The error notification component is responsible for watching all CLIR tool components and alerting the user to any failure and its causation. For example, when there is no translation available from the dictionary, for some term/terms, the error notification component will notify the user why his/her query is not translated. As is shown in Figure 3, the user submits the German query "ehemaligen Universität Student" , the tool notified the user that there is no translation available for the term "ehemaligen" from the bilingual dictionary and at the same time the tool provides an automatic translation for the rest of the terms which is "university student".

5 Language Specific Prototype Evaluation

In addition to the more general usability evaluations presented above, we discuss in the following an evaluation of the accuracy of the disambiguation algorithm for specific languages. Furthermore, we were interested in evaluating whether the support provided by our CLIR tool is appropriate to guide the user in improving the translation and thus to improve the performance of the CLIR retrieval process in general. Here we focus on English/German. For evaluations using Arabic/English see [3, 5].

5.1 English/German Evaluation

Different from the evaluation performed in [3, 5], where the test queries have multiple quite different meanings, here, in order to have a challenged evaluation, we selected 100 test instances of five polysemous words (occupation, plant, movement, passage and bank) for each 20 test instances from one of the most popular Word Sense Disambiguation evaluation data sets (SemEval 2010)[5] [17]. It is very difficult to disambiguate polysemous words as they have separate different meanings that are related to one another. For example, the English polysemous word "plant" can have these related meanings in German "gewächs", "pflanze", "vegetation" etc. Another example is the English polysemous word "passage" that has these separate related meanings in German "Durchgang", "Durchtritt", "Durchfahrt", "Durchlass", "Überfahrt", "Verlauf" , etc. Furthermore, disambiguating polysemous words is a very difficult task with scores being very close to the baseline measure [15]. This evaluation is performed in particular, to check whether the proposed CLIR tool, in this chapter is appropriate to select the correct translation corresponding to the given polysemous word in the source language.

5.1.1 Experiment Setup

For the CLIR word sense disambiguation task in [17], numbers of English nouns were given. For each English noun 20 test instances were provided. For each test instance, possible translations in the target languages were also provided (hand-tagged Gold standards translations). There were two types of scoring the translations, one based on scoring the best translation and the other based on scoring the best 5 translations in the target languages. For our experiment, we used only the first, where we conducted the test only on selecting the best translation of the ambiguous word in the target language. There were two types of tests; one is a bilingual test, where the ambiguous word is translated to one target language or a multilingual test where the ambiguous word is translated into five languages (Dutch, French, German, Spanish and Italian). For our test, we selected the bilingual test where the

[5] http://webs.hogent.be/~elef464/lt3_SemEval.html#_subtasks

test instances (queries) are in English and the translations are in German. The test instances are long sentences where some of them are greater than 63 words in length, (see Figure 5) which do not fit into a real life scenario CLIR, where the search engine queries are usually between 2.4 and 2.7 in length [14]. In order to deal with this and have significant evaluation for our disambiguation algorithm, we adapted the test sentences and extracted only important words from each test instance. After removing stop words, for each test instance, important words were selected. This task has been performed by the users, 10 users, each has 10 test instances. They constructed their queries by selecting a few words which describe their needs in the test instances context. For the test instances shown in Figure 5, these words are selected: "physical", "cash", "movement". The users were requested to select as few words as possible to express their need. The new test instances (queries) ranged from 2 to 7 words in length with the average being 4.1.

```
<instance id="12">
    <context>
        (4) Account should also be taken of complementary activities carried out in other international fora, in particular
        those of the Financial Action Task Force on Money Laundering (FATF), which was established by the G7 Summit
        held in Paris in 1989. Special Recommendation IX of 22 October 2004 of the FATF calls on governments to take
        measures to detect physical cash <head>movements</head>, including a declaration system or other disclosure obligation.
    </context>
</instance>
```

Fig. 5 Test instance number "12" for the ambiguous word "movement"

The Gold standards translations (in 5 languages – e.g., see Figure 6) were extracted from the Europarl parallel corpora[6] [16]. Europarl parallel corpus is a collection of documents for 21 languages. These parallel corpora were extracted from the proceedings of the European Parliament. To construct the Gold standard translations, human annotators were requested to select one of the automatically provided translations from the corpora. Each number in front of each possible translation reflects the number of times this translation was picked up by the human annotators. Figure 6 shows the Gold standard translations for the first 10 test instances for the polysemous word "occupation".

The test sentences (in English) were selected from the JRC-ACQUIS multilingual parallel corpora[7]. The JRC-ACQUIS multilingual parallel corpora is the total body of European law that are applicable in European members states. Currently this corpora is a collection of text written from 1950 up to now, however this text is growing continuously. There were 2 test data: first, development test data which contains 5 polysemous nouns (occupation, passage, movement, plant and bank) each were provided with 20 test instances. Second, test data which contain 50 English nouns for each 20 test instances were provided. For our experiment, due to the adaption of this test data to

[6] http://www.statmt.org/europarl/

[7] http://wt.jrc.it/lt/Acquis/

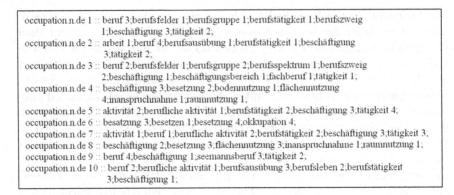

```
occupation.n.de 1 :: beruf 3;berufsfelder 1;berufsgruppe 1;berufstätigkeit 1;berufszweig
                     1;beschäftigung 3;tätigkeit 2;
occupation.n.de 2 :: arbeit 1;beruf 4;berufsausübung 1;berufstätigkeit 1;beschäftigung
                     3;tätigkeit 2;
occupation.n.de 3 :: beruf 2;berufsfelder 1;berufsgruppe 2;berufsspektrum 1;berufszweig
                     2;beschäftigung 1;beschäftigungsbereich 1;fachberuf 1;tätigkeit 1;
occupation.n.de 4 :: beschäftigung 3;besetzung 2;bodennutzung 1;flächennutzung
                     4;inanspruchnahme 1;raumnutzung 1;
occupation.n.de 5 :: aktivität 2;berufliche aktivität 1;berufstätigkeit 2;beschäftigung 3;tätigkeit 4;
occupation.n.de 6 :: besatzung 3;besetzen 1;besetzung 4;okkupation 4;
occupation.n.de 7 :: aktivität 1;beruf 1;berufliche aktivität 2;berufstätigkeit 2;beschäftigung 3;tätigkeit 3;
occupation.n.de 8 :: beschäftigung 2;besetzung 3;flächennutzung 3;inanspruchnahme 1;raumnutzung 1;
occupation.n.de 9 :: beruf 4;beschäftigung 1;seemannsberuf 3;tätigkeit 2;
occupation.n.de 10 :: beruf 2;berufliche aktivität 1;berufsausübung 3;berufsleben 2;berufstätigkeit
                      3;beschäftigung 1;
```

Fig. 6 Gold standard translation for the polysemous word "occupation" based on human annotators

fit for an actual CLIR scenario and due to the unavailability of the full test data, we could use only the development test data, so at the end we evaluated our disambiguation algorithm based on 100 test instances (queries).

5.1.2 Evaluation of Disambiguation Algorithm

The goal of the evaluation was based on two perspectives, first to evaluate the performance of the disambiguation algorithm in German and English languages. In order to achieve this goal, we smoothly integrated more languages into the proposed tool in this chapter (the first prototype included only Arabic and English language pair). The integration was a trivial task where only two steps were needed. We obtained an English-German dictionary and English-German parallel corpora from Europarl parallel corpora [16]. No modification for the disambiguation algorithm as well as for the contextual information algorithm was required. Second was to evaluate whether user interaction, could improve the performance of the disambiguation algorithm by selecting relevant term/terms proposed by the tool.

In order to evaluate the performance of the disambiguation algorithm, we used the precision measurement which is proposed by many researchers for a word sense disambiguation task e.g., [10, 12]. Precision is the proportion of the correctly disambiguated senses for the ambiguous word. In the gold standard translations, the translation that has a larger number associated with it compared to other possible translations are ranked first (see Figure 6). We compared the result of the disambiguation algorithm only with this translation i.e., if the first ranked translation was selected by 4 human annotators and the second was selected by only 3 human annotators, we consider only the first as correct even if for algorithm proposes the second ranked translation as the correct one. In order to give the user wide possibilities to interact with the tool, the tool provides the user with 5 ranked translations along with their contextual information. Furthermore, a list of possible interactive related terms, to the user query, is presented to the user. We assumed the tool translation correct when it is displayed within the 5 ranked translations provided by the tool. Figure 7 shows one of the test instances "health plant animal" for the polysemous word "plant". The tool successfully presented to the user for the polysemous word "plant", the two correct senses (based on human annotators disambiguation) "pflanze" in fist rank, and "Gewächs" in third rank.

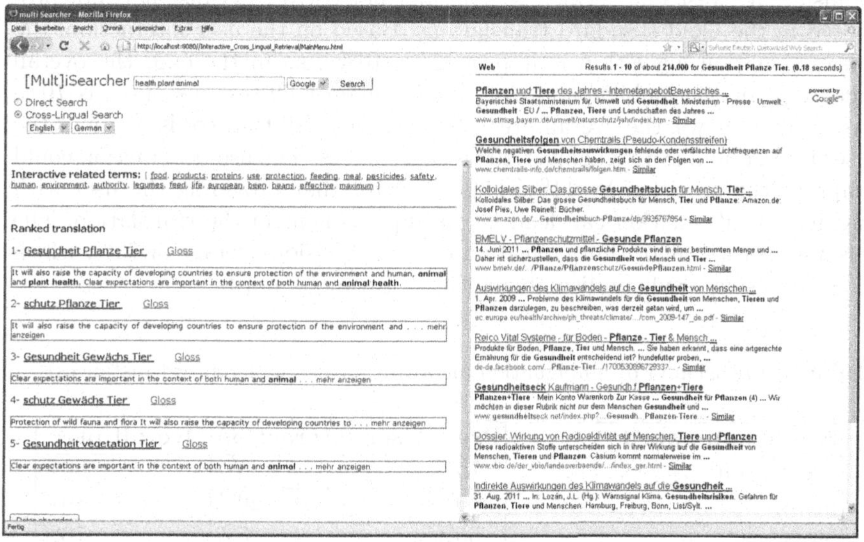

Fig. 7 The cross-lingual retrieval for one of the test instances "health plant animal" for polysemous word "plant"

For the ambiguous word (occupation), we find that the disambiguation algorithm gained up to 75%. With the user interaction, the disambiguation algorithm improved by 5%. For example, the algorithm failed to provide the correct translation for the user query "high occupation rate" but after the user interaction and the selection of the proposed relevant term "unemployment", the disambiguation algorithm could provide the user with the correct translation "besetzung". For the ambiguous word "plant", the algorithm without the user interaction could provide the correct translation for 11 test instances out of 20 and gained accuracy up to 60%. However, with the user interaction, the disambiguation algorithm is improved by 20 % and gained an overall accuracy average of 80%. For the ambiguous word "movement" the algorithm gained up to 65% accuracy and with the user interaction, the disambiguation algorithm improved by 5% and gained an overall average of 70%. The disambiguation algorithm for the ambiguous word "passage" gained up to 50% and with the user interaction, the disambiguation algorithm improved by 15% and gained an overall average of 65%. For the ambiguous word "bank", the disambiguation algorithm could disambiguate 12 out of 20 test instances with accuracy being 60% The accuracy is an underestimation of the real accuracy since some of the Gold standard translations are not direct translations for the given query terms. For example, we consider the query "palestinian people west bank", the proposed translation in the Gold standard for the "west bank" is "westjordanufer", the word "jordan" does not exist in the query. Therefore, our algorithm could propose only the translation "west ufer" which we consider not correct because it is not proposed in the Gold standard translations. Based on the user interaction, the disambiguation algorithm is improved by 10% , so in the end, the overall accuracy for the ambiguous word "bank" against 20 test instances is 70%. As Table 1 shows, the overall precision average of all test words, without the user interaction, is 62% against 100 test instances. The user interaction could improve the precision by 11%. This indicates that providing the user with significant information can lead to an improvement in the translation. The disambiguation algorithm gained an overall precision average of 73%, which is a promising result in disambiguating polysemous words.

Table 1 Precision of disambiguation algorithm (average of 100 test instances)

Ambiguous word	**Prec.**	**User interaction**	**Overall prec.**
occupation	75%	5%	80%
plant	60%	20%	80%
movement	65%	5%	70%
passage	50%	15%	65%
Bank	60%	10%	70%
Overall average Precision	**62%**	**11%**	**73%**

6 Conclusion

We proposed an interactive CLIR approach in order to investigate the feasibility and the validity of utilizing interactive translation methods for disambiguation tasks in CLIR processes. To ensure that a user has a certain confidence in selecting a translation, which he/she possibly cannot even read or understand, the approach provides appropriate information about translation alternatives and their meaning, so that a user has a certain degree of confidence in the translation. Based on a literature review of current CLIR tools, we furthermore identified several common issues and shortcomings, which we tried to tackle in the approach proposed in this chapter. The interface components are integrated in order to perform the CLIR task smoothly, from submitting the query till getting the relevant documents. The smooth design of the approach is on the one hand obtained by performant back-end components and on the other hand by providing the user with some control over the query translation process. Furthermore, the proposed CLIR approach considers the user as an integral part of the CLIR process, in that the user can interact with the tool in a way that allows him/her improve the translation and thus to improve the overall performance of the CLIR process.

References

1. Aggarwal, C.C., Yu, P.S.: On effective conceptual indexing and similarity search in text data. In: Proc. of 2001 IEEE Intl. Conf. on Data Mining, ICDM 2001, pp. 3–10 (2001)
2. Ahmed, F., Nürnberger, A.: Evaluation of n-gram conflation approaches for arabic text retrieval. Journal of the American Society for Information Science and Technology (JASIST) 60(7), 1448–1465 (2009)
3. Ahmed, F., Nürnberger, A.: multi searcher: can we support people to get information from text they can't read or understand? In: Proc. of 33rd Intl. ACM SIGIR Conf. (SIGIR 2010), pp. 837–838. ACM, NY (2010)
4. Ahmed, F., Nürnberger, A., De Luca, E.W.: Revised n-gram based automatic spelling correction tool to improve retrieval effectiveness. Research J. on Computer Science and Computer Engineering with Applications (Polibits) 40, 39–48 (2009)
5. Ahmed, F., Nürnberger, A., Nitsche, M.: Supporting Arabic Cross-Lingual Retrieval Using Contextual Information. In: Hanbury, A., Rauber, A., de Vries, A.P. (eds.) IRFC 2011. LNCS, vol. 6653, pp. 30–45. Springer, Heidelberg (2011)
6. Bian, G.W., Teng, S.Y.: Integrating query translation and text classification in a cross-language patent access system. In: Proc. of NTCIR-7 Workshop Meeting, pp. 16–19 (2008)
7. Braschler, M., Peters, C., Schäuble, P.: Cross-language information retrieval (clir) track overview. In: Proc. of 8th Text Retrieval Conference, pp. 25–33 (2000)
8. Capstick, J., Diagne, A.K., Erbach, G., Uszkoreit, H., Leisenberg, A., Leisenberg, M.: A system for supporting cross-lingual information retrieval. Information Processing and Management: an International 36(2), 275–289 (2000)
9. Carbonell, J.G., Yang, Y., Frederking, R.E., Brown, R.D., Geng, Y., Lee, D.: Translingual information retrieval: A comparative evaluation. In: Proc. of 15th Intl. Joint Conf. on Artificial Intelligence, pp. 708–714 (1997)
10. Dagan, I., Itai, A.: Word sense disambiguation using a second language monolingual corpus. Computational Linguistics 20(4), 563–596 (1994)
11. De Luca, E.W., Hauke, S., Nürnberger, A., Schlechtweg, S.: MultiLexExplorer - combining multilingual web search with multilingual lexical resources. In: Combined Works. on Language-Enhanced Educat. Techn. and Devel. and Eval. of Robust Spoken Dialogue Sys., pp. 17–21 (2006)
12. Fakhrahmad, S., Rezapour, A., Jahromi, M.Z., Sadreddini, M.: A new word sense disambiguation system based on deduction. In: Proceedings of the World Congress on Engineering (WCE 2011), pp. 1271–1281 (2011)

13. Fernandez-Amoros, D., Gil, R.H., Somolinos, J.A.C., Somolinos, C.C.: Automatic Word Sense Disambiguation Using Cooccurrence and Hierarchical Information. In: Hopfe, C.J., Rezgui, Y., Métais, E., Preece, A., Li, H. (eds.) NLDB 2010. LNCS, vol. 6177, pp. 60–67. Springer, Heidelberg (2010)
14. Gabrilovich, E., Broder, A., Fontoura, M., Joshi, A., Josifovski, V., Riedel, L., Zhang, T.: Classifying search queries using the web as a source of knowledge. ACM Transactions on the Web 3(2), 1–28 (2009)
15. van Gompel, M.: Uvt-wsd1: A cross-lingual word sense disambiguation system. In: Proceedings of the 5th International Workshop on Semantic Evaluation, SemEval 2010, pp. 238–241 (2010)
16. Koehn, P.: Europarl: A parallel corpus for statistical machine translation. In: Proceedings of the Tenth Machine Translation Summit, pp. 79–86. AAMT, Phuket (2005)
17. Lefever, E., Hoste, V.: Semeval-2010 task 3: Cross-lingual word sense disambiguation. In: Proceedings of the 5th International Workshop on Semantic Evaluation, SemEval 2010, pp. 15–20. Association for Computational Linguistics (2010)
18. Lopez-Ostenero, F., Gonzalo, J., Penas, A., Verdejo, F.: Interactive Cross-Language Searching: Phrases are Better than Terms of Query Formulation and Refinement. In: Peters, C., Braschler, M., Gonzalo, J. (eds.) CLEF 2002. LNCS, vol. 2785, pp. 416–429. Springer, Heidelberg (2003)
19. Oard, D.W.: Cross-language text retrieval research in the usa. In: Proc. of 3rd DELOS Workshop; Cross-Language Information Retrieval, No. 97-W003 in Ercim Workshop Proc. European Research Consortium for Informatics and Mathematics, pp. 7–16 (1997)
20. Oard, D.W., He, D., Wang, J.: User-assisted query translation for interactive cross-language information retrieval. Information Processing and Management: an International Journal 44(1), 181–211 (2008)
21. Ogden, W.C., Davis, M.W.: Improving cross-language text retrieval with human interactions. In: Proceedings of the 33rd Hawaii International Conference on System Sciences, p. 3044. IEEE Computer Society, Washington, DC (2000)
22. Petrelli, D., Beaulieu, M., Sanderson, M., Demetriou, G., Herring, P., Hansen, P.: Observing users, designing clarity: A case study on the user-centered design of a cross-language information retrieval system. Journal of the American Society for Information Science and Technology (JASIST) 55(10), 923–934 (2004)

Machine Translation at Work

Aljoscha Burchardt, Cindy Tscherwinka, Eleftherios Avramidis,
and Hans Uszkoreit

Abstract. Machine translation (MT) is – not only historically – a prime application of language technology. After years of seeming stagnation, the price pressure on language service providers (LSPs) and the increased translation need have led to new momentum for the inclusion of MT in industrial translation workflows. On the research side, this trend is backed by improvements in translation performance, especially in the area of hybrid MT approaches. Nevertheless, it is clear that translation quality is far from perfect in many applications. Therefore, human post-editing today seems the only way to go. This chapter reports on a system that is being developed as part of taraXŰ, an ongoing joint project between industry and research partners. By combining state-of-the-art language technology applications, developing informed selection mechanisms using the outputs of different MT engines, and incorporating qualified translator feedback throughout the development process, the project aims to make MT economically feasible and technically usable.

1 Introduction

Just a few years ago, English was considered the lingua franca of the future, at least in business contexts. Today, the situation has drastically changed, particularly in light of the developments in web communication and publishing. The amount of online content in other languages has exploded. According to some estimates, the European market volume for translation and interpretation, including software localisation and website globalisation, was €5.7 billion in 2008 and was expected to grow by 10% per annum.[1] Yet this existing capacity based mostly on human translation is by far not enough to satisfy current and future translation needs. The integration of MT would seem to be the most promising way of managing translation in a cost-effective and timely fashion in the future but, sur-

Aljoscha Burchardt · Eleftherios Avramidis · Hans Uszkoreit
DFKI – Language Technology Lab, Alt-Moabit 91c, 10559 Berlin, Germany
e-mail: aljoscha.burchardt@dfki.de, eleftherios.avramidis@dfki.de,
hans.uszkoreit@dfki.de

Cindy Tscherwinka
euroscript Deutschland, Alt-Moabit 91, 10559 Berlin, Germany
e-mail: Cindy.Tscherwinka@euroscript.de

[1] European Commission Directorate-General for Translation, Size of the language industry in the EU, Kingston Upon Thames, 2009.

A. Przepiórkowski et al. (Eds.): *Computational Linguistics*, SCI 458, pp. 241–261, 2013.
DOI: 10.1007/978-3-642-34399-5_13

prisingly, neither the economic feasibility of MT nor the correlation with the real-world needs of professional translators and LSPs have been analysed in depth up to now. From the LSP's point of view, MT lacks basic functionality to be a true help in professional translation processes. Therefore, despite continuous improvements to MT quality, this technology is economically viable only in very specific translation scenarios.

Within research, the move from rule-based to statistical MT systems and more recently to hybrid ones has led to translation results one would not have dreamt of just a few years ago. Still, there is no one-size-fits-all MT engine available. The various MT paradigms have different strengths and shortcomings – not only in terms of quality. For example, rule-based MT (RBMT) offers good control of the overall translation process, but setting up and maintaining such a system is very costly as it requires trained specialists. Statistical MT (SMT) is cheap, but it requires huge amounts of computing power and training data, which may not be available so that new languages and domains can be included. Translation Memory Systems (TMS) provide human translation quality, but are limited in coverage due to their underlying design. Selecting the right system (combination) for the right task is an open question.

This paper reports on a system that is being developed as part of the taraXŰ project, which aims to find answers to questions such as:

1. Can MT help human translators reduce translation costs without sacrificing quality?
2. How can MT be integrated into professional translation processes?
3. How can a hybrid system utilise input properties, metadata about system behaviour, linguistic analysis of input and output text, etc.?
4. When is post-editing most effective?
5. How can the result of human post-editing be fed back into the system to help improve it?

To answer these questions, integrated and human-centric research is needed. Some related work is available on supporting human translators in using MT and post-editing (e.g., Casacuberta et al., 2009; Koehn, 2009; Specia 2011), but for the most part, no language professionals were involved in the design and calibration of such systems. The question of how professional translators can optimally be supported in their translation workflow is still open. In this article we focus on questions 2-3 and 5. Questions 1 and 4 will be addressed in the further course of the project.

As the first part of its analytic process, the taraXŰ system makes a selection within a frame of hybrid MT, including RBMT, TMS, and SMT. Then, a self-calibration component is applied, supplemented by controlled language technology and human post-processing results from previous translations in order to match real-world translation concerns. A novel feature of this project is that human translators are integrated into the development process from the very beginning: after several rounds of collecting and implementing user feedback, the selection and calibration mechanisms will be refined and iteratively improved.

In this chapter we will consider the motivations behind the work from two different angles: We will give an overview of the language service provider's general conditions when working with MT in a professional translation environment and the scientific motivation for working on hybrid MT in the taraXŰ project. After that, we will describe the actual system being built in the project and show how it takes account of the prerequisites for using MT in a professional translation workflow. The final part of the chapter provides initial observations on the user feedback and shows the feasibility of a sample selection mechanism. Some notes on further work and conclusions close the chapter.

2 Translation Quality and Economic Pressure – The Language Service Provider's Motivation

The provision of high quality translations – preferably at a low cost – is a matter of balancing a maximum level of automation in the translation process on the one hand with carefully designed quality assurance measures on the other hand. The ever-growing translation volumes and increasingly tight schedules cry out for the integration of machine translation technology in professional translation workflows. Though, for a variety of reasons, this is only economically feasible under very specific conditions. This section provides a background on LSPs' requirements concerning translation workflows and in particular Translation Memory Systems (TMS). This background is helpful for understanding the design of our system, which we describe later on in this chapter.

2.1 Re-using Language Resources

For a long time machine translation technology has been developed and improved with the clear focus on providing understandable translations for end users (inbound translation). Its primary goal has been to put people in a position to understand the meaning of some foreign language sentence (information gisting) using machine translation that preserves the source language meaning and complies with general target language requirements such as spelling and grammar. While this is a helpful procedure in a zero-resources scenario, the typical LSP scenario is that of outbound translation for a specific customer (content producer). Usually, language resources from past translations are available, having been revised and approved. The contained material should not undergo MT again; instead, the resources should be accounted for, during the translation process, in order to reduce redundancy in the manual effort of post-editing and proofreading.

Professional translators needed a comfortable translation environment that enables maximum reuse of past translations, in order to minimise translation effort and ensure maximum consistency. This need resulted in the development of Translation Memory (TM) systems, which are now standard in professional translation workflows.

In TM systems, language resources do not only exist in the form of aligned source and target language sentences in the memory. Usually, a client-specific terminology database is available for translation as well. Such a termbase stores multilingual terminology and has usually been carefully compiled and validated during product development. It is a very important part of corporate identity and binding for translation. Apart from terms, it also stores additional information that helps the translator, such as definitions, sample sentences or usage information. Most TM systems support term recognition, which integrates the termbase into the editor and helps the translator by displaying relevant terminology for the current segment. Terminology verifiers check the translation for compliance with the termbase and warn the translator if source language terms are not translated according to the information stored in the termbase.

By comparison, in MT terminological assets can be factored into translation, but compliance with given terminology still has to be manually checked by the translator. All lexical choices that do not comply with a given terminology have to be identified and manually modified by the human translator. That is true even if the machine translation result would be acceptable from a semantic point of view.

When employing MT, one has to keep in mind that target language requirements may also go beyond the lexical level and concern phrasing at the sentence level. From this aspect, even semantically suitable and grammatical translations may be rejected if stylistic requirements of the target language are not met. On the contrary, when working with a TM System the translations suggested to the translator are human-made, client-specific, validated and therefore expected to be grammatically and orthographically correct and in line with the target language requirements concerning style and terminology.

The matching algorithms of the TM systems provide an estimate of the similarity of a proposed translation derived from the memory with the correct translation of a given sentence. As a rule of thumb, 100% matches – where the sentence to be translated exactly matches a source language sentence in the memory including formatting information – do not need to be modified; high fuzzy matches require minor post-editing effort and low fuzzy matches require a new translation. These estimates enable a calculation of translation effort and are the basis of pricing. The more translations can be re-used, the lower the price.

2.2 Machine Translation and Post-editing Effort

The advantage of MT over Translation Memories is that it can deliver translations even if there are no past translation projects that can be reused for a new translation job. In its current form, however, it is missing a valuable feature of TMs: It is not possible to estimate machine translation quality or the post-editing effort required to create a high quality translation without reference to the post-edited result. An MT translation may have the value of a 100% match as well as a zero match. That poses a real problem for the LSP. On what basis does it make a fair offer?

Assessing the time and effort needed to post-edit MT output may be possible in a very restricted scenario, in which source language document characteristics are known, the MT engine is optimised according to the expected source, and when there is already some experience of MT results in that scenario. But even this is an assessment at the document level and subject to change. For the economic application of MT in any given scenario, methods for automatically assessing MT quality at the sentence level are needed. As mentioned earlier, such an assessment includes not only an evaluation of orthography and grammar of the source language content, but also how well target language requirements, such as terminology and style, are met. In the context of software documentation, there may also be more general criteria such as the maximum permissible length of a sentence.

2.3 *Translation Workflow Requirements*

Apart from these content-related requirements, there are also workflow requirements that have to be taken into account when deciding how MT can be deployed in an economically feasible way, the most important being

- the processability of various file formats apart from plain text,
- the possibility of accessing MT from the familiar and functionally mature working environment of the professional translator
- the possibility of making use of the post-edited machine translation results for future translation jobs
- the continuous optimisation of the MT engines with minimum manual effort.

The professional translator generally works with a TM system that consists of a translation memory, a terminology database and an editor. The TM system performs the task of stripping formatting information from the translatable content. Most TM systems come with filters that can handle a variety of file formats. Translatable content is then compared with the translation memory and filled up with target language content according to a defined threshold value in a bilingual file. The translator then uses the editor, from which he or she has access to the translation memory and the terminology database for translation. The proposals, if there are any, from the translation memory are fed into the editor where they are manually edited. After translation, the new source and target language pairs are stored in the translation memory and the bilingual file is saved as a target language file in the original file format.

In an effort to avoid the need to work with multiple software tools and to provide the translator with MT in his or her familiar working environment, a trend for integrating MT into the translation memory has emerged. The major TM system vendors have already integrated MT engines into their products and innovative LSPs have built their own solutions to integrate MT.

Following this approach, the translator's working environment does not change, except that there is one more proposed translation – marked as being MT – to check for suitability for post-editing, meaning that the effort of post-editing is smaller than that of creating a new translation. Even the remaining problem of saving the post-edited machine translations for future translation jobs in the memory is handled by the TMS.

2.4 *Optimisation of MT Engines*

For rule-based systems, substantial quality leaps can be achieved by means of dictionary enhancement. System improvement is a targeted measure and applies mainly to a specific document to be translated. In an optimal translation workflow, terminology is provided and translated before document translation and stored in the terminology database to allow for active terminology recognition during translation via the TMS and thus ensure consistent translation. However, because there are usually tight deadlines, terminology often has to be collected during document translation. If the source language documents always concern a certain domain, subsequent dictionary enhancement will most likely increase the machine translation quality of future translation jobs. Otherwise, subsequent dictionary enhancement does not justify the effort, as it is a costly process and requires the skills of a trained specialist.

Concerning statistical systems, improvement is currently achieved by training the system with ever more bilingual data, and weighting certain training content for domain specificity. As opposed to the optimisation of rule-based systems, it is impossible to optimise statistical MT systems on the basis of the content to be translated. Instead, comparable material is used in the optimisation process to avoid overfitting.

In summary, the LSP is confronted with a variety of issues when introducing MT into existing translation workflows. Some can be resolved easily, while others certainly require some development effort, and it still remains completely unclear how others are best dealt with. Questions such as how to determine MT quality and post-editing effort before translation takes place, how to automatically choose the best machine translation given a specific translation scenario, and how to economically optimise MT quality are the subject of research and development in the taraXŰ project.

3 Hybrid Machine Translation – Scientific Motivation

This section explains the scientific motivation for including a hybrid MT system in our project by briefly discussing strengths and weaknesses of the main MT paradigms. We will also discuss different modes of MT support that are offered by our system.

The idea of using computers to translate between languages can be traced back to the late 1940s and was followed by substantial funding for research during the 1950s and again in the 1980s. At its most basic level, MT simply substitutes words in one natural language with words in another language. This naive approach can be useful in subject domains that have a very restricted, formulaic language such as weather reports. However, to obtain a good translation of less standardised texts, larger text units (phrases, sentences, or even whole passages) must be matched to their closest counterparts in the target language. One way of approaching this is based on linguistic rules. Research on rule-based MT goes back to the early days of Artificial Intelligence in the 1960s, and some systems have reached a high level of sophistication (e.g., Schwall & Thurmair, 1997). While this symbolic approach allows close control of the translation process, maintenance is expensive as it relies on the availability of large sets of grammar rules carefully designed by a skilled linguist.

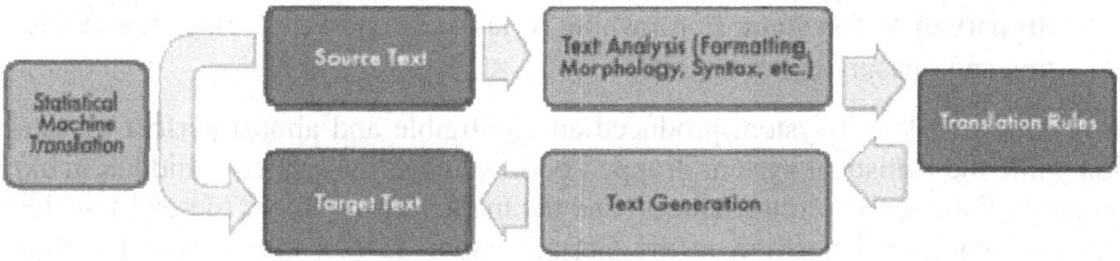

Fig. 1 Statistical (left) and rule-based (right) MT (from: Burchardt et. al. 2012)

In the late 1980s when computational power increased and became cheaper, interest in statistical models for MT began to grow. Statistical translation models are derived "automatically" by analysing parallel, bilingual text corpora such as the Europarl parallel corpus, which contains the proceedings of the European Parliament in 21 European languages. Given enough data, statistical MT works well enough to derive approximate translations of foreign language texts. Since the mid-1990s, statistical MT has become the prevalent MT approach in the research community (e.g., Koehn et al., 2007; Li et al., 2010). Well-known large free online translation services also rely on statistical MT. Compared with rule-based systems, statistical MT more often generates ungrammatical output. Fig. 1 schematically illustrates a statistical and a rule-based MT pipeline.

Hybrid MT is a recent trend (e.g., Federmann et al., 2009; Chen et al., 2009) for leveraging the quality of MT. Based on the observation that different MT systems often have complementary strengths and weaknesses, different methods for hybridisation are investigated that aim to "fuse" an improved translation out of the good parts of several translation possibilities.

3.1 Complementary Errors

Typical difficulties for statistical MT are morphology, sentence structure, long-range re-ordering and missing words, while strengths are disambiguation and lexical choice. Rule-based MT systems are typically strong in morphology, sentence structure, and have the ability to handle long-range phenomena. Weaknesses arise from parsing errors and wrong lexical choice. The following examples illustrate the complementary nature of such system errors.

- (1) **Input**: Wir sollten ihn auf keinen Fall heute zerstören.
  - **Human translation:** We definitely shouldn't destroy it/him today.
  - **Rule-based MT system**: *We should not destroy it in any case today.*
  - **Statistical MT system**: *We should, in any case, it today.*
- (2) **Input**: Für eine ganze Reihe von Bereichen bringt dies die Drosselung und schrittweise Einstellung von Aktivitäten mit sich.
  - **Human translation**: For a wide range of sectors this means the reduction and phasing out of activities.
  - **Rule-based MT system**: For a whole series of fields this brings the restriction and step-by-step attitude from activities with themselves.

– **Statistical MT system**: For a whole series of areas of this brings the restriction and phasing out of activities.

In (1), the rule-based system produced an intelligible and almost perfect translation while the statistical system dropped both the negation, which is hidden in *auf keinen Fall* (under no circumstances) and the main verb *zerstören* (destroy), which has a sentence-final position in the German input. In (2), the statistical system came up with a good translation. The rule-based system chose the wrong sense of *Einstellung* (attitude instead of discontinuation) and mistranslated the German *bringt ... mit sich* (brings) literally into *brings ... with themselves.*

In the design of a hybrid system, a fundamental conceptual decision is to try and merge the translation results of different systems into a new sentence, a process known as system combination. An alternative is to select the best sentence from different translation results. The taraXŰ system described in the next section follows the latter approach in selecting system outputs from all major MT paradigms in a hybrid architecture.

3.2 Measuring Translation Quality

In the development of MT engines, a great deal of effort has been spent finding measures that correlate well with human judgements when comparing translation systems for quality. Commonly used measures of MT quality such as BLEU (Papineni et al., 2001) depend on the availability of human reference translations, which are only available in artificial development scenarios and cannot be taken as absolute because of the inherent subjectivity of the human judgement. Furthermore, the correlation of these metrics to human judgments has been mainly measured on ranking between different machine translation systems (e.g., Callison-Burch et al., 2006). While ranking systems is an important first step, it does not provide many scientific insights for their improvement.

On the other side, a number of common translation quality standards are used to assess translations produced by professional human translators that mostly assess surface criteria at the sentence level such as the well-formedness of the translation (e.g., SAE J2450) and sometimes more abstract criteria at the document level such as terminology, style, or consistency (e.g., LISA QA). The notion of translation quality, however, is also relative to factors such as the purpose of the translation (e.g., providing information vs. giving instructions) or the expectation of the recipient (e.g., elaboration vs. brevity).

Overall, research and development in MT is confronted with heterogeneous requirements from human translation workflows, a variety of different MT paradigms, and quality criteria that partially rely on human judgement. Therefore, the taraXŰ project follows a human-centric hybrid approach.

3.3 Post-editing vs. Standalone MT

Two different translation scenarios are studied in the project and thus have to be handled by the project's system:

- **Standalone MT.** As the name suggests, this is a pure MT scenario, where the hybrid system selection mechanism has to find the translation that best preserves the meaning of the source language sentence.
- **Human post-editing.** In this scenario, a human translator post-edits the translation result in order to reach the level of a high-quality human translation. The system should thus select and offer the translation that is most easy to edit. Ideally, it would even indicate if none of the outputs is good enough for post-editing and creating a translation manually would require the least effort.

The sentences humans select as the "best translation" and "easiest-to-post-edit" in fact differ. To anticipate one result from a MT post-editing task, described later, here is the result of a human expert ranking of four different results of MT engines and the sentence that was chosen for post-editing:

- **Rank 1:** *Our experience shows that the majority of the customers in the three department stores views not more at all on the prices.*
- **Rank 2:** *Our experience shows that the majority of the customers doesn't look on the prices in the three department stores any more.*
- **Rank 3:** *Our experience shows that the majority of the customers does not look at the prices anymore at all in the three department stores.*
- **Rank 4:** *Our experience shows that the majority of customers in the three Warenhäusern do not look more on prices.*
- **Editing result** (on Rank 4): *Our experience shows that the majority of customers in the three department stores no longer look at the prices.*

The rationale behind this is clear. Due to the untranslated word *Warenhäusern* (department stores), this translation is ranked worst. But by translating one word and making a few other changes, it can be made acceptable.

4 The taraXŰ MT System at Work

4.1 Translation Workflow

In this section, we provide a high-level description of the core parts of the taraXŰ system. The results of experiments and evaluations are presented in the next section. Our system follows the approach of deploying several MT engines in one infrastructure and selecting the best of all machine translations for further use. To allow for maximum usability in professional translation environments we embedded the process of requesting the MT results, selecting the best translation and post-editing in a TMS workflow. In such a workflow, MT is applied to new content only and input can be processed in a variety of file formats. The translator uses his or her familiar working environment and various quality assurance measures that come with the TMS can be applied to the bilingual content. Once post-edited, the machine translation becomes part of the memory content and is available for reuse.

In taraXŰ the translator is offered the most suitable machine translation out of all available machine translations. The target language document is therefore a patchwork of the best translations selected on a sentence level, i.e. the best translation for each one of the original sentences. That is at least true for a gisting scenario, in which the target language document is not subject to post-editing and needs to be fully translated. In the case of post-editing there still can be source language content in the translated file if the expected quality of the machine translation is so low that it would not speed up the translation process and would possibly irritate the translator.

4.2 System Architecture

The taraXŰ framework integrates rule-based and statistical machine translation engines, a system for automatic source and target language quality assessment and a translation memory system.

Each of the systems we use in our experiments is proven and well established in its own closed field of application. Given that, and as a practical extension to current attempts at combining several MT engines, taraXŰ offers an informed selection mechanism that chooses the best out of a choice of machine translations.

The selection mechanism evaluates all available system-immanent information that could be derived from the systems. Most of the systems generate information that give insight into the translation process. We are convinced that at least part of that information can be used to estimate translation quality and possibly post-editing effort. Additional workflow characteristics for choosing among the available translations are incorporated into the selection mechanism. The optimal translation output is then used.

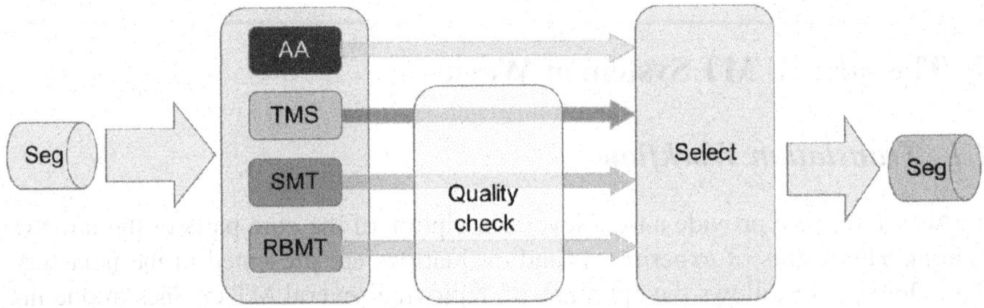

Fig. 2 System architecture showing main steps: input analysis/translation, output checking, and selection

Fig. 2 depicts the components of the taraXŰ approach and the way they interact. As can be seen, the segmented source language content is routed through all the embedded systems. The source language quality assessment component (AA) provides the selection mechanism (Select) with categorised information about spelling, grammar and style. The TMS (TMS) and all MT (RBMT, SMT) engines produce target language output and provide system-immanent metadata, which is also fed to the selection mechanism (Select). The machine translation results are subject to target language quality assessment (AA) and the same categorised information as for source language quality assessment is provided to the selection mechanism. Fig. 3 illustrates the process of generating a target language file.

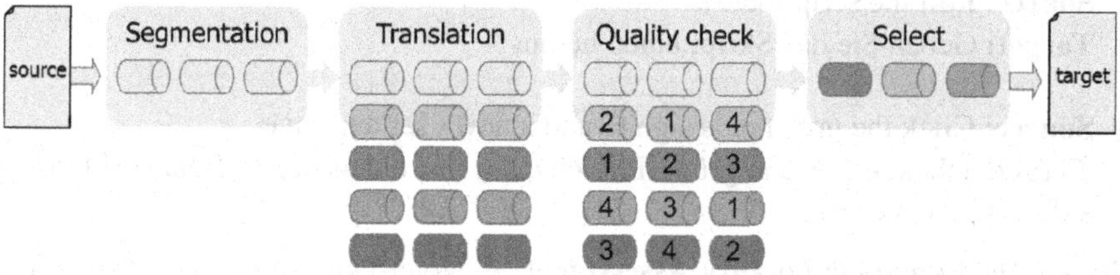

Fig. 3 Overview of the selection process

4.3 Selection Mechanism

The Selection Mechanism takes over the task of automatically assessing the quality of machine translation, given the field of application. To be of real practical value we are convinced that the selection mechanism must closely model human selection behaviour. Our approach is to model the correlation between machine accessible information and human selection, using Machine Learning techniques for fine-tuning and extracting patterns. The incorporation of pragmatic and system-immanent information that is derived from the

- translation workflow (information gisting or post-editing)
- source language quality assessment
- target language quality assessment and
- behaviour of the MT engines

is relevant to this. Following our approach, the following aspects are taken into consideration:

(i) Source Language Quality Assessment. Regarding the complementary strengths and weaknesses of the different MT paradigms, evaluating the source content characteristics on a sentence level should have a positive effect on MT selection. Moreover, in a practical scenario not only well-formed and grammatically correct source language content can be expected. Non-native writers may produce erroneous content with respect to spelling and grammar that is handled differently by statistical and rule-based MT engines. The following examples show how source language quality affects machine translation quality.

- **Source:** Edit the script bellow:
- **Target:** Überarbeiten Sie das Schriftgebrüll:
-
- **Source:** The Greens plus click to add a new user and save.
- **Target:** Das Grünzeug plus Klick, um einen neuen Bediener hinzuzufügen und zu sichern.

The German translations are incomprehensible nonsense. Correction of the source language sentences results in understandable machine translations that can be made perfect with a few post-edits.

- **Source:** Edit the script below:
- **Target:** Geben Sie das Skript unten heraus:

- **Source:** Click the green plus sign to add a new user and save.
- **Target:** Klicken Sie das grüne Pluszeichen, um einen neuen Benutzer hinzuzufügen, und sichern.

(ii) Target Language Quality Assessment. A target-language quality assurance functionality can obviously be of help for the selection. Primarily the degree of compliance with target language requirements (and in particular with terminology, grammar and style) allows the graded assessment of post-editing effort.

(iii) Behaviour of the MT Engines. Instead of treating the machine translation engines as a "black box", we look further into their internal functioning. As the translation process consists of several "decoding" steps, such an observation should lead to useful conclusions regarding the quality of its result. The simplest indication is the inability of a system to find the correct translation of a word. More advanced indications depend on the respective translation method. For example, a statistical system's scoring mechanism may signify too many ambiguities regarding phrasal/lexical selection, whereas rule-based systems may flag their inability to fully analyse part of the input text.

5 Observations From System Operation and Further Development

TaraXŰ, aiming to offer a full solution for translation workflows as outlined in Section 2, is subject to ongoing research and development. This is accomplished within an iterative process, where the operation of the first modules produces results that are used as a driving force for the development of extensions for the system. As we have now outlined the current structure of the system, we will go on to discuss the results and observations from its operation in our controlled development environment.

5.1 Observing Translator Preferences and MT Behaviour

As the goals of the system are human-centric, the iterative process of evaluation and further development includes the involvement of actual human users, namely professional translators. In this section we provide a prototype instance of the aforementioned selection mechanism (i.e. with no actual automatic functionality) and observe the translator preferences on the MT outputs offered. These preferences are analysed to show the relative performance of each MT system, in particular errors and hints as to the degree to which they complement each other.

Experiment Structure. Two tasks were performed by the professional translators, mirroring the two modes of our selection mechanism:

1. In the **first task**, annotators ranked the output of the four systems, according to how well these preserve the meaning of the source sentence. In a subsequent step, they classified the two main types of errors (if any) of the best translation. We used a subset of the error types suggested by Vilar et al., (2006).
2. In the **second task**, the translators selected the translation that is easiest to post-edit and then edited it. They were asked to perform only the minimal post-editing necessary to achieve acceptable translation quality. No target language requirements were set.

Technical Details. In order to accomplish the tasks, we employed external Language Service Providers that offer professional translation services by humans, hoping to get as close to the target user group as possible. Annotation guidelines describing the evaluation tasks as well as examples on the correct usage of the error classification scheme and minimal post-editing were provided to the translators. The evaluation interface was set up taking advantage of the *Appraise* evaluation tool (Federmann 2010), which through its web-based platform allows for remote work and interoperability. A sample screenshot of the interface can be seen in Fig. 4.

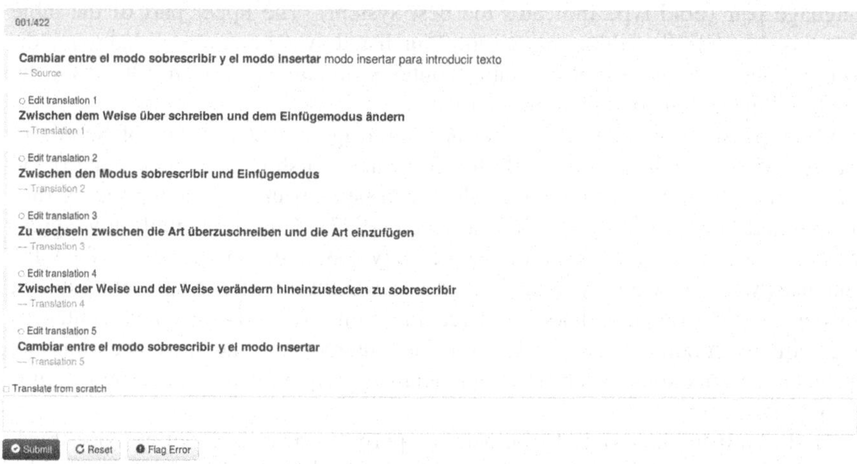
001/422
Cambiar entre el modo sobrescribir y el modo insertar modo insertar para introducir texto
— Source
Edit translation 1
Zwischen dem Weise über schreiben und dem Einfügemodus ändern
— Translation 1
Edit translation 2
Zwischen den Modus sobrescribir und Einfügemodus
— Translation 2
Edit translation 3
Zu wechseln zwischen die Art überzuschreiben und die Art einzufügen
— Translation 3
Edit translation 4
Zwischen der Weise und der Weise verändern hineinzustecken zu sobrescribir
— Translation 4
Edit translation 5
Cambiar entre el modo sobrescribir y el modo insertar
— Translation 5
Translate from scratch
Submit
Reset
Flag Error

Fig. 4 The Appraise interface used within taraXŰ

Data. The experiment was performed based on the language directions German-English, English-German and Spanish-German. The corpus size was about 50,000 words in total or 2,000 sentences per language pair.

The test data was domain-oriented. The "news" part of the test set consisted of 1,030 test sentences from two WMT shared tasks (Callison-Burch et al, 2008 and Callison-Burch et al, 2010), subsampled proportionally to each one of the documents contained). The "technical documentation" part contained 400 sentences extracted from the community-based translations of the OpenOffice2 documentation

(Tiedemann, 2009), post-processed manually in order to remove wrong alignments and translation mismatches.

The translation outputs were provided by four distinct state-of-the-art MT/TM implementations: The statistical MT-systems *Moses* (Koehn, 2005), trained using Europarl and News Corpus (Callison-Burch et al, 2010), and *Google Translate*, the rule-based system *Lucy* (Alonso & Thurmair, 2003) and the translation memory system *Trados* (Carroll, 2000). The latter was included in order to investigate a baseline scenario where no MT is involved. It was filled with the same bilingual material that was used for training Moses. In order to enforce high fuzzy matches for part of the evaluation data, we took some sentences from the bilingual material and modified them minimally as regards grammar, spelling and negation – mirroring a real-life translation scenario with existing language resources. There were no training data constraints for Google Translate and Lucy. None of the MT engines were optimized before the evaluation.

Observations. The main observations and conclusions from the ranking and error classification task are described below.

i) User preference. An overview of the users' preference to each one of the systems is depicted in Table 1, where system ranks are averaged for each domain and language pair (bold type indicates the best system). The upper part of the table contains the overall ranking among the four listed systems for each language direction. The bottom part of the table contains domain-specific results from the news and the technical domain respectively.

When possible, we measured inter-annotator agreement in terms of Scott's pi (Scott, 1955). Pi values were 0.607 for German-English (2 annotators) and 0.356 for German-English (3 annotators), which can be respectively interpreted as substantial and fair (following Landis and Koch, 1977). This comparably low agreement seems to stem partly from the deliberately simple design of the interface: the sentences were presented without context and we did not offer a special treatment for ties, i.e. the evaluators had to enforce a ranking even if two or more sentences were equally good/bad. A second reason for disagreement may have been missing "project specifications" such as target language requirements concerning terminology, style, etc.

Trados is listed here only for orientation purposes as it does not produce translations, but is more of a storage system for existing bilingual material. Its translation results provide translations of already known content whose source was similar to the original test sentences.

One observation is that the MT system ranks are comparably close to each other. The narrow margin between the ranks indicates that no system is better than another in all cases, which supports the hypothesis that their performance is complementary and that their combination can thus be beneficial.

ii) Error classification. As described above, the evaluators were asked to choose the two most important errors of the best translation. The results of this classification are presented in Table 2. The table reads as follows: 3.2% of errors made by system Lucy

Table 1 Human ranking results, as the average position of each system in each task

	Lucy	Moses	Google	Trados
Overall	2.00	2.38	**1.86**	3.74
DE-EN	2.01	2.46	**1.73**	3.80
ES-DE	**1.85**	2.42	1.99	3.72
EN-DE	2.12	2.28	**1.89**	3.71
WMT10	**2.52**	2.59	2.69	2.21
OpenOffice	1.72	2.77	**1.56**	3.95

Table 2 Human error classification: error distribution for each translation system (error classes are based on Vilar et al., 2006)

	Lucy	Moses	Trados
Missing content word(s)	3.2%	16.8%	12.6%
Wrong content word(s)	**34.6%**	**24.6%**	**33.2%**
Wrong functional word(s)	**18.6%**	11.8%	11.0%
Incorrect word form(s)	13.1%	14.6%	9.1%
Incorrect word order	16.1%	**22.0%**	13.4%
Incorrect punctuation	3.7%	3.4%	2.1%
Other error	10.7%	6.8%	**18.6%**

were missing content words, 34.6% wrong content words, etc. It can be seen that the most common errors in all systems are wrong content words. The next most frequent error type is incorrect word order, followed by wrong functional words and incorrect word form. This indicates the need for improvement of reordering and lexical choice techniques for all translation systems.

Table 3 Five types of automatically classified edits for three translation systems as a distribution over all translated words per system. Percentages averaged over the total number of errors of each system

	correcting word form	correcting word order	adding missing word	deleting extra word	correcting lexical choice
Lucy	4.3%	7.0%	**4.4%**	6.2%	**23.7%**
Moses	4.9%	9.0%	**7.5%**	4.9%	**21.8%**
Trados	2.6%	**4.9%**	8.1%	6.5%	**47.7%**

iii) Post-editing. A similar analysis has been carried out on the outputs that were chosen to be post-edited, which are not necessarily the best ranked ones. In fact, only 33% of post-edited outputs were ranked as the best. More experiments are, however, needed to check whether the post-editors' intuitive choices really mirror post-editing difficulty.

It should be noted that the Google Translate system was not considered as an option for editing. We took this decision because we have no way of influencing the system and wanted to avoid futile efforts.

The post-edited output was compared with the original MT output, in order to conclude which types of editing are most frequent for each system. The following five types of edits (Popović and Burchardt, 2011) are taken into account: correcting word form (morphology), correcting word order, adding missing word, deleting extra word and correcting lexical choice.

Table 3 presents numbers for each of the five correction types for the three systems. It reads as follows: In 4.3% of the words translated by Lucy, the human translator corrected the word form, for 7% of the words, the order was corrected, etc. The most frequent correction for all systems is lexical choice, however for the Trados system the number of lexical corrections is significantly higher than for the other systems. This is explained by the fact that the choice of the target language sentence proposed for editing is not based on a semantic level but on the level of string similarity of the source sentences. The next most frequent type of correction is the word order. The Moses system shows slightly more incorrectly positioned words (order errors) than other systems. These results confirm the human error classification: the main weakness for all systems is incorrect lexical choice, i.e. wrong content words, and incorrect word order. These two aspects should be taken into account for further improvement of the systems.

5.2 *Selection Mechanism*

In order to show the feasibility of our approach, we are including here the results of the first experimental implementation of the Selection Mechanism (Avramidis, 2011). It has been trained using 1,000 Spanish sentences, each of them accompanied by five system outputs in English. The goal of the training was that, for each input sentence, the mechanism selects the system output which is the closest to the known translation, using Levenshtein distance (Levenshtein, 1966). The model was trained with a Support Vector Machine algorithm (Tsochantaridis et. al.), which was given system-immanent information by the five systems such as the use of segment combination in the case of RBMT (see Avramidis, 2011 for details).

The resulting mechanism was tested on a test set of 1,000 sentences, translated by the same systems, also including the system-immanent information from the translation process. We collected only the best chosen output for each sentence and evaluated the quality of the output using the automatic evaluation metric of BLEU.

Our model reported having successfully compared the quality of the sentences in 63% of the cases. Although this is relatively low, the overall translation quality is encouraging: as can be seen in Table 4, we achieved a BLEU score that is at least comparable with the best systems in the task. We consider this to be an encouraging result, indicating some feasibility for future efforts on the selection mechanism. A more suitable metric for the purpose of post-editing is Word Error Rate, which shows an average of the editing distance between the suggested MT output and the goal translation, which we managed to decrease significantly. Finally, our selection mechanism performed better than state-of-the-art system combination approaches in ML4HMT-2011 (Federmann et al, 2012; Table 5).

Finally, we want to report on speed improvement as one indicator for improved efficiency of the translation process. Table 6 contrasts average processing time for translation done from scratch versus post-editing of MT output on a collection of documents including 1806 sentences sampled from the most recent evaluation campaign in taraXŰ. These results indicate that post-editing the machine-translation output is significantly faster than translating the same sentences from scratch. Our comparative usage statistics show an average speed-up of about 16 seconds per sentence, which sums up to about 8 working hours for the entire set.

Table 4 The translation performance of a sample implementation of the selection mechanism for Spanish-English using only system-immanent features trained using Levenshtein distance

System	BLEU	WER
Hierarchical SMT	19.68	62.37
Lucy RBMT	**23.37**	64.78
Metis SMT	12.62	77.62
Apertium RBMT	22.30	64.91
Moses-based SMT	23.14	60.66
Selection Mechanism	**23.54**	**46.13**

Table 5 Results of human ranking, when comparing our selection Mechanism with other system combination methods

System	Overall rank
Syscomb DCU	2.52
Lucy RBMT enhanced	2.05
Syscomb LIUM	2.87
Selection Mechanism	**2.50**

Table 6 Average processing time for translation done from scratch versus post-editing of machine translation output

Sentence length	Average time (sec)	
(words)	from scratch	post-editing
0 – 19	45.94	33.71
20 – 39	96.80	74.59
40 – 59	155.84	99.09
60 – 79	73.83	18.00
all	**63.32**	**46.93**

6 Further Work

The experiments described were performed with "off-the-shelf" core systems, which means using general corpora to train the SMT engine and general vocabulary for RBMT. Since then, different methods of optimisation have been applied to the engines to achieve better overall translation results. Future experiments will allow to assess potential quality improvements with respect to the characteristics of the different test sets. We are convinced that source language characteristics have a major effect on translation quality and optimisation potential.

This gradual approach allows for the comparison of MT quality with respect to the three intertwining levels: domain adaptation, source language characteristics and more detailed error analysis on a word basis. Future experiments will also include more language directions.

Optimisation will be handled pragmatically, as would be feasible in a professional translation workflow. The focus will be on domain adaptation, which means re-training the SMT engines with domain-specific material and adding domain-specific terminology to the RBMT dictionaries. Moreover, it is planned to give the translators more guidance regarding the target language requirements, e.g., regarding style and expected translation quality.

More features that we expect to have a definite impact on quality assessment will be used for the selection mechanism. They will be generated primarily from TMS functionality, like the matching algorithms and quality assurance functionality mentioned above. We expect that a deviation from terminological requirements may have more weight on the semantic level than on post-editing effort, but that grammatical misconstructions will most likely have a greater impact on post-editing effort than on semantic closeness to the original meaning.

Lastly, the threshold for post-editing has not yet been defined in the context of our larger experimental study. This threshold strongly depends on the individual memory, but we hope to find some guidance for automatic assessment in the course of the next evaluation studies.

7 Conclusions

In this chapter, we have shed light on the economic and scientific requirements for integrating machine translation (MT) into professional human translation workflows. We have described the taraXŰ system prototype that is being developed in an ongoing joint project between language service providers and MT research experts.

By pairing a hybrid MT system with a translation memory (TM) architecture, we believe that we have brought together the best of both worlds – the state of the art in MT and professional translation workflows. So far the evaluation results are in line with our presumptions that different systems have complementary strength and weaknesses. Unsurprisingly, all systems leave room for improvements, which will be examined in the next phase of the project.

As regards the assessment of translation quality, the divergences in ranking and post-editing results between several annotators indicate that more flexible automatic measures are needed to assess quality realistically. This subjective factor in translation also runs counter to the use of reference translations for automatically assessing translation quality.

What is definitely needed for effective post-editing is support with respect to the type of divergence from target language requirements. This is particularly the case when the MT is correct in terms of spelling and grammar, but needs to be adapted regarding lexical choice or style.

Acknowledgments. The work presented here is a joint work. The authors would like to thank Patrick Bessler, Christian Federmann, Horst Liebscher, Maja Popović, Marcus Watts, and David Vilar for their contributions. Many thanks also go to the anonymous reviewers for helping clarify certain points. The taraXŰ project is financed by TSB Technologiestiftung Berlin – Zukunftsfonds Berlin, co-financed by the European Union – European Regional Development Fund.

References

Alonso, J., Thurmair, G., Deutschland, C.: The Comprendium Translator System. In: Proceedings of the Ninth Machine Translation Summit, New Orleans (2003)

Avramidis, E.: DFKI System Combination with Sentence Ranking at ML4HMT-2011. In: Proceedings of the International Workshop on Using Linguistic Information for Hybrid Machine Translation and of the Shared Task on Applying Machine Learning Techniques to Optimising the Division of Labour in Hybrid Machine Translation, San Francisco (2011)

Burchardt, A., Egg, M., Eichler, K., Krenn, B., Kreutel, J., Leßmöllmann, A., Rehm, G., Stede, M., Uszkoreit, H.: The German Language in the Digital Age. Springer (2012)

Callison-Burch, C., Fordyce, C., Koehn, P., et al.: Further Meta-Evaluation of Machine Translation. In: Proceedings of the Third Workshop on Statistical Machine Translation, pp. 70–106. Association for Computational Linguistics, Columbus (2008)

Callison-Burch, C., Koehn, P., Monz, C., et al.: Findings of the 2010 Joint Workshop on Statistical Machine Translation and Metrics for Machine Translation. Proceedings of the Joint Fifth Workshop on Statistical Machine Translation and Metrics, pp. 17–53. Association for Computational Linguistics, Uppsala (2010)

Callison-Burch, C., Osborne, M., Koehn, P.: Re-evaluating the Role of Bleu in Machine Translation Research. In: Proceedings of the 11th Conference of the European Chapter of the Association for Computational Linguistics, Trento, pp. 249–256 (2006)

Carroll, S.: Introducing the TRADOS workflow development. Translating and the Computer 22. Aslib Proceedings, London (2000)

Casacuberta, F., Civera, J., Cubel, E., et al.: Human Interaction for High Quality Machine Translation. Communications of the ACM 52, 135–138 (2007)

Chen, Y., Jellinghaus, M., Eisele, A., et al.: Combining Multi-Engine Translations with Moses. In: Proceedings of the Fourth Workshop on Statistical Machine Translation, pp. 42–46. Association for Computational Linguistics, Athens (2009)

Federmann, C.: Appraise: An Open-Source Toolkit for Manual Phrase-Based Evaluation of Translations. In: Proceedings of the Seventh International Conference on Language Resources and Evaluation, Valletta (2010)

Federmann, C., Avramidis, E., Costa-Jussa, M.R., et al.: The ML4HMT Workshop on Optimising the Division of Labour in Hybrid Machine Translation. In: Proceedings of the Twelfths International Conference on Language Resources and Evaluation, Istanbul (2012)

Federmann, C., Theison, S., Eisele, A., et al.: Translation Combination using Factored Word Substitution. In: Proceedings of the Fourth Workshop on Statistical Machine Translation, pp. 70–74. Association for Computational Linguistics, Athens (2009)

Koehn, P.: Europarl: A Parallel Corpus for Statistical Machine Translation. In: Proceedings of the Tenth Machine Translation Summit, Phuket (2005)

Koehn, P.: A Process Study of Computer-aided Translation. Machine Translation 23, 241–263 (2009)

Koehn, P., Hoang, H., Birch, A., et al.: Moses: Open Source Toolkit for Statistical Machine Translation. In: Proceedings of the Forty-Fifth Annual Meeting of the Association for Computational Linguistics, pp. 177–180. Association for Computational Linguistics, Prague (2007)

Landis, J.R., Koch, G.G.: The Measurement of Observer Agreement for Categorical Data. Biometrics 33, 159–174 (1977)

Levenshtein, V.: Binary Codes Capable of Correcting Deletions and Insertions and Reversals. Soviet Physics Doklady 10, 707–710 (1966)

Li, Z., Callison-Burch, C., Dyer, C., et al.: Joshua 2.0: A Toolkit for Parsing-Based Machine Translation with Syntax, Semirings, Discriminative Training and Other Goodies. In: Proceedings of the Joint Fifth Workshop on Statistical Machine Translation and Metrics, pp. 133–137. Association for Computational Linguistics, Uppsala (2010)

Papineni, K., Roukos, S., Ward, T., Zhu, W.-J.: BLEU: A Method for Automatic Evaluation of Machine Translation. In: Proceedings of the Fortieths Annual Meeting of the Association for Computational Linguistics, pp. 311–318. Association for Computational Linguistics, Pennsylvania (2002)

Popovic, M., Burchardt, A.: From Human to Automatic Error Classification for Machine Translation Output. In: Proceedings of the Fifteenth International Conference of the European Association for Machine Translation, Leuven (2011)

Schwall, U., Thurmair, G.: From METAL to T1: Systems and Components for Machine Translation Applications. In: Proceedings of the Sixth Machine Translation Summit, pp. 180–190 (1997)

Scott, W.A.: Reliability of Content Analysis: The Case of Nominal Scale Coding. Public Opinion Quarterly 19, 321–325 (1955), doi:10.1086/266577

Specia, L.: Exploiting Objective Annotations for Measuring Translation Post-editing Effort. In: Proceedings of the Fifteenth International Conference of the European Association for Machine Translation, Leuven (2011)

Tiedemann, J.: News from OPUS—A Collection of Multilingual Parallel Corpora with Tools and Interfaces. Recent Advances in Natural Language Processing 5, 237–248 (2009)

Tsochantaridis, I., Hofmann, T., Joachims, T., Altun, Y.: Support Vector Machine Learning for Interdependent and Structured Output Spaces. In: Proceedings of the Twenty-First International Conference on Machine Learning, Banff, Alberta (2004)

Vilar, D., Xu, J., D'Haro, L.F., Ney, H.: Error Analysis of Machine Translation Output. In: Proceedings of the Fifth International Conference on Language Resources and Evaluation, Genoa, pp. 697–702 (2006)

Anubis – Speeding Up Computer-Aided Translation

Rafał Jaworski

Abstract. In this paper, the idea of Computer-Aided Translation is first introduced and a modern approach to CAT is then presented. Next, we provide a more detailed description of one of the state-of-art CAT systems – memoQ. Then, the author's approach to the idea – the Anubis system – is described and evaluated. While Anubis is comparable to memoQ in terms of the precision of provided translation memory matches, it outperforms memoQ when it comes to the speed of searching the translation memory. In the experiments carried out, Anubis turned out to be over 430% faster than memoQ. The result was achieved thanks to the author's algorithm of minimizing the size of translation memory index, so it can be stored in computer's RAM memory.

The paper is divided in two parts: the first describes the field of CAT, where the Anubis system can be applied. The second gives a detailed description of Anubis itself, proving its potential usefulness.

1 Introduction

In the modern world, we observe an increasing use of AI techniques. Can these techniques prove useful in practical applications? This paper focuses on one specific AI mechanism, associated with the domain of machine translation – Computer Aided Translation (abbreviated CAT). It is designed to faciliate the work of a human translator whose task is to translate a document from one language (called the source language) to another (target language). The machine's task in the process is providing suggestions for translation of each sentence to be translated. Such suggestions are then reviewed by the human translator and are used to produce the final translation. From the technical

Rafał Jaworski
Adam Mickiewicz University
Poznań, Poland
e-mail: rjawor@amu.edu.pl

A. Przepiórkowski et al. (Eds.): *Computational Linguistics*, SCI 458, pp. 263–280, 2013.
DOI: 10.1007/978-3-642-34399-5_14

point of view, the main difficulty of the problem lies in the process of generating suggestions for translation of a sentence. To deal with it, most CAT systems incorporate so called translation memories – databases containing previously translated sentences. More specifically, a translation memory is a set of pairs of sentences, where the first sentence is in the source language and the second – in the target language. During the translation process, given an input sentence in source language, the system looks for a similar one in the translation memory. If such a sentence is found, its translation is returned as a suggestion for translation of the input sentence. If not, the sentence is translated by the human translator and added to the translation memory to enrich it.

2 CAT History

The early beginnings of what is today called Computer-Aided Translation date back to 1980s ([3]) when systems of this class were developed in Japan. Japanese computer companies (Fujitsu, Hitachi, NEC, Sharp, Toshiba) worked on software faciliating the process of translation primarily in the directions: Japanese-English and English-Japanese (though other languages were also taken into consideration). The systems relied on automatic translations which were corrected by human translators in the process of post-editing. Machine translation carried out by the systems was based either on a direct word-to-word transfer or on a very superficial lexicographical analysis. Interestingly, these systems tended to focus on a specific domain of texts. The benefits of this focus included lower costs of lexical resources preparation (due to smaller volume of dictionaries), faster translation (for the same reason) and higher precision of translations. The most popular domains the systems focused on were computer science and information technology.

A notable milestone in the history of computer-aided translation was the creation of the ALPS system in 1981 (as described in [3]). ALPS was the first CAT system designed for personal computers and released to the market. It offered the following functionalities:

- multilingual word-processing
- automatic dictionary
- terminology consultation
- interactive translation
- repetitions extraction

Especially the last feature is worth mentioning as it was a very early concept of a translation memory. All translations carried out by a translator were stored in a so called "repetitions file". While working on a new translation, a translator could compare the new sentence to those in the repetitions file. The process was facilited by automatic search of sentence fragments.

Sadly, ALPS did not turn out to be profitable. Nevertheless, several other systems were developed shortly after. By the end of 1980s, the translators had realised how much they could benefit from using a computer as a translation tool. Not only did the CAT systems provide tools for the translation process, but also faciliated word processing and managament of the work. The class of such systems is now called "Translation workstations". The earliest vendors of translation workstations, as stated in [3], were:

- Trados (Translator's Workbench, still developed)
- STAR AG (Transit)
- IBM (the TranslationManager, no longer marketed)
- the Eurolang Optimizer (also no longer available).

More recently, in the 1990s and 2000s many more appeared:

- Atril (Déjà Vu)
- SDL (the SDLX system)
- Xerox (XMS)
- Terminotix (LogiTerm)
- MultiCorpora (MultiTrans)
- Champollion (WordFast)
- MetaTexis
- ProMemoria
- Kilgray Translation Technologies (memoQ system)

Nowadays CAT tools are widely popular among freelance translators and translation agencies.

3 Key Concepts

The field of Computer-Aided Translation incorporates a variety of concepts and techniques. It is difficult to find standards regarding the design and functionalites of CAT systems. Nevertheless, it is important to distinguish CAT from other related fields, such as machine translation (MT), natural language processing (NLP) or more general field of artificial intelligence.

This section lists key concepts and definitions that clarify what should and what should not be called CAT. The definitions are inspired by those found in [8].

Computer-Aided Translation (also called Computer-Assisted Translation) is a term used to describe computer techniques used to faciliate the process of translation.

Machine Assisted Human Translation (MAHT) in CAT is the work of a human translator on the translation process. The human translator is the performer of translations while the computer plays a supportive role. This relation is a crucial characteristic of MAHT. There exists an approach where these roles are reversed – Human Assisted Machine Translation (HAMT) in which the human helps the computer in carrying out the translations. This, however, is closely related to machine translation, and is not a part of CAT.

Machine Translation (MT) is an action of fully automatic text translation. The translation is carried out entirely by the computer with no help of human translators whatsoever. Even though MT is not a proper part of CAT, MT systems are sometimes used in CAT systems to provide rough suggestions of translation. Human translator is then responsible for carrying out the post-editing. Such a hybrid technique can be regarded as a CAT technique.

Translation Workbench also known as MAHT Workbench or Integrated Translation System is a piece of computer software offering a variety of CAT techniques along with utilities faciliating the work with text in general.

Translation Memory (TM) is a database of previously carried out translations. It is assumed that TM contains only high-quality translations which can be reused in future. The reuse of translations by means of Translation Memory is the most widely recognized and appreciated feature of CAT. It reflects the natural work process of a translator before the era of computerization, when instead of using databases, translators took notes of phrases and sentences to use them later in their work. Apart from cost saving (once translated sentence does not need to be translated again), Translation Memories grant the consistency of translations, which in some cases is essential.

Terminology consultation is a mechanism of automated dictionary lookups during text translation. It is a widely popular CAT mechanism applied in a majority of CAT systems. During translation of a sentence, a translator is provided with dictionary matches of words or phrases that appeared in the sentence. Typically, multiple dictionaries are searched for terminology matches. These dictionaries are divided into two categories: built-in dictionaries and user-created glossaries. The first are usually vast and comprehensive general dictionaries while the latter serve for storing more specialistic terms. User-created glossaries have similar effect as translation memories – they allow to reduce the cost of translation as well as to ensure consistency. The dictionaries and glossaries are most useful when the translation memory fails to provide a good suggestion.

Text aligning is a process of creating translation memories out of previously translated documents. Professional translators who do not use a CAT system often store their translations in the form of pairs of documents (either in electronic form or on paper). In order to use these translations as a translation memory, the documents need to be aligned. The first stage of the text alignment procedure is importing two monolingual documents into the computer and splitting them into sentences. This process can be automated as automatic sentence splitting procedures usually prove effective. The next step is sentence alignment, i.e. determining, which sentences are each other's translation. There are automatic procedures to perform this task but their results often need correction. Therefore, computer software faciliating manual alignment correction has been developed and included in some CAT systems.

As the latter sections will be dealing with details regarding the new CAT algorithm, some concepts from the field of computer science and natural language processing will also be defined in this section.

Hashing (in computer science) is a procedure carried out on a larger set of data in order to produce its shorter version. This short, simple dataset is called a hash. Hash is typically used to identify the original data.

Lemmatization (in natural language processing) is a process of substituting a word with its base dictionary form, called lemma. Lemmatization must be done with the use of a dictionary.

Stemming (in natural language processing) is a process of substituting a word with its shorter version, called stem. Stem does not necessarily has to be a dictionary word. Stemming is often done without the use of a dictionary, by simply removing word's flexion suffixes.

4 CAT Usability Studies

4.1 CAT on the Market

In recent years, CAT tools have revolutionized the market of translations (see for instance [11] or [1]). As they allow for reducing the cost of translation, translation agencies using CAT can offer their services at lower prices. Moreover, consistency of translations granted by CAT became not only a desired but demanded quality. As a result, CAT allows the agencies to offer better translations for lower prices and be far more competitive than companies preferring the traditional work model.

However, many translators are still resentful towards CAT. Some agencies need to enforce the usage of CAT tools by obligating the translators to that. Similarily, it is not true that all the freelance translators use CAT. They argue that CAT can generate additional work as post-editing of a sentence with simultaneous terminology consultation and possible other MAHT procedures can take longer than translating this sentence "from scratch" without any help. Where does the truth lie? How helpful are CAT systems on average?

Several studies have been made to find answers to these questions. The popularity of CAT tools and the fact that agencies using them are more competitive on the market suggests that CAT indeed faciliates the process of translation.

4.2 Individual CAT Productivity Study

As for freelance translators, results and conclusions of the analysis described in [12] can provide answers to the questions posted above. The author of [12] is a professional translator who had been working in this profession before CAT tools became widely popular. During the CAT revolution she became familiar with several tools, as different clients demanded the use of different systems. The author agrees that using CAT tools increases productivity but decided to measure the profits precisely. This is an essential calculation as the majority of clients demand prices reduction for translations with translation memories. It must be known weather the profit from using CAT tools can make up for the reduction of translation prices.

The author of the article was using three different CAT tools: DejaVu (version X), Trados (version 6) and a custom made tool developed by one of her clients. In DejaVu and Trados she was using her own translation memories, accumulated over the years, totalling approximately 150 000 entries each. The jobs carried out by the author were split into two categories: full and reduce priced. In the first case the client did not demand price reduction for using a translation memory and the translator was using her own resources. In the second case, the client provided the translator with a specialized translation memory, required its use and demanded price reduction for sentences found in the memory. Consequently, with respect to the CAT tool used and price reduction policy, the jobs were split into the following categories:

1. Trados full price
2. Trados reduced price
3. DejaVu full price
4. DejaVu reduced price
5. Custom tool full price
6. Custom tool reduced price
7. No CAT used

The amount of translation work taken into consideration in this study is shown in Table 1. Translation productivity was measured in the unit of words per hour. The baseline speed of translation with no CAT tool used was **250** words/hour. The calculated productivity for each translation tools is shown in Table 2. The results are very impressive and show a considerable profit

Table 1 The amount of work in productivity study

CAT tool	No of projects	Total word count	Total time (h)
Trados	36	158940	192.8
DejaVu	25	42525	74.35
Custom CAT tool	26	155023	271.2
No CAT tool	3	2617	6.5

Table 2 Translation productivity for different CAT tools

CAT tool	Productivity (words/hour)
Trados (full and reduced)	824.3
Trados (full price)	424.5
Trados (reduced price)	1104.3
DejaVu (full and reduced)	571.9
Custom CAT tool (only reduced)	571.6
No CAT tool	250

from using CAT tools. It is understandable that the productivity is higher when the client provides a specialized translation memory. The quality of a translation memory is equally important as the power of a CAT tool. Specialized TMs, dedicated for a specific translation task, have a good chance of providing good translation suggestions. In the light of these facts, the client's demand for lower price is justified.

The study also showed that using a CAT tool with a translation memory consisting of translations accumulating over time has an advantage over translating with no help of such a tool.

What follows from this study is that Computer-Aided Translation is a practical, useful technique, worth focusing on.

5 The memoQ CAT System

In this section we describe the memoQ system [4] by Kilgray Translation Technologies. Its translation memory searching module will serve as a baseline for the evaluation of Anubis system.

5.1 Overview

MemoQ is a modern Computer-Aided Translation tool which can be considered as a full translator's workbench. The system was first introduced in 2006. Thanks to its robustness and customizability, the system has gained much popularity and is still in the process of development. Its features include:

- Translation memory
- Terminology base
- Automatic quality checks (based on TM, term bases, spellcheckers and many others)
- Translation editor interface
- Translation memory editor
- Formatting tags analysis
- Text aligner
- Real-time preview for .doc, .docx, .ppt, .pptx, .html and XML file formats

- Handling TMX, XLIFF, bilingual DOC/RTF document formats
- Compatibility with Trados, WordFast and STAR Transit document workflows
- Customizable add-in architecture

Out of the above features, one of the most important factors that build the strength of memoQ is the integration with other popular CAT tools. It faciliates the transition from these tools to memoQ and thus helps to acquire former Trados or WordFast users.

5.2 The memoQ Translation Memory

Just as in the majority of CAT tools on the market, the key functionality of the memoQ system is a translation memory. It is searched for so called "100% matches" (sentences in the TM identical to the one being translated) as well as for "fuzzy matches" (TM sentences similar to the translated sentence in terms of a fuzzy similarity measure). The system provides a percentage similarity score along with every TM match. The user can set a similarity threshold for TM matches. The matches with scores below the threshold will not be returned as suggestions for translation.

A distinguishing feature of memoQ translation memory are so called "101% matches". The artificial score of 101% is assigned to 100% matches which are found in the same context in TM and in the translated document. For those matches the translator is guaranteed that they require the minimal amount of reviewing. This context matching is referred to as "ICE matching".

Another interesting technique used in memoQ is "SPICE matching". It provides another possibility for a TM match to achieve 101% score. SPICE matching is executed when the translated document is in the XML format. The "over-perfect" score is given to 100% matching segments that refer to the same information in terms of XML attributes and their values. This situation is common e.g. in localisation documents.

The memoQ translation memory has, however, a drawback – it is not well optimized for search speed. This is mainly due to the fuzzy sentence similarity measure which is not the fastest known algorithm for sentence searching. In practice, CAT technique's speed is one of the key factors that decide if it is usable. A translator has to receive the translation suggestions and perform post-editing within the time he or she would have translated the sentence manually. Hence, the speed of sentence searching is crucial, especially when having to deal with translation memories of a considerable size.

6 Anubis Sentence Search Algorithm

In order to speed up searching for sentences in a translation memory, we developed the Anubis system. Its goal is to find in a TM all sentences similar to a given sentence in the shortest possible time. As in other TM search systems, each TM match is assigned a percentage similarity score.

When designing the new algorithm, we decided to take advantage of the achievements in the fields of natural language processing and approximate string matching. This section describes the search algorithm, revealing its main idea – the use of custom designed, compact index.

6.1 The Anubis System

The algorithm was implemented in the Anubis system. Anubis is used to manage a translation memory for the use of Computer-Aided Translation and to perform TM searches. Technically, Anubis is based on the described suffix-array-based index stored in RAM memory. The system carries out the following operations:

- Adding a sentence pair (example) to the RAM-based index.
- Storing the RAM index to the hard disk.
- Restoring the RAM index from the hard disk.
- Searching the index for examples whose source sentence is similar to a given input sentence.

The typical use case of the system is divided into three stages: preparation, system initialization and translation. The preparation consists of the following steps:

1. Building the index from previously collected translation memory (stored for example in the TMX format).
2. Storing the index into hard disk.

Once the translation memory is stored on the hard disk, the system is able to load it on demand. Loading of the translation memory takes place during the initialization of the Anubis system. The initialization is done in the following steps:

1. Restore the index from hard disk.
2. Initialize system's NLP tools (such as the stemmer, mentioned in the Section 2.1).

Because of the possibility of sharing resources (such as translation memory, stemmer, etc.), Anubis is especially efficient when designed in a client-server architecture. In such architecture, Anubis enables multiple clients to search for sentences in the translation memory as well as add new sentences to it. The typical translation stage (run on a client by a human translator) would be:

1. Get translation suggestions for current sentence.
2. Carry out the translation with the help of suggestions.
3. Add the newly translated sentence to the translation memory for future reuse.
4. Repeat steps 1-3 until the end of the document.

The following subsections will give a detailed description of the procedures that the translation memory search algorithm consists of.

6.2 Sentence Hash

The procedure **hash** is invoked on a sentence and returns its hashed version. The first step of the procedure is removal of all punctuation and technical characters from the stentence. In the next step, stop words (such as functional words) are discarded using predefined stop words list. Finally, all the sentence words are stemmed (by the means of a stemmer dedicated for the sentence language). After these operations, a sequence of word stems (tokens) containing only the most significant information is obtained.

6.3 Suffix Array

A suffix array is a data structure widely used in the area of approximate string matching (see [7] and [6]). For a given set of strings (or sentences) suffix array is an index, consisting of all suffixes that can be obtained from the strings. For example, the following sentences:

1. Operation is finished
2. Sun is shining
3. I am tall

would produce the following suffixes:

- operation, is, finished
- is, finished
- finished
- sun, is, shining
- is, shining
- shining
- i, am, tall
- am, tall
- tall

The suffixes are then sorted in lexicographical order and put into a so called suffix array, along with the id of the sentence the suffix originates from and the offset of the suffix. The resulting suffix array for the above example is shown in Table 3. The algorithm described in this paper uses its own method of coding such a suffix array in order to reduce memory usage. The method uses the general idea of Huffman's coding algorithm (described in [2]). It assignes numeric codes to each sentence and maintaines a code-token dictionary. This technique proves to be effective, as the ratio of distinct tokens divided by total number of tokens in a corpus is often small (see [10] for ratio of 0.033). This ratio gets even smaller when each token is stemmed before

Table 3 Example suffix array

Suffix	Sentence id	Offset
am, tall	3	1
finished	1	2
i, am, tall	3	0
is, finished	1	1
is, shining	2	1
operation, is, finished	1	0
shining	2	2
sun, is, shining	2	0
tall	3	2

processing. In an experimental corpus of 3 593 227 words, the number of distinct stemmed tokens was 17 001, resulting in a ratio of 0.005. Because integers are less memory consuming than strings representing the tokens and the dictionary takes up a relatively small amount of memory, this method leads to a significant memory usage decrease.

6.4 Adding to the Index

Another procedure that has to be carried out before searching is adding a sentence to the index. Procedure **indexAdd** takes two parameters: the sentence and its unique id. It takes advantage of a simple hash-based **dictionary**, capable of storing, retrieving and creating codes for tokens. The index is represented by the object **array**.

The procedure first generates the hashed version of the sentence. Then, every token from the hash is substituted with a code. Finally, the procedure generates all the suffixes from the modified hash and adds them to the index with the sentence's id and offset parameters. These additions preserves the suffix array sorting. The procedure is described in Figure 1.

6.5 Searching the Index

The algorithm for searching the index uses a procedure called **getLongest-CommonPrefixes** and an object **OverlayMatch**. Both of them will be defined in this section before introducing the main search algorithm.

The procedure **getLongestCommonPrefixes**, described in Figure 2, takes one parameter – a series of tokens – and returns a set of suffixes from the array having the longest common prefix with the input series. It takes advantage of the array's method **subArray** which returns the set of suffixes from the array beginning with a given series of tokens. The method subArray is optimized (by means of the binary search).

Algorithm 1: adding to the index

```
procedure indexAdd(s,id)
    h := hash(s)
    for all (Token t in h)
        code := dictionary.get(t)
        if (code == null)
            code := dictionary.createNewCode(t)
        t := code  //substitute a token with its code
    for (i = 0 to length(h))
        array.addSuffix(h.subsequence(i,length(h)),id, i)
end procedure
```

Fig. 1 Algorithm for adding a single sentence to the index

Algorithm 2: getLongestCommonPrefixes procedure

```
procedure getLongestCommonPrefixes(h)
    longestPrefixesSet := empty set
    pos := 0
    currentScope := array
    while(not empty(currentScope) and pos < length(h))
        currentScope := currentScope.subArray(h.subSequence(0,pos))
        if (size(currentScope) > 0)
            longestPrefixesSet := currentScope
        pos := pos + 1
    return longestPrefixesSet
end procedure
```

Fig. 2 The getLongestCommonPrefixes procedure

The **OverlayMatch** object holds information about the degree of similarity between the searched sentence and one of the sentences in the suffix-array-based index. Each sentence candidate found in the index has its own OverlayMatch object which is used to assess its similarity to the searched sentence. The object is described in Figure 3. For example, if the searched sentence (pattern) is: "There is a small book on the table" and the candidate sentence (example) is: "I know there is a small pen on the desk", the OverlayMatch object would be:

patternMatches : { [0,3]; [5,6] }
exampleMatches : { [2,5]; [7,8] }

The main search algorithm – the procedure **search** – takes one parameter: a series of tokens and returns a map of candidate sentence ids and their OverlayMatch objects. It is described in Figure 4.

The OverlayMatch object definition

```
object OverlayMatch {
    patternMatches - a list of disjunctive intervals,
                     representing the overlay of the searched sentence
    exampleMatches - a list of disjunctive intervals,
                     representing the overlay of the candidate sentence
}
```

Fig. 3 The OverlayMatch object

Algorithm 3: The main search procedure

```
procedure search(h)
   for(i = 0 to length(h))
      longestPrefixes := getLongestPrefixes(h.subSequence(i,length(h)))
      for all (Suffix suffix in longestPrefixes)
         prefixLength := longestPrefixes.getPrefixLength()
         currentMatch := matchesMap.get(suffix.id)
         currentMatch.addExampleMatch(suffix.offset,
                                      suffix.offset+prefixLength)
         currentMatch.addPatternMatch(i, i+prefixLength)
end procedure
```

Fig. 4 The search procedure

6.6 Computing Score

For a given input sentence (pattern) and a candidate sentence (example) having an OverlayMatch object, the similarity score is computed using the following formula:

$$score = \frac{\sum_{i=0}^{n} patternMatches[i].length + \sum_{i=0}^{m} exampleMatches[i].length}{length(pattern) + length(example)}$$

where:

- $patternMatches[k].length$ is the length (in tokens) of k-th interval
- $length(pattern)$ is the length of the pattern

For the example used to present the OverlayMatch object in section 2.4, the computed score would be:

$$\frac{(4+2)+(4+2)}{8+10} \approx 66.6\%$$

7 Evaluation

Anubis was evaluated in order to measure the precision of its translation suggestions and the speed of TM searching. For comparison, the system memoQ, described in Section 5 was selected.

7.1 Precision Evaluation

Precision evaluation procedure was aimed at determining wheather Anubis is able to provide as valuable translation suggestions as memoQ. As there is no good way of scoring the value of suggestions automatically, human translators were involved in the evaluation process.

The experiment was carried out on a genuine translation memory, coming from a translator who built it during his work. Using this kind of TM makes this evaluation significantly more credible than that using bilingual corpora acquired from the Internet. These corpora are often the results of text alignment, not sentence-by-sentence translation. For that reason, they differ from real-life translation memories (e.g. bilingual corpora tend to contain longer sentences).

The statistics of the translation memory used in the experiment are presented in Table 4.

Table 4 Experiment's translation memory statistics

Attribute	**Value**
Source language	Polish
Target language	English
Translation unit count	215 904
Polish word count	3 194 713
English word count	3 571 598
Total word count	6 766 311

The evaluation procedure was split into three phases: preparation, translation and annotation. The preparation phase involved selecting at random 1500 Polish sentences from the translation memory (let's denote it as set $TEST$). During the translation phase, Anubis and memoQ provided translation suggestions for every sentence from $TEST$ using the procedure described in Figure 5.

In this procedure, each sentence is translated using the translation memory it came from, so it is expected to appear as one of possibly many suggestions. If a translation system returns an empty set of suggestions, it will signal the pathological situation in which a 100% match is not found by the system. If the returned *suggestions* set contains only one element, the perfect match, it will be intepreted as the situation where the system could not find a good

Translation procedure

```
for all sentence in TEST
    suggestions = getTranslationSuggestions(sentence)
    if (size(suggestions) == 0)
        report('error!')
    else if (size(suggestions) == 1)
        report('no translation found')
    else
        suggestions.remove(0)
    print(sentence,suggestions[0])
```

Fig. 5 Translation procedure

match. Finally, if the *suggestions* set contains more than one suggestion, they are (except for the first one, which is the 100% match) the proper suggestions. In the report only the first proper suggestion is included along with its similarity score.

The translation procedure was run both by Anubis and memoQ. The results of these runs are shown in Table 5. "Reported errors" correspond to the number of error reports generated by the Translation procedure described in Figure 5. Both systems did not commit any error and successfully detected all the 100% matches.

"Translated sentences" count indicates the number of sentences for which at least one proper suggestion has been found (i.e. the *suggestions* set contained more than 1 element). The results were comparable, though memoQ translated 20% more sentences. This was due to the fact that Anubis had a 50% similarity threshold set, whereas memoQ did not.

A situation in which system A translated a sentence and system B did not translate this sentence was called system's A "knock-out". In the light

Table 5 Translation phase statistics

	memoQ	Anubis
Sentences analyzed	1500	1500
Reported errors	0	0
Translated sentences	1156	962
Knock-outs	285	91
Awarded better scores	491	374
Common translations	871	
including:		
- identical translations	444	
- different translations	427	
Scores correlation (Pearson)	0.541	
Scores correlation (Spearman)	0.553	

Table 6 "Knock-out" statistics

	Annotator 1	Annotator 2
Total memoQ knock-outs	285	
Valuable memoQ knock-outs	162	246
Total Anubis knock-outs	91	
Valuable Anubis knock-outs	34	45

of the previous figure, indicating the total number of translations, it is not surprising that memoQ scored more "knock-outs" than Anubis.

These "knock-outs", however, were further analyzed to determine if the translation suggestions corresponding to them were valuable. This analysis was performed by two human translators, who were given a set of sentence pairs. Each pair constituted of the source sentence and the suggestion provided by a CAT system. For each pair a translator was to answer a yes/no question: would the provided suggestion be helpful in translating the source sentence into the target language? This procedure was the first step of the experiment's annotation phase. The results of the "knock-outs" analysis are shown in Table 6. This study indicates that the similarity threshold for Anubis can be set to a lower value.

In the translation phase, the scores of suggestions provided by the two systems were also taken into consideration. The correlation of these scores (treated as random variables) was measured using statistical methods (see [9]). The coefficient values around 0.5 indicate some correlation between the scores. MemoQ was the system that was awarding better scores in general for the same suggestions.

The two systems provided identical translation suggestions for 444 out of total 871 common translations. The remaining 427 translations were judged by two human translators in the key step of the annotation phase. The translators were given a set of 3-tuples: source sentence, translation from system A, translation from system B. In order to obtain more balanced results, the order of translations was changed randomly. For each tuple, the annotators were to assign one of the possible scores

- first system's suggestion is better
- second system's suggestion is better
- both translations are of equal value

The criterion of the suggestion's value was: "how helpful is the suggestion in translating the source sentence into the target language?"

The results of this annotation are shown in Table 7. The inter-annotator agreement was: the annotators agreed in 262 cases, disagreed 165 times out of which only 24 were strong disagreements (one annotator assigned a win to one system and the other annotator to the other system).

These results show that suggestions generated by the two systems are roughly comparable.

Table 7 Suggestion precision statistics

	Annotator 1	Annotator 2
Total translations	427	
Anubis wins	**156**	**103**
MemoQ wins	150	96
Draws	121	228

7.2 Speed Evaluation Compared to memoQ

Apart from the precision, the most important characteristic of a CAT system is its speed. During the translation phase of the above experiment, the translation speed was measured. The tests were run on a machine with Intel Core 2 Duo 2.0 GHz CPU and 3GB RAM memory. The results of the speed test are shown in Table 8. These results show a considerable advantage of Anubis. The speed of suggestion generation can be a crucial characteristic of a CAT system in the context of translation memory servers. As the volume of collected translation memories grows, translation agencies try to realize an idea of centered TM, shared among their translators. On the other hand, freelance translators often team up in communities and share their translation resources (an example of such a project is Wordfast VLTM – Very Large Translation Memory [5]).

Table 8 MemoQ speed test results

	memoQ	Anubis
Total translations	1500	
Translation time [s]	414.6	**94.7**
Average speed [sent/s]	3.618	**15.846**

These tendencies lead to creation of vast translation memories, shared among multiple users. In this architecture, it is essential to have an effective translation memory search algorithm. The algorithm will have to be able to deal with a large data set in a short time, as multiple users working simultaneously will query the TM often. Thus, the idea of speeding up translation memory searching is a key idea for next generation CAT tools.

8 Conclusions

This paper presented the idea od Computer-Aided Translation, which is gaining more and more popularity in today's market of translations. The technique is known to improve the efficiency of human translator's work. A good CAT system provides precise translation suggestions based on the translation memory. The other key factor that determines the system's usability is the speed of suggestions generation. Although nowadays this CAT technique is often wrapped up in a whole translator's workbench (like in the memoQ system), translation memory searching is still the key mechanism that is used to help the translator.

A novel approach to translation memory building, storing and searching was presented in this paper. We propose the system Anubis, based on state-of-art techniques of approximate string searching and natural language processing. The results of the first tests show that the precision of translation suggestions generated by Anubis can be compared to the memoQ system. However, Anubis offers much higher speed of translations, especially for large translation memories which are the future of CAT.

The system Anubis is still in the process of development. The study on Computer-Aided Translation along with the system's evaluation indicate that Anubis has a chance to become a usable, helpful tool for translators.

References

1. Craciunescu, O., Gerding-Salas, C., Stringer-Keeffe, S.: Machine translation and computer-assisted translation: a new way of translating? The Translation Journal 8(3) (2004)
2. Huffman, D.: A method for the construction of minimum-redundancy codes. Proceedings of the I.R.E., 1098–1102 (1952)
3. Hutchins, J.: Machine translation: a concise history. In: Wai, C.S. (ed.) Computer Aided Translation: Theory and Practice. Chinese University of Hong Kong (2007)
4. Multiple: Kilgray translation technologies: memoq translator pro, `http://kilgray.com/products/memoq/`
5. Multiple: Wordfast community: Very large translation memory project, `http://www.wordfast.com/`
6. Navarro, G.: A guided tour to approximate string matching. ACM Computing Surveys (CSUR) 33(1) (March 2001)
7. Navarro, G., Baeza-Yates, R., Sutinen, E., Tarhio, J.: Indexing methods for approximate string matching. IEEE Data Engineering Bulletin 24, 2001 (2000)
8. Palacz, B.: A comparative study of cat tools (maht workbenches) with translation memory components. Master thesis written under guidance of prof. Włodzimierz Sobkowiak, Adam Mickiewicz University (2003)
9. Stigler, S.M.: Francis galton's account of the invention of correlation. Statistical Science (1989)
10. Tufiş, D., Irimia, E.: Roco-news: A hand validated journalistic corpus of romanian (2006)
11. Twiss, G.: A comparative study of cat tools (maht workbenches) with translation memory components. Proz.com The Translator Workspace (2006)
12. Vallianatou, F.: Cat tools and productivity: Tracking words and hours. The Translation Journal Volume 9(4) (2005)

Incorporating Subject Areas into the Apertium Machine Translation System

Jordi Duran, Lluís Villarejo, Mireia Farrús, Sergio Ortiz, and Gema Ramírez

Abstract. The Universitat Oberta de Catalunya (Open University of Catalonia, UOC), is a public university based in Barcelona. The UOC is characterised by three main factors: (a) it is a virtual university based in an e-Learning model, (b) it is based in a strongly Spanish-Catalan bilingual region, and (c) students come from around the world, so that linguistic and cultural diversity is a crucial factor.

Within this context, it becomes essential to meet the UOC's linguistic needs taking into account its particular characteristics. One of the tools created to this end is the adaptation of Apertium, a free/open-source rule-based machine translation platform, which can be found under http://apertium.uoc.edu/, customised to the translation needs of the institution in order to offer the best possible service to their user community.

In order to continue adapting and adding value to the existing tool for generalisable large-scale applications, the UOC's translation system has recently implemented a semantic filter based on subject fields aimed at improving the translation quality and at better fitting the university needs. The paper will explain all the steps of this adaptive process, as well as a demonstration of the resulting tool: (a) the choice of the subject fields according to the university studies, (b) the design and implementation of the dictionaries used to extract the required information to filter and disambiguate homonym and polysemous terms, including source code in the dictionaries, and (c) the design and implementation of the corresponding web interface.

1 Introduction

During the last decades, machine translation (MT) tools have proved highly useful in human translation work, since they save human translators a great deal of time

Jordi Duran · Lluís Villarejo · Mireia Farrús
Universitat Oberta de Catalunya, Barcelona, Spain
e-mail: lvillarejo@uoc.edu, mfarrusc@uoc.edu, jdurancal@uoc.edu

Sergio Ortiz · Gema Ramírez
Prompsit Language Engineering, Alacant, Spain
e-mail: sortiz@prompsit.com, gramirez@prompsit.com

A. Przepiórkowski et al. (Eds.): *Computational Linguistics*, SCI 458, pp. 281–292, 2013.
DOI: 10.1007/978-3-642-34399-5_15

by quickly providing initial draft translations (Craciunescu et al. 2004; Kirchhoff et al. 2004; Villarejo et al. 2009a).

The Language Technologies group from the Office of Learning Technologies at the Universitat Oberta de Catalunya (Open University of Catalonia, UOC) and the university's Language Service have recently been working on improvements to the UOC's Apertium-based translation (http://apertium.uoc.edu). Apertium is an open-source MT platform that can be easily modified and adapted to users' needs (Forcada et al. 2009).

At the time of writing, the UOC has created an Apertium-based web application incorporating specific terminology used by the university. It has also contributed to the development of the system multiple-format support (Word, PowerPoint, Excel, Writer, Calc, Impress, HTML files, compressed files, etc.) for document translation service, which includes the option of creating and exporting translation memories that allow for the customisation of translations (Villarejo et al. 2009a; Villarejo et al. 2009b; Villarejo et al. 2010). The advantage of working with Apertium is that all linguistic improvements made to the system are saved in a common repository so that its entire community of users can benefit from them.

The Apertium application was placed at the disposal of the UOC's internal community (the university's teaching staff and staff who manage and edit teaching material) in December 2009, and was well received by the users. Soon we realised that a user translating legal texts and another one translating computing texts required from different translations. With a view to obtaining high quality translations fully adapted to the university, the UOC subsequently set about to implement the one sense per discourse assumption to improve the quality of Apertium's translations.

The one sense per discourse assumption states that if a polysemous word is used more than once in a coherent discourse, it is very likely that all its occurrences refer to the same sense. This assumption has been proved to be effective to improve results in tasks like Word Sense Disambiguation (Gale et. al. 1992) and Statistical Machine Translation (Carpuat 2009). If we could detect that some translation pairs (source and target word) are specifically used in some of the university subject areas, we could build up a dictionary with them, ask the user to manually introduce the subject area for the source text which is to be translated and implement this way the one sense per discourse assumption. In this paper we describe how we incorporated subject areas related to the university fields of knowledge and the subjects taught into Apertium in order to improve the translation quality. The inclusion of subject areas in the MT system, where they act as semantic filters, allows for the disambiguation of homonyms and polysemes. The translation process thus takes place on the basis of the topic of the source text.

This chapter describes the work carried out in relation to incorporating subject areas into Apertium. Sections 1 to 6 of the article deal with the planning of the project, and sections 7 to 9 with its execution and conclusions, plus future lines of work.

2 Relevance of the Project

The new Apertium interface enables any user to translate documents on the basis of the subject area corresponding to each source text. The function in question can also be deactivated by selecting the 'General vocabulary' option instead of a specific subject area.

Additionally, the subject area analysis process makes it possible to identify new specific terminology and incorporate it into the MT system, as well as to produce glossaries of terms to complement universities' teaching material and academic and administrative documentation.

As Apertium is an open-source platform, all linguistic improvements made to it are stored in a freely accessible general repository. Any user can thus benefit from the semantic filters implemented in this project. Given that the subject areas have been established in relation to the UOC's fields of knowledge, the improvements made to Apertium in this project will be extremely useful in any sphere related to the university.

3 Goals

The incorporation of subject areas into the Apertium machine translation system within the context of the Universitat Oberta de Catalunya sought to accomplish some objectives. In the first place, a specific analysis and selection of the most useful subject areas for the UOC's community should be performed. As a consequence, the subject areas will serve as filters for the MT system with a view to improving its translated output. Secondly, the creation of a glossary of polysemes and homonyms which can be disambiguated by assigning subject areas to them will be fulfilled. And finally, the implementation of the subject filters in the MT system's interface so that users can choose a subject area and generate a much more accurate translation featuring the terminology used in the field in question will be carried out.

4 Beneficiaries

First of all, it is of great importance for the success of the project to identify those target people that will be potential users of the new machine translation. The main groups set to benefit from the project are the following ones:

(1) Administrative staff looking to circulate versions of documents or communiqués in various languages and who want to use MT to speed up their work.
(2) Linguists who correct and translate universities' teaching, academic and administrative documentation.
(3) External language professionals who work with university language services.

(4) Teaching staff seeking to use MT to prepare the teaching material with which they provide students.
(5) Students who want to use MT to obtain rough translations to enable them to understand material or carry out work in languages in which they lack proficiency.
(6) In general, institutions, organisations, businesses, the media and members of the public interested in using MT in their day-to-day activities.

5 Project Stages and Groups Involved

Language technology projects require collaboration between two types of experts, namely linguists, with their mastery of language, and IT engineers specialising in natural language processing, who provide access to the technology necessary for the implementation of such projects. In that regard, the UOC boasts a multidisciplinary team that combines the linguistic expertise of the staff of its Language Service and the skills of the language technology specialists from its Office of Learning Technologies.

Broadly speaking, the project consisted of three stages. Each stage and the groups it involved are briefly outlined below.

Stage 1 included subject area analysis and description, as well as the compilation of university material from which specific vocabulary could be extracted. The activities involved in this stage were carried out by the Language Service and the Office of Learning Technologies.

Stage 2 included producing a glossary of semantically ambiguous terms and adding tags to dictionary entries to identify their subject area. The activities involved in this stage were carried out by the Language Service together with the Office of Learning Technologies and Prompsit, a company that had already participated in technical work geared to MT in similar projects. In the case in hand, Prompsit designed the data files for subject areas format and the integration within Apertium and collaborated in the generation of linguistic data.

Stage 3 involved the implementation of subject areas in the new Apertium interface. The Language Service and the Office of Learning Technologies designed the interface, assessed it from a linguistic perspective and disseminated it, while Prompsit carried out the actual implementation work as well as subsequent technical testing.

6 General Tool Description

At present, the UOC has an internal translation service based on Apertium (available at http://apertium.uoc.edu). Anyone wishing to do so may download the same version of the system from a freely accessible general code repository (http://sourceforge.net/projects/apertium/).

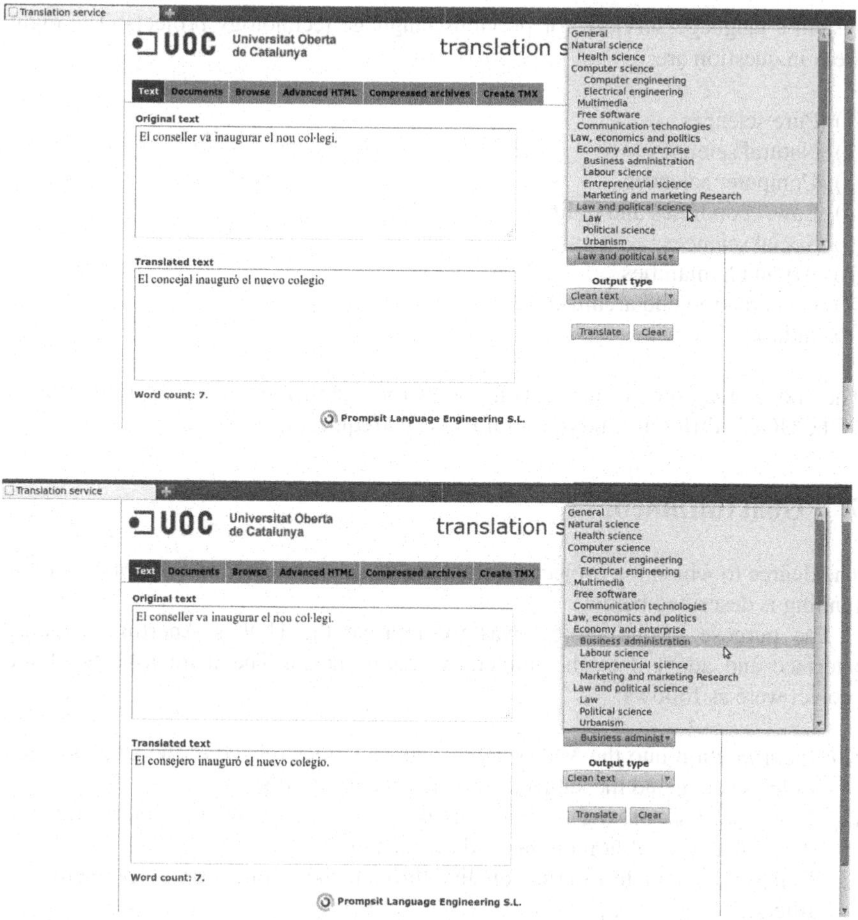

Fig. 1 Proposed new interface. In the example, word "conseller" in Catalan is translated into "concejal" (councillor) in Spanish in the Law and Political Science domain but into "consejero" (advisor) in the Bussiness Administration domain.

The MT system can be used to translate texts entered directly in the platform's text box, documents in multiple formats, web pages (retaining the full original structure), HTML files and compressed files. It also allows for the creation and use of TMX-format (Translation Memory eXchange) translation memories, which help to generate much more accurate translations with a greater degree of customisation.

The proposed new interface developed in this project features a drop-down list of subject areas from which users can choose in line with the topic of the translation to be performed (See Figure 1). The list of subject areas is based on a general basic classification of eight areas established by the Catalan university system's language services for previous language technology projects. The eight areas in question are:

(1) Pure sciences
(2) Natural sciences
(3) Computer sciences
(4) Law, economics and politics
(5) Social sciences
(6) Art and humanities
(7) Construction and architecture
(8) Industry

The above areas served as a starting point for establishing sub-areas on the basis of the UOC's different study programmes and requirements.

7 Goal fulfilment

The degree to which the project fulfilled the goals established in its application for funding is described below.

The purpose of the project was to improve the UOC's Apertium existing interface and adapt it to the university requirements. The main aspects of the project were as follows:

(1) Incorporation into the MT system of subject areas related to the UOC's fields of knowledge and the subjects taught at the university.
(2) Configuration of the subject areas to act as semantic filters for the disambiguation of homonyms and polysemes.
(3) Enhancement of translation quality through the application of the semantic filters.
(4) Open access for the system's community of users, who can apply and benefit from the improvements made.
(5) Ease of use, even for those unfamiliar with MT.

8 Stages and Execution of the Project

As stated in section 5, the work was carried out in three different stages, which are described below in detail.

Stage 1

a) Selecting subject areas

As mentioned previously, the classification established by the Catalan university system's language services (see Table 1) was used as a starting point for selecting

Table 1 Subject areas established by the Catalan university system's language services (2001)

	Area	Sub-areas
1	Pure sciences	Chemistry, physics, mathematics, statistics
2	Natural sciences	Earth science, life sciences, health sciences, veterinary science, agriculture, stockbreeding, fishing, environmental science
3	Computer science	Computer science, telecommunications, electronics, electrical engineering
4	Law, economics and politics	Law, public administration, defence, economics, home economics, administration, work and employment, politics
5	Social sciences	Sociology, anthropology, psychology, pedagogy, communication, information, sports, games, tourism, hotel industry, food
6	Art and humanities	Geography, history, philology, philosophy, religion, art, music
7	Construction and architecture	Construction, architecture
8	Industry	Industrial engineering, various industries

subject areas to act as semantic filters. The classification was further developed to include the subject areas contained in the UOC's institutional repository and analysed in this project (see Table 2).

b) Compiling material

Parallel documents (i.e. documents in Catalan with translations into Spanish or English) corresponding to different subject areas (see Table 1) were used for the purpose of identifying ambiguous terms. The documents in question were taken from the volume of texts corrected and translated by the UOC's Language Service from July to December 2009.

Stage 2

a) Designing dictionaries

Apertium has a monolingual dictionary for each of its languages and a bilingual dictionary for each of its language pairs. Each dictionary, be it monolingual or bilingual, is a separate file that can be modified and used on the basis of the language pair involved in a translation.

The same system of separate files was used for the purpose of assigning subject areas to ambiguous terms. More specifically, two different files were used, one for

Table 2 Subject areas analysed in this project

0.	**General**
2.	**Natural sciences**
2.1.	Health sciences
3.	**Computer science**
3.1.1.	Computer science engineering
3.1.3.	Technical engineering in computer systems
3.2.	Multimedia
3.3.	Free software
3.4.	Communication technologies
4.	**Law, economics and politics**
4.1.	Economics and business
4.1.1.	Business administration and management
4.1.2.	Labour sciences
4.1.3.	Business studies
4.1.5.	Marketing and market research
4.2.1.	Law
4.2.2.	Political sciences and public management
4.2.3.	City management and urban planning
5.	**Social sciences**
5.1.	Information and knowledge society
5.2.2.	Teaching
5.2.3.	Psychology
5.2.4.	Educational psychology
5.2.	Psychology and educational sciences
5.3.	E-learning
5.4.1.	Communication
5.4.2.	Audiovisual communication
5.4.3.	Information and documentation
5.5.1.	Tourism
5.5.2.	Travel programme
5.6.	Food systems, society and culture
5.7.	Humanitarian cooperation, peace and sustainability
6.	**Art and humanities**
6.1.	Humanities
6.1.1.	Philosophy
6.2.	Languages and cultures
6.3.1.	Catalan philology
6.3.2.	Translation and technologies
6.3.4.	Arab and Islamic studies
6.3.5.	Cultural management
8.	**Industry**

specifying subject areas and the other for indicating the grammatical and semantic information required for disambiguating terms:

- *Subject area file.* This file contains a list of subject areas, each with an identifying code. For instance, the e-learning subject area, which has been assigned the code 5.3, is represented as follows:

  <d n="5.3">e-learning</d>

- *Disambiguation file.* This file indicates the different subject areas that can be assigned to a particular term. The Spanish term *mesa*, for example, can be assigned to the general corpus area (code 0), the law, economics and politics area (code 4) or the humanities area (code 6.1). The corresponding Spanish-Catalan dictionary entries are as follows:

  <e d="0"><p><l>mesa<s n="n"/></l><r>taula<s n="n"/>.
  <e d="4"><p><l>mesa<s n="n"/></l><r>mesa<s n="n"/>...
  <e d="6.1"><p><l>mesa<s n="n"/></l><r>tabula<s n="n"/>...

  The following rules must be observed where the disambiguation file is concerned:

  (1) At least the grammatical category must be indicated to take advantage of the Apertium part-of-speech disambiguation module based on hidden Markov models (Rabiner 1989), which automatically selects the most probable part-of-speech for homographs.
  (2) The specific subject area must always be indicated.
  (3) The terms that appear in the file must be incorporated into the corresponding monolingual and bilingual dictionaries.
  (4) If a single term has two translations within a given subject area, an exclamation mark must be used to denote the translation to be prioritised. The criterion used to prioritise a translation over another is explained in the following section.

b) Extracting ambiguous terms

The first step in analysing ambiguous terms on the basis of the parallel corpus consisted of extracting relevant terms for each subject area, including the general area. Term relevance within an area was determined on the basis of its keyness, i.e. its quality of being key in its context (Scott and Tribble 2006; Bondi & Scott 2010). If, proportionately speaking, a term appears with significantly greater frequency in a text corresponding to a specific area than in an extensive general corpus (a reference corpus), it is said to have keyness in the area in question. The higher a term's keyness value in an area, the more relevant it is within that area.

The extraction of ambiguous terms was carried out in six steps, applying the concept of keyness:

Step 1. Calculating keyness values for each area

The keyness values of the terms appearing in each specific area were calculated, giving rise to a list (let's name it List 1), which is an ordered list showing each term's relevance in the area in question.

Step 2. Looking up terms in different areas (I)

In each subject area, a search was performed for terms also appearing in at least one other area and which thus had a certain probability of being ambiguous. This gave rise to another list (let's name it List 2) consisting of pairs of words and its keyness value. (Examples of the eight words with highest keyness values in the e-learning area in Catalan are: "*ser* (to be) 1513"; "*haver* (to have) 1437"; "*aprenentatge* (learning) 1199"; "*uoc*: 918"; "*anar* (to go) 713"; "*xarxa* (net) 670"; *curs* (course) 659; *ensenyament* (teaching) 659"; etc..

Step 3. Filtering using the reference corpus

The thousand most relevant terms, according to the keyness value, in the general reference corpus were extracted. Given that terms that are highly relevant in the reference corpus are held to be of little relevance in specific areas, those extracted from the reference corpus were deleted from List 2. Generic verbs (e.g. to be, to have, to make, to do and to go) were also deleted. The result was List 3, consisting of pairs of words and its keyness value. (Examples of the nine words with highest keyness values in the e-learning area in Catalan are: "*aprenentatge* (learning) 1199"; "*uoc* 918"; "*xarxa* (net) 670"; "*curs* (course) 659"; "*ensenyament* (teaching) 659"; "*nou* (new) 594"; "*social* (social) 572"; "*obert* (open) 561"; "*estudiant* (student) 561"; etc.

Step 4. Looking up terms in different areas (II)

With the initial lists having been altered, a new search for terms appearing in two or more areas was performed, giving rise to List 4.

Step 5. Filtering terms

Terms from languages other than that under analysis were deleted from List 4. Terms other than verbs, nouns or adjectives were subsequently deleted. The result was List 5.

Step 6. Checking for ambiguity

Having obtained a filtered list of terms and the subject areas in which they appear, a search was performed for terms with more than one translation in other languages and in other subject areas. Firstly, to that end, the texts of the Catalan-Spanish and Catalan-English documents corresponding to each area were aligned by using the Hunaling alignment tool (version 1.0). Next, the aligned sentences containing the relevant terms were extracted, classified by area. That information was used to manually determine which terms have different translations in different subject areas. A final list of relevant terms by subject area having different translations in different subject areas was obtained.

Stage 3

The list of terms with different translation by subject areas was incorporated into the translation engine's disambiguation files. Apertium's workflow was modified to cope with this information during translation and, finally, the UOC's Apertium user interface was modified to allow subject area selection. At the time of writing (April, 2012), the *apertium.uoc.edu* translation service is awaiting the corresponding update.

9 Conclusions and Future Lines of Work

By way of conclusion, it must be emphasised that the project's original goals were fulfilled. The UOC's Office of Learning Technologies worked with the university's Language Service and Prompsit to ensure that the incorporation of subject fields into the Apertium MT system was carried out with the utmost linguistic and technological rigour.

Nevertheless, the UOC aims at constantly improve the needs of the academic community, and to this end, new lines of work are planned as future developments related to the Apertium MT system.

First of all, one the future objectives is to extend the range of potentially ambiguous UOC terminology incorporated into the new interface featuring subject areas. Second and related to the first objective, another goal is to extend the range of subject areas that serve as semantic filters as the UOC adds further study programmes to the courses it offers. Third, the UOC aims at systematising the ambiguous terminology extraction process as more parallel documents become available from the Language Service and to automatise polysemy disambiguation by exploring the most recent work inside Apertium related with automatic lexical selection (Tyers 2012). Finally, an evaluation of the usefulness of each term identified in this project is also in the scope ot the future work and will be done through the analysis of new documents corresponding to each subject area covered.

Acknowledgements. The work in question was made possible by funding that the article's authors obtained from the Office of the Vice President for Research and Innovation of the UOC, through the APLICA call for innovation projects geared to management made by the university's Open Innovation Office.

References

Bondi, M., Scott, M.: Keyness in Texts. Benjamins, Amsterdam (2010)

Carpuat, M.: One sense per discourse. In: Proceedings of the NAACL HLT Workshop on Semantic Evaluations: Recent Achievements and Future Directions, Boulder, Colorado, pp. 19–27 (June 2009)

Craciunescu, O., Gerding-Salas, C., Stringer-O'Keeffe, S.: Machine Translation and Computer-Assisted Translation: a New Way of Translating? The Translation Journal 8(3) (2005)

Forcada, M.L., Tyers, F.M., Ramírez-Sánchez, G.: The Free/Open-Source Machine Translation Platform Apertium: Five Years on. In: Proceedings of the First International Workshop on Free/Open-Source Rule-based Machine Translation FreeRBMT, Alacant, Spain (2009)

Gale, W., Church, K., Yarowski, D.: One sense per discourse. In: HLT 1991: Proceedings of the Workshop on Speech and Natural Language, Morristown, USA, pp. 233–237. Association for Computational Linguistics (1992)

Kirchhoff, K., Turner, A.M., Axelrod, A., Saavedra, F.: Application of Statistical Machine Translation to Public Health Information: a Feasibility Study. Journal of the American Medical Informatics Association 18, 473–478 (2011), doi:10.1136/amiajnl-2011-000176

Rabiner, L.R.: A tutorial on hidden Markov models and selected applications in speech recognition. Proceedings of the IEEE 77(2), 257–286 (1989)

Scott, M., Tribble, C.: Textual Patterns: keyword and corpus analysis in language education. Benjamins, Amsterdam (2006)

Tyers, F.M., Sánchez-Martínez, F., Forcada, M.L.: Flexible finite-state lexical selection for rule-based machine translation. Research paper accepted at the EAMT (to be released, 2012)

Villarejo, L., Cullen, D., Corral, A.: La integració de les tecnologies de la llengua en el flux de treball del Servei Lingüístic de la UOC. Llengua i ús, Revista tècnica de Política Lingüística 46 (2009a)

Villarejo, L., Ortiz, S., Ginestí, M.: Joint efforts to further develop and incorporate Apertium into the document management flow at Universitat Oberta de Catalunya. In: Proceedings of the First International Workshop on Free/Open-Source Rule-based Machine Translation (2009b)

Villarejo, L., Farrús, M., Ortiz, S., Ramírez, G.: A web-based translation service at the UOC based on Apertium. In: Proc. of the International Multiconference on Computer Science and Information Technology, Wisla, Poland, pp. 525–520 (2010)

Author Index